Smart Innovation, Systems and Technologies 352

Series Editors

The Smart Innovation, Systems and Technologies book series encompasses the topics of knowledge, intelligence, innovation and sustainability. The aim of the series is to make available a platform for the publication of books on all aspects of single and multi-disciplinary research on these themes in order to make the latest results available in a readily-accessible form. Volumes on interdisciplinary research combining two or more of these areas is particularly sought.

The series covers systems and paradigms that employ knowledge and intelligence in a broad sense. Its scope is systems having embedded knowledge and intelligence, which may be applied to the solution of world problems in industry, the environment and the community. It also focusses on the knowledge-transfer methodologies and innovation strategies employed to make this happen effectively. The combination of intelligent systems tools and a broad range of applications introduces a need for a synergy of disciplines from science, technology, business and the humanities. The series will include conference proceedings, edited collections, monographs, handbooks, reference books, and other relevant types of book in areas of science and technology where smart systems and technologies can offer innovative solutions.

High quality content is an essential feature for all book proposals accepted for the series. It is expected that editors of all accepted volumes will ensure that contributions are subjected to an appropriate level of reviewing process and adhere to KES quality principles.

Indexed by SCOPUS, EI Compendex, INSPEC, WTI Frankfurt eG, zbMATH, Japanese Science and Technology Agency (JST), SCImago, DBLP.

All books published in the series are submitted for consideration in Web of Science.

Ireneusz Czarnowski · R. J. Howlett ·
Lakhmi C. Jain
Editors

Intelligent Decision Technologies

Proceedings of the 15th KES-IDT 2023 Conference

Editors
Ireneusz Czarnowski
Gdynia Maritime University
Gdynia, Poland

R. J. Howlett
KES International Research
Shoreham-by-sea, UK

Lakhmi C. Jain
KES International
Selby, UK

ISSN 2190-3018 ISSN 2190-3026 (electronic)
Smart Innovation, Systems and Technologies
ISBN 978-981-99-2971-9 ISBN 978-981-99-2969-6 (eBook)
https://doi.org/10.1007/978-981-99-2969-6

This Springer imprint is published by the registered company Springer Nature Singapore Pte Ltd.
The registered company address is: 152 Beach Road, #21-01/04 Gateway East, Singapore 189721, Singapore

Preface

This volume contains the proceedings of the 15th International KES Conference on Intelligent Decision Technologies (KES-IDT 2023). The conference was held in Rome, from 14 to 16 of June, 2023.

The KES-IDT is an international annual conference organized by KES International. The KES Conference on Intelligent Decision Technologies belongs to a sub-series of the KES Conference series.

The KES-IDT provides space for the presentation and discussion of new research results under the common title "Intelligent Decision Technologies". The conference has an interdisciplinary character, giving opportunities to researchers from different scientific areas and different application areas to show how intelligent methods and tools can support the decision making processes.

This year the submitted papers has been allocated to the main track and 6 special sessions. Each submitted paper has been reviewed by 2–3 members of the International Program Committee and International Reviewer Board. 28 papers were accepted for presentation during the conference and inclusion in the KES-IDT 2023 proceedings.

We are very satisfied with the quality of the papers and would like to thank the authors for choosing KES-IDT as the forum for the presentation of their work. We also would like to thanks Prof. Bożena Kostek and Prof. Andrzej Czyżewski for their interesting and valuable keynote speech during the conference. Their keynotes have been also included in the conference proceedings.

We also gratefully acknowledge the hard work of the KES-IDT international program committee members and the additional reviewers for taking the time to review the submitted papers and selecting the best among them for presentation at the conference and inclusion in the proceedings.

We hope that KES-IDT 2023 significantly contributes to the fulfilment of academic excellence and leads to even greater successes of KES-IDT events in the future.

June 2023

Ireneusz Czarnowski
R. J. Howlett
Lakhmi C. Jain

KES-IDT 2023 Conference Organization

Honorary Chairs

Lakhmi C. Jain	KES International, UK
Gloria Wren-Phillips	Loyola University, USA

General Chair

Ireneusz Czarnowski	Gdynia Maritime University, Poland

Executive Chair

Robert J. Howlett	KES International Research, UK

Program Chairs

Jose L. Salmeron	University Pablo de Olavide, Seville, Spain
Antonio J. Tallón-Ballesteros	University of Huelva, Spain

Publicity Chairs

Izabela Wierzbowska	Gdynia Maritime University, Poland
Alfonso Mateos Caballero	Universidad Politécnica de Madrid, Spain

Keynote Speeches

Bożena Kostek	Gdańsk University of Technology, Poland
Andrzej Czyżewski	Gdańsk University of Technology, Poland

Special Sessions

Decision Making Theory for Economics

Takao Ohya	Kokushikan University, Japan
Takafumi Mizuno	Meijo University, Japan

Large-Scale Systems for Intelligent Decision Making and Knowledge Engineering

Sergey V. Zykov	HSE University, Russia
Hadi M. Saleh	HSE University, Russia

Artificial Intelligence Innovation in Daily Life

Hadi Saleh	HSE University, Russia

Reasoning-Based Intelligent Applied Systems

Jair Minoro Abe	Paulista University, Brazil

Recent Development of Multivariate Analysis and Model Selection

Rei Monden	Hiroshima University, Japan
Mariko Yamamura	Department of Radiation Effects Research Foundation, Japan

Applied Methods of Machine and Deep Learning

Nikita Andriyanov	Financial University under the Government of the Russian Federation

International Program Committee and Reviewers

Jair Minoro Abe	Paulista University, Brazil
Witold Abramowicz	Poznan University of Economics and Business, Poland
Miltos Alamaniotis	University of Texas at San Antonio, USA
Nikita Andriyanov	Financial University, Russia
Ahmad Taher Azar	Prince Sultan University, Saudi Arabia
Valentina Balas	Aurel Vlaicu University of Arad, Romania

Dariusz Barbucha	Gdynia Maritime University, Poland
Monica Bianchini	University di Siena, Italy
Leszek Borzemski	Wrocław University of Science and Technology, Poland
Oliver Bossert	McKinsey & Co. Inc., Germany
Adriana Burlea-Schiopoiu	University of Craiova, Romania
Prof. Frantisek Capkovic	Slovak Academy of Sciences, Slovakia
Amitava Chatterjee	Jadavpur University, India
Amine Chohra	Paris East University Criteil (UPEC), France
Marco Cococcioni	University of Pisa, Italy
Paolo Crippa	University Politecnica delle Marche, Italy
Gloria Cerasela Crisan	Vasile Alecsandri University of Bacau, Romania
Matteo Cristani	University of Verona, Italy
Ireneusz Czarnowski	Gdynia Maritime University, Poland
Vladimir Dimitrieski	University of Novi Sad, Serbia
Dinu Dragan	University of Novi Sad, Serbia
Eman El-Sheikh	University of West Florida, USA
Laura Falaschetti	University Politecnica delle Marche, Italy
Margarita N. Favorskaya	Reshetnev Siberian State University of Science and Technology, Russia
Claudia Frydman	Aix Marseille University, France
Mauro Gaggero	National Research Council of Italy
Mauro Gaspari	University of Bologna, Italy
Christos Grecos	Arkansas State University, USA
Ioannis Hatzilygeroudis	University of Patras, Greece
Dawn Holmes	University of California, USA
Katsuhiro Honda	Osaka Metropolitan University, Japan
Tzung-Pei Hong	National University of Kaohsiung, Taiwan
Xiangpei Hu	Dalian University of Technology, China
Anca Ignat	University Alexandru Ioan Cuza of Iasi, Romania
Yuji Iwahori	Chubu University, Japan
Nikita Jain	Poornima College of Engineering, India
Piotr Jedrzejowicz	Gdynia Maritime Univeristy, Poland
Dragan Jevtic	University of Zagreb, Croatia
Nikos Karacapilidis	University of Patras, Greece
Prof. Pawel Kasprowski	Silesian University of Technology, Poland
Petia Koprinkova-Hristova	Bulgarian Academy of Sciences
Aleksandar Kovačević	University of Novi Sad, Serbia
Boris Kovalerchuk	Central Washington University, USA
Marek Kretowski	Bialystok University of Technology, Poland
Kazuhiro Kuwabara	Ritsumeikan University, Japan
Georgy Lebedev	Sechenov University, Russia

Pei-Chun Lin	Feng Chia University, Taiwan
Ivan Lukovic	University of Belgrade, Serbia
Fiammetta Marulli	University della Campania L. Vanvitelli, Italy
Michele Mastroianni	University of Salerno, Italy
Alfonso Mateos	Universidad Politecnica de Madrid, Spain
Lyudmila Mihaylova	University of Sheffield, UK
Yasser Mohammad	Assiut University, Egypt
Rei Monden	Hiroshima University, Japan
Mikhail Moshkov	King Abdullah University of Science and Technology, Saudi Arabia
Vesa A. Niskanen	University of Helsinki, Finland
Marek R. Ogiela	AGH University of Science and Technology, Poland
Takao Ohya	Kokushikan University, Japan
Mario F. Pavone	University of Catania, Italy
Isidoros Perikos	University of Patras, Greece
Petra Perner	Institute of Computer Vision and Applied Computer Sciences, Germany
Gloria Phillips-Wren	Loyola University, USA
Camelia Pintea	Technical University of Cluj-Napoca, Romania
Bhanu Prasad	Florida A&M University, USA
Dilip Kumar Pratihar	Indian Institute of Technology Kharagpur, India
Radu-Emil Precup	Politehnica University of Timisoara, Romania
Jim Prentzas	Democritus University of Thrace, Greece
Prof. Małgorzata Przybyła-Kasperek	University of Silesia in Katowice, Poland
Marcos G. Quiles	Federal University of Sao Paulo, Brazil
Marina Resta	University of Genova, Italy
lvaro Rocha	University of Lisbon, Portugal
Wojciech Sałabun	West Pomeranian University of Technology in Szczecin, Poland
Hadi Saleh	HSE University, Russia
Mika Sato-Ilic	University of Tsukuba, Japan
Milos Savic	University of Novi Sad, Serbia
Bharat Singh	Big Data Labs, Germany
Aleksander Skakovski	Gdynia Maritime University, Poland
Urszula Stańczyk	Silesian University of Technology, Poland
Ruxandra Stoean	University of Craiova, Romania
Piotr Szczepaniak	Lodz University of Technology, Poland
Masakazu Takahashi	Yamaguchi University, Japan
Choo Jun Tan	Wawasan Open University, Malaysia
Shing Chiang Tan	Multimedia University, Malaysia

Claudio Turchetti	University Politecnica delle Marche, Italy
Eiji Uchino	Yamaguchi University, Japan
Fen Wang	Central Washington University, USA
Junzo Watada	Waseda University, Japan
Yoshiyuki Yabuuchi	Shimonoseki City University, Japan
Mariko Yamamura	Department of Radiation Effects Research Foundation, Japan
Hiroyuki Yoshida	Harvard Medical School, USA
Beata Zielosko	University of Silesia in Katowice, Poland
Alfred Zimmermann	Reutlingen University, Germany
Carl Vogel	Trinity College Dublin, Ireland
Sergey V. Zykov	HSE University, Russia

Contents

Keynote Speeches

Data, Information, Knowledge, Wisdom Pyramid Concept Revisited in the Context of Deep Learning ... 3
Bożena Kostek

Multimedia Industrial and Medical Applications Supported by Machine Learning ... 13
Andrzej Czyżewski

Main Track

Accessibility Measures and Indicators: A Basis for Dynamic Simulations to Improve Regional Planning ... 25
Victoria Kazieva, Christine Große, and Aron Larsson

Image-Multimodal Data Analysis for Defect Classification: Case Study of Industrial Printing ... 35
Hiroki Itou, Kyo Watanabe, and Sumika Arima

Image-Multimodal Data Analysis for Defect Classification: Case Study of Semiconductor Defect Patterns ... 48
Daisuke Takada, Hiroki Itou, Ryo Ohta, Takumi Maeda, Kyo Watanabe, and Sumika Arima

Decision-Making Model for Updating Geographical Information Systems for Polish Municipalities Using the Fuzzy TOPSIS Method ... 62
Oskar Sęk and Ireneusz Czarnowski

An Ontology-Based Collaborative Assessment Analytics Framework to Predict Groups' Disengagement ... 74
Asma Hadyaoui and Lilia Cheniti-Belcadhi

Artificial Intelligence Innovation in Daily Life

On Sensing Non-visual Symptoms of Northern Leaf Blight Inoculated Maize for Early Disease Detection Using IoT/AI ... 87
Theofrida Julius Maginga, Deogracious Protas Massawe, Hellen Elias Kanyagha, Jackson Nahson, and Jimmy Nsenga

Inference Analysis of Lightweight CNNs for Monocular Depth Estimation 97
Shadi Saleh, Pooya Naserifar, and Wolfram Hardt

Arabic Text-to-Speech Service with Syrian Dialect 109
Hadi Saleh, Ali Mohammad, Kamel Jafar, Monaf Solieman, Bashar Ahmad, and Samer Hasan

A Systematic Review of Sentiment Analysis in Arabizi 128
Sana Gayed, Souheyl Mallat, and Mounir Zrigui

Reasoning-Based Intelligent Applied Systems

Optimize a Contingency Testing Using Paraconsistent Logic 137
Liliam Sayuri Sakamoto, Jair Minoro Abe, Aparecido Carlos Duarte, and José Rodrigo Cabral

Age-Group Estimation of Facial Images Using Multi-task Ranking CNN 147
Margarita N. Favorskaya and Andrey I. Pakhirka

Evaluation Instrument for Pre-implementation of Lean Manufacturing in SMEs Using the Paraconsistent Annotated Evidential Logic Eτ Evaluation Method 157
Nilton Cesar França Teles, Jair Minoro Abe, Samira Sestari do Nascimento, and Cristina Corrêa de Oliveira

Recent Development of Multivariate Analysis and Model Selection

Modified C_p Criterion in Widely Applicable Models 173
Hirokazu Yanagihara, Isamu Nagai, Keisuke Fukui, and Yuta Hijikawa

Geographically Weighted Sparse Group Lasso: Local and Global Variable Selections for GWR 183
Mineaki Ohishi, Koki Kirishima, Kensuke Okamura, Yoshimichi Itoh, and Hirokazu Yanagihara

Kick-One-Out-Based Variable Selection Method Using Ridge-Type C_p Criterion in High-Dimensional Multi-response Linear Regression Models 193
Ryoya Oda

Estimation Algorithms for MLE of Three-Mode GMANOVA Model with Kronecker Product Covariance Matrix 203
Keito Horikawa, Isamu Nagai, Rei Monden, and Hirokazu Yanagihara

Implications of the Usage of Three-Mode Principal Component Analysis with a Fixed Polynomial Basis 214
Rei Monden, Isamu Nagai, and Hirokazu Yanagihara

Spatio-Temporal Analysis of Rates Derived from Count Data Using Generalized Fused Lasso Poisson Model 225
Mariko Yamamura, Mineaki Ohishi, and Hirokazu Yanagihara

Large-Scale Systems for Intelligent Decision Making and Knowledge Engineering

An Ontology Model to Facilitate Sharing Risk Information and Analysis in Construction Projects 237
Heba Aldbs, Fayez Jrad, Lama Saoud, and Hadi Saleh

Designing Sustainable Digitalization: Crisisology-Based Tradeoff Optimization in Sociotechnical Systems 250
Sergey V. Zykov, Eduard Babkin, Boris Ulitin, and Alexander Demidovskiy

Decision Making Theory for Economics

Calculations by Several Methods for MDAHP Including Hierarchical Criteria 263
Takao Ohya

Utilization of Big Data in the Financial Sector, Construction of Data Governance and Data Management 273
Shunei Norikumo

A Block Chart Visualizing Positive Pairwise Comparison Matrices 282
Takafumi Mizuno

Applied Methods of Machine and Deep Learning

Neural Networks Combinations for Detecting and Highlighting Defects in Steel and Reinforced Concrete Products 293
Nikita Andriyanov, Vitaly Dementiev, and Marat Suetin

Modern Methods of Traffic Flow Modeling: A Graph Load Calculation Model Based on Real-Time Data 302
Roman Ekhlakov

Application of Machine Learning Methods for the Analysis of X-ray Images of Luggage and Hand Luggage 310
Nikita Andriyanov

Author Index 317

About the Editors

Ireneusz Czarnowski is a professor at the Gdynia Maritime University. He holds B.Sc. and M.Sc. degrees in Electronics and Communication Systems from the same University. He gained the doctoral degree in the field of computer science in 2004 at Faculty of Computer Science and Management of Poznan University of Technology. In 2012, he earned a postdoctoral degree in the field of computer science in technical sciences at Wroclaw University of Science and Technology. His research interests include artificial intelligence, machine learning, evolutionary computations, multi-agent systems, data mining and data science. He is an associate editor of the Journal of Knowledge-Based and Intelligent Engineering Systems, published by the IOS Press, and a reviewer for several scientific journals.

Dr. Robert Howlett is the executive chair of KES International, a non-profit organization that facilitates knowledge transfer and the dissemination of research results in areas including intelligent systems, sustainability and knowledge transfer. He is a visiting professor at Bournemouth University in the UK. His technical expertise is in the use of intelligent systems to solve industrial problems. He has been successful in applying artificial intelligence, machine learning and related technologies to sustainability and renewable energy systems; condition monitoring, diagnostic tools and systems; and automotive electronics and engine management systems. His current research work is focused on the use of smart microgrids to achieve reduced energy costs and lower carbon emissions in areas such as housing and protected horticulture.

Dr. Lakhmi C. Jain, Ph.D., M.E., B.E.(Hons) fellow (Engineers Australia) is with the University of Technology Sydney, Australia, and Liverpool Hope University, UK.

Professor Jain serves the KES International for providing a professional community the opportunities for publications, knowledge exchange, cooperation and teaming. Involving around 5,000 researchers drawn from universities and companies worldwide, KES facilitates international cooperation and generates synergy in teaching and research. KES regularly provides networking opportunities for professional community through one of the largest conferences of its kind in the area of KES.

Keynote Speeches

Data, Information, Knowledge, Wisdom Pyramid Concept Revisited in the Context of Deep Learning

Bożena Kostek(✉)

ETI Faculty, Audio Acoustics Laboratory, Gdańsk University of Technology, Narutowicza 11/12, 80-232 Gdańsk, Poland
bokostek@audioakustyka.org

Abstract. In this paper, the data, information, knowledge, and wisdom (DIKW) pyramid is revisited in the context of deep learning applied to machine learning-based audio signal processing. A discussion on the DIKW schema is carried out, resulting in a proposal that may supplement the original concept. Parallels between DIWK pertaining to audio processing are presented based on examples of the case studies performed by the author and her collaborators. The studies shown refer to the challenge concerning the notion that classification performed by machine learning (ML) is/or should be better than human-based expertise. Conclusions are also delivered.

Keywords: Data · Information · Knowledge · Wisdom (DIKW) Pyramid · audio signals · machine learning (ML)

1 Introduction

Data, information, knowledge, and wisdom (DIKW) are typically presented in the form of a pyramid [1–7]. When looking at this schema, several thoughts may occur. The first refers to what is not included in it, i.e., intuition and perception. These two notions are certainly needed in acquiring knowledge and wisdom. Moreover, the Cambridge dictionary definition of wisdom shows: "the ability to use your knowledge and experience to make good and judgments." So, other missing factors are experience and judgment. Then, what about courage, pushing the boundaries of believing in what one is doing, perseverance, and expertise, which is not an exact synonym for knowledge or wisdom but competence, proficiency, or aptitude. Then, one may look for humility in wisdom to know one's limits. What about understanding and intelligence? Where do they fit in this schema?

Furthermore, we cannot treat the pyramid as an equation in which a sum of data, information, and knowledge equals wisdom. So, maybe this concept should not be presented as a pyramid but rather as a chaotic mixture of all those factors mentioned blending into each other (see Fig. 1). Figure 1 illustrates some of the notions recalled above. Of course, there is sufficient ocean-like space to contain more ideas and beliefs to supplement such a visualization concept.

I. Czarnowski et al. (Eds.): KESIDT 2023, SIST 352, pp. 3–12, 2023.
https://doi.org/10.1007/978-981-99-2969-6_1

On the contrary, the hierarchical presentation has advantages, as it shows the direction from raw data through information and knowledge toward wisdom [6, 7], measured as a level of insight. Even though there is an observation that the levels of DIKW hierarchy wisdom are fuzzy, moreover, the path from data to wisdom is not direct. This is further expanded in a strong thread on DIKW within the rough set community [8, 9]. A wistech (wisdom technology) was introduced, defined as a salient computing and reasoning paradigm for intelligent systems [8].

Fig. 1. Word cloud translating data, information, knowledge, and wisdom (DIKW) pyramid to the chaotic concept of acquiring knowledge and wisdom (created with the use of https://www.wordclouds.com/).

In that respect, all these questions pertain to machine learning (ML) and artificial intelligence (AI). Nowadays, the first attempts that employ learning algorithms are called conventional or baseline. Still, when an artificial neuron was modeled by McCulloch and Pitts [10], and Rosenblatt later proposed a perceptron to classify pictures, there were already ambitious plans for what could be done with such an approach. A well-known Rosenblatt's statement envisioned that perceptron would be able to "recognize people and call out their name," "instantly translate speech in one language to speech or writing in another language," "be fired to the planets as mechanical space explorers," but also "reproduce itself" and be self-conscious in the future [10, 11]. However, before this belief came true, several decades passed, and technology had to change to employ graphical units instead of CPU (central processing unit), resulting in data-hungry deep learning [12]. A very apt statement of Vandeput on the last decade's machine learning evolvement refers to the "deep learning tsunami" [10]. However, already in the'50s of the previous century, a need for data was recognized [10]. Even though one may discern a difference between machine learning and deep learning, the first regarded as prediction, classification, etc., and the second considered as "algorithms inspired by the structure and function of the brain called artificial neural networks" [11]; they both need data.

In most cases, data should be structured and annotated by experts. The latter notion, however, may no longer apply as synthesized data may substitute carefully crafted datasets [13]. Moreover, a knowledge-based approach to machine learning may not be necessary as relevant features are extracted automatically in some deep model structures [13, 14]. Contrary, the notion of imperfect data in the sense of incomplete, too small a size, unbalanced, biased, or unrepresentative [15] is still valid.

In this paper, examples of work performed by the author and her collaborators are shown further on. This short review of study encompasses intelligent audio processing.

2 Intelligent Music and Speech Processing

Even though music information retrieval (MIR) is a well-established area encompassing musical sound, melody and music genre recognition, music separation and transcription, music annotating, automatic music mixing, music composing, etc. [16–18], there is a void between human- and machine-based processing, which is sometimes referred to as a semantic gap or bridging a semantic gap [19], i.e., finding interconnections between human knowledge, content collections, and low-level signal features. There are two layers to music services, i.e., a general map of the relation between songs – interconnections between the users and songs, and a personalization layer – information from the above analysis is confronted with the user's music preferences, mood, emotions, and not only what the particular user listens to but what songs they like to combine. In the dictionary of terms related to MIR, one should include music representation, which may be obtained by automatic tagging using metadata (ID3v2), included in, e.g., Gracenote or FreeDB databases; manual tagging by experts or social tagging; content-based; low-level description of music (feature vectors based on MPEG-7 standard, Mel-Frequency Cepstral Coefficients (MFCC), and dedicated descriptors), or 2D maps as features (e.g., spectrograms, mel-spectrograms, cepstrograms, chromagrams, MFCCgrams, designated for deep learning [20]. One should not forget collaborative filtering in Music Recommendation Systems (MRS) that creates maps based on neighbors or taste compatibility.

None of the mentioned representations is devoid of problems. For example, if there are millions of songs in a music service, then even very active users cannot listen to 1% of the music sources; thus, this may result in an unreliable recommendation if the co-occurrence-based method is considered. Contrary to the above consideration, low-level descriptors seem a straightforward representation. However, when comparing time- or time-frequency representation of music/speech signals, one may notice that sounds of the same instrument differ regardless of their representation (see Fig. 2). Obviously, male and female voices uttering the same sentence also differ (see Fig. 2). This may cause identification problems.

From the derivation of signal representation (as shown in Fig. 2) to much more sophisticated tasks is not so far. This may be illustrated based on the identification process of mixed and often overlapped instruments in a music piece to decide which classes are contained in the audio signal [21]. This task was performed on Slakh dataset [22], designated for audio sources separation and multi-track automatic transcription, consisting of 2,100 songs. In most cases, there are four instruments in a song in Slakh. This concerns piano, bass, guitar, and drums.

In Fig. 3, a block diagram of the deep model designated for musical instrument identification is shown [21].

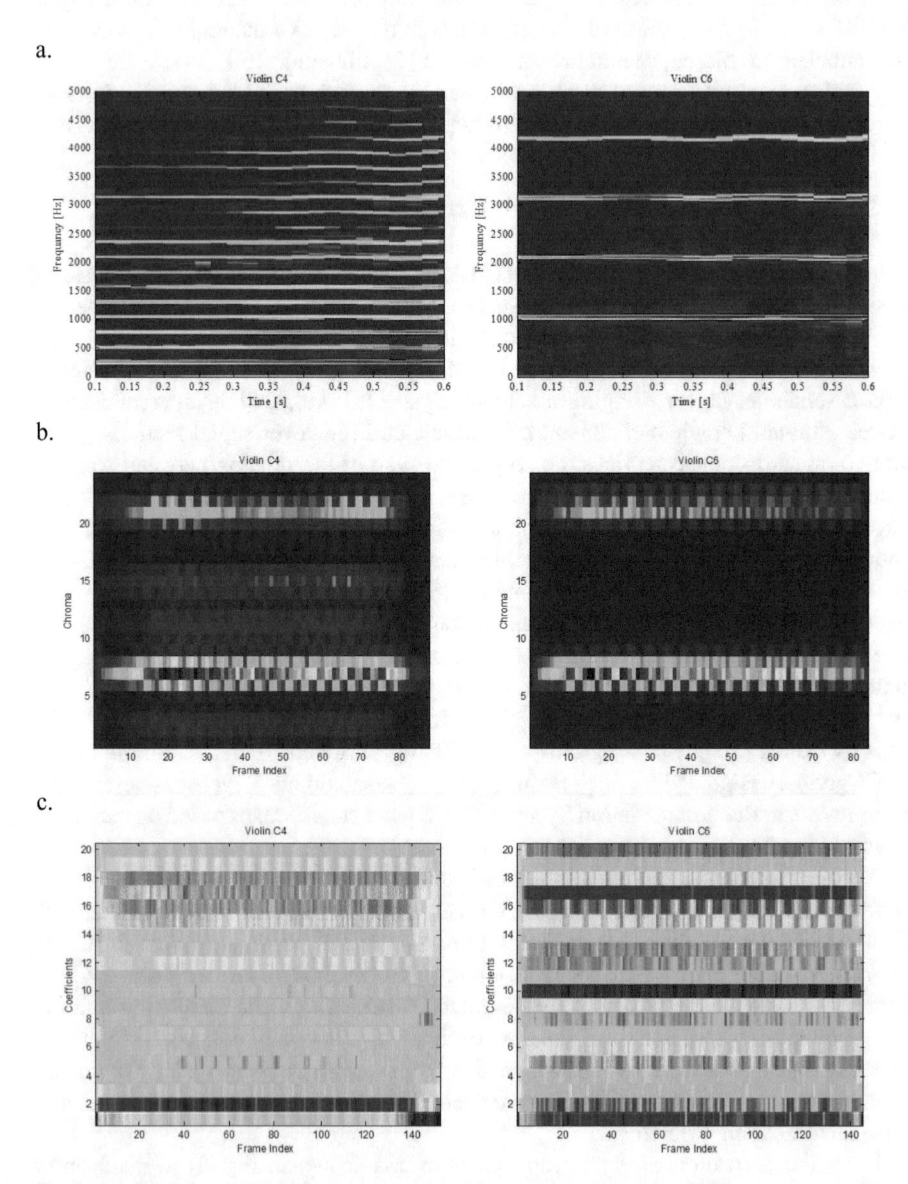

Fig. 2. Violin C4 and C6: (a) spectrograms, (b) chromagrams, (c) MFCCgrams; male/female utterance: (d) spectrograms, (e) chromagrams, (f) MFCCgrams.

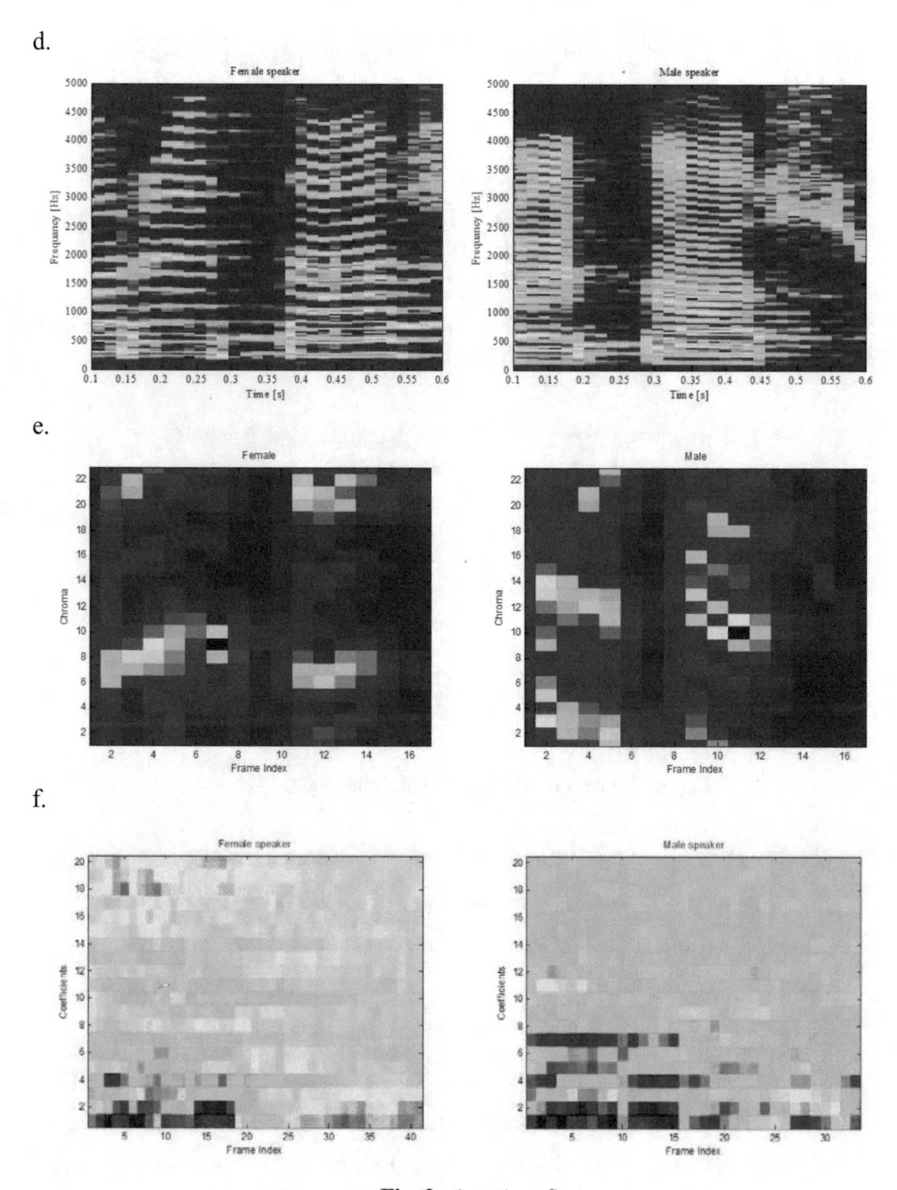

Fig. 2. (*continued*)

The dataset for music identification was divided into three parts: training set – 116,369 examples; validation set – 6,533; evaluation set – 6,464. Figure 4 refers to the metrics obtained during the training and validation processes [21]. In Fig. 5, a histogram of instruments contained in a music piece, identified by the deep model, is presented.

Overall results were as follows: precision – 0.95; recall – 0.94; AUC ROC (area under the ROC curve) – 0.94; true positive – 21,064; true negative – 2,470; false positive – 1,283; false negative – 1,039.

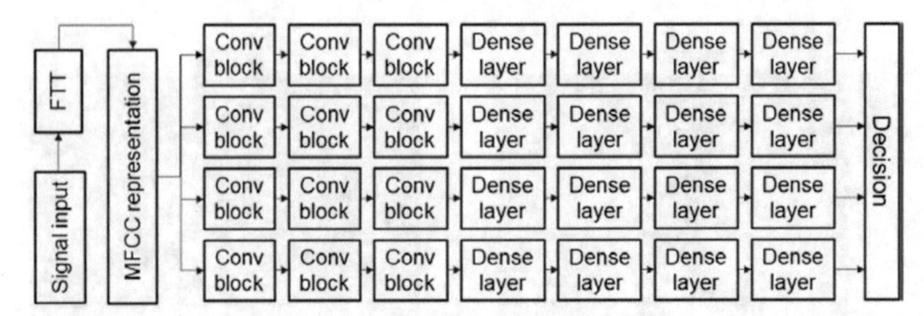

Fig. 3. A block diagram of the deep model designated for musical instrument identification based on Mel Frequency Cepstral Coefficients (MFCCs).

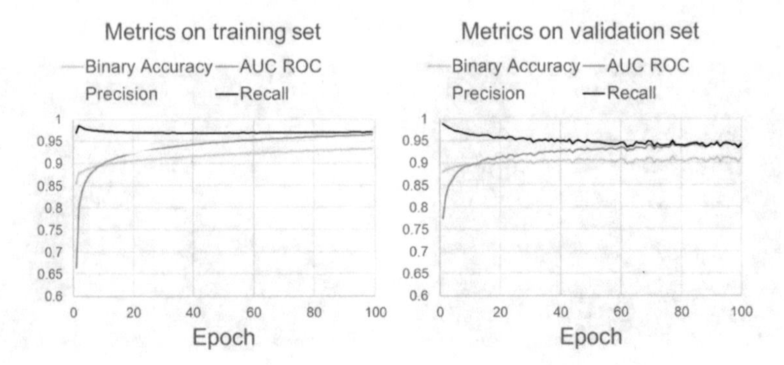

Fig. 4. Metrics on validation and training sets [21].

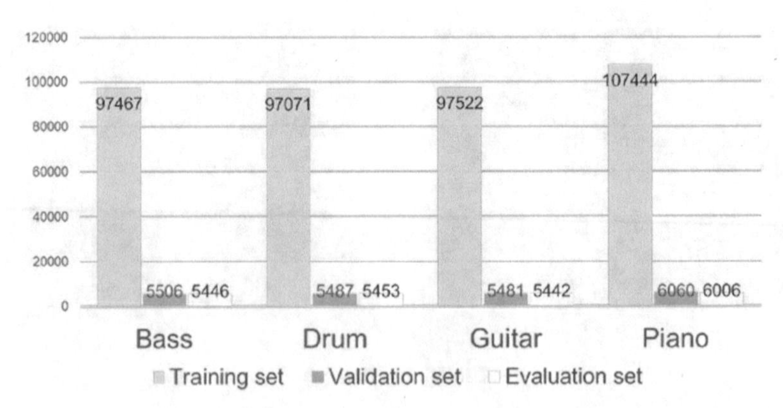

Fig. 5. Histogram of instruments contained in a music piece [21].

Another example concerns the autonomous audio mixing using the wave U-Net deep model [23]. The signal waveform and music genre label are provided at the net input. Individual models are mixed to a stem (stem-mixing is a method of mixing audio material based on creating groups of audio tracks and processing them separately prior to combining them into a final master mix). Then, stems are mixed within the given genre to the entire mix.

In Fig. 6, spectrograms resulting from a mix prepared by a professional mixer and that of the deep U-Net model are shown. One can see that these signal representations are visually indistinguishable, which was further confirmed by the outcome of listening tests (see Fig. 7).

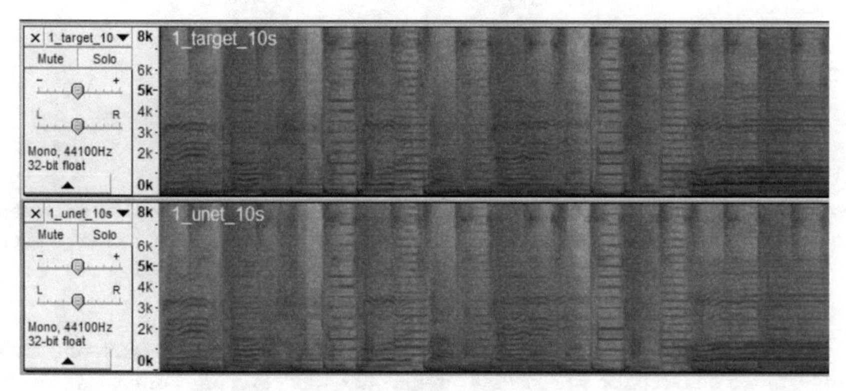

Fig. 6. Spectrograms resulted from autonomous audio mixing using mixes prepared by a professional mixer and the U-Net deep model.

In Fig. 7, the results of the subjective tests checking the quality of mixes prepared by autonomous deep model, technology-based (Izotope), anchor (filtered, low-quality sound), and reference mixes, the last one referring to professionally created mix [23], are shown. Listeners correctly identified both the reference and anchor signals. The U-Net model, in the listeners' opinion, is almost as good as the reference signal and is much better than state-of-the-art-based technology [23].

In music processing – information is often provided by tagging music and its user's behavior and actions (i.e., creating an ecosystem); it is contained in music services (MRS) within the frameworks of music ecosystem (music + users of music services); music content is analyzed at the low-level features, or there is a mixture of approaches.

In speech processing – datasets are collected by, e.g., automatically extracting speech and conversations from TV, radio, Facebook, YouTube, and other resources, as well as listened to and recorded by Alexa, Siri, Google, WhatsApp, etc. Indeed, there exist (and are still created) resources prepared manually dedicated to a particular problem [20]; however, as already mentioned, synthesized data may fill in these needs [13].

In the speech area, several applications may be discerned, e.g., speech recognition enabling communication, healthcare assistance, etc.; voice recognition/authentication systems; emotion recognition in speech and singing; voice cloning (testing vulnerability to attack speaker verification system); automatic aging of biometric voice; pronunciation learning by 2L (second language) speakers; automatic diagnosis or computer-aided diagnosis based on speech characteristics retrieved from the patient's voice (voice, speech and articulation disorders, Parkinson disease, dementia, dysarthria, etc.).

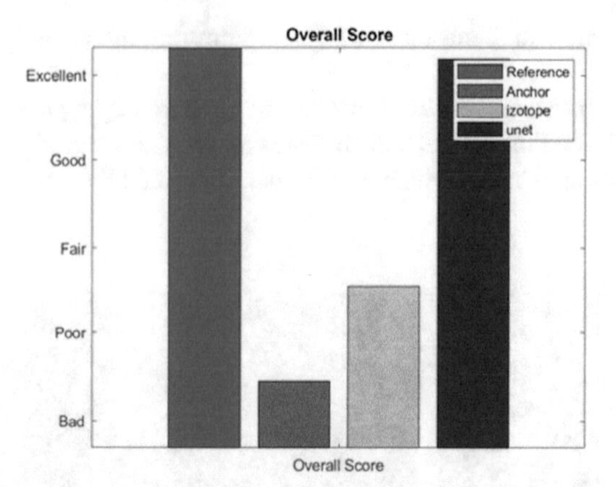

Fig. 7. Outcome of the subjective tests checking the quality of mixes prepared by autonomous deep mode, technology-based (Izotope), anchor (low-quality filtered signal), and the reference mixes, the last one referring to professionally created mix [23].

Voice authentication (VA), i.e., testing vulnerability to attack speaker verification system, based on DeepSpeaker-inspired architecture models using various parametrization approaches, brought high values of accuracy and a low level of equal error rate. The outcome of such a study for voice authentication based on the DeepSpeaker-inspired model, along with various representations, such as VC (vocoder), MFCC (Mel Frequency Cepstral Coefficients); GFCC (Gammatone Frequency Cepstral Coefficients), and LPC (Linear Predictive Coding Coefficients) is shown in Table 1. Depending on what criterion is more important, i.e., equal error rate (EER) or the number of epochs (each epoch took between 7 and 30 min), one may optimize the approach to VA.

Table 1. Outcomes of a DeepSpeaker-inspired model, along with various representations, such as VC (vocoder), MFCC (Mel Frequency Cepstral Coefficients); GFCC (Gammatone Frequency Cepstral Coefficients), and LPC (Linear Predictive Coding Coefficients).

Representation/model	F-score	Accuracy	EER	Epochs
VC model (MFCC)	0.875	0.997	0.0208	895
MFCC	0.784	0.8641	0.0829	400
GFCC	0.732	0.8132	0.1378	400
LPC	0.741	0.8216	0.0936	400

3 Conclusions

Challenges that could be identified within audio technology are related to the role of human factors such as, for example, the user's personality and experience, emotions in the user's models, and personalized services. Emotions are one of the most important aspects of interpersonal communication, as spoken words often – in addition to their content – contain additional, more deeply hidden meanings. Recognizing emotions, therefore, plays a crucial role in accurately understanding the interlocutor's intentions in all human-computer (and vice versa) technology. When searching for the keyword "emotion recognition in speech" on Google in December, the number shown was 17,500,000 results; today, as of January 8th, the value increased by almost 2 million. This shows the extremely high and growing importance of this issue, which can also be observed within the scientific community [24, 25].

Moreover, speech signal contains phonemic variation, temporal structure, prosody, timbre, and voice quality. It also includes various aspects of the speaker's profile. State-of-the-art methods employ deep learning to recognize all these components in audio signals. One may say that what is easily discerned and analyzed by a human may no longer escape an ML-based approach, as this is already happening.

Finally, the author hopes that this paper is another voice in the discussion regarding whether this is already the stage when algorithms gain wisdom on their own.

References

1. Liew, A.: DIKIW: data, information, knowledge, intelligence, wisdom and their interrelationships. Bus. Manage. Dyn. **2**, 49–62 (2013)
2. Rowley, J.: The wisdom hierarchy: representations of the DIKW hierarchy. J. Inf. Sci. **33**(2), 163–180 (2007). https://doi.org/10.1177/0165551506070706
3. Tuomi, I.: Data is more than knowledge: Implications of the reversed knowledge hierarchy for knowledge management and organizational memory. J. Manage. Inf. Syst. **16**(3), 103–117 (1999)
4. Wood, A.M.: The wisdom hierarchy: From signals to artificial intelligence and beyond. A framework for moving from data to wisdom. https://www.oreilly.com/content/the-wisdom-hierarchy-from-signals-to-artificial-intelligence-and-beyond/.Accessed 29 Dec 2022
5. Barlow, M.: Learning to Love Data Science, 2015, O'Reilly Media, Inc., ISBN: 9781491936580. Accessed 29 Dec 2022
6. Mahmood, I., Abdullah, H.: WisdomModel: convert data into wisdom. Appl. Comput. Inform. (2021). https://doi.org/10.1108/ACI-06-2021-0155 https://www.emerald.com/insight/content/doi/https://doi.org/10.1108/ACI-06-2021-0155/full/html
7. Van Meter, H.J.: Revising the DIKW pyramid and the real relationship between data. Inf. Knowl. Wisdom, Law, Technol. Humans **2**, 69–80 (2020). https://doi.org/10.5204/lthj.1470
8. Jankowski, A., Skowron, A., Swiniarski, R.: Interactive rough-granular computing in wisdom technology. In: Yoshida, T., Kou, G., Skowron, A., Cao, J., Hacid, H., Zhong, N. (eds.) AMT 2013. LNCS, vol. 8210, pp. 1–13. Springer, Cham (2013). https://doi.org/10.1007/978-3-319-02750-0_1
9. Skowron, A., Jankowski, A.: Towards W2T foundations: interactive granular computing and adaptive judgement. In: Zhong, N., Ma, J., Liu, J., Huang, R., Tao, X. (eds.) Wisdom Web of Things. WISEITBS, pp. 47–71. Springer, Cham (2016). https://doi.org/10.1007/978-3-319-44198-6_3
10. Vandeput N.: A Brief History of Neural Networks from Data Science for Supply Chain Forecasting. https://medium.com/analytics-vidhya/a-brief-history-of-neural-networks-c234639a43f1. Accessed 29 Dec 2022
11. New Navy Device Learns By Doing; Psychologist Shows Embryo of Computer Designed to Read and Grow Wiser. https://www.nytimes.com/1958/07/08/archives/new-navy-device-learns-by-doing-psychologist-shows-embryo-of.html. Accessed 29 Dec 2022

12. Leung, K.: How to Easily Draw Neural Network Architecture Diagrams. https://towardsdatascience.com/how-to-easily-draw-neural-network-architecture-diagrams-a6b6138ed875. Accessed 29 Dec 2022
13. Korzekwa, D., Lorenzo-Trueba, J., Drugman, T., Calamaro, S., Kostek, B.: Weakly-supervised word-level pronunciation error detection in non-native English speech. In: INTERSPEECH (2021). https://doi.org/10.21437/interspeech.2021-38
14. Leung, W.-K., Liu, X., Meng, H.: CNN-RNN-CTC based end-to-end mispronunciation detection and diagnosis. In: ICASSP, pp. 8132–8136 (2019)
15. Unquestioned assumptions to imperfect data. https://heyday.xyz/blog/research-project-challenges. Accessed 29 Dec 2022
16. Moffat, D., Sandler, M. B.: Approaches in intelligent music production. Arts (8), 5, 14, September (2019)
17. De Man, B., Reiss, J.D.: A knowledge-engineered autonomous mixing system. In: Audio Engineering Society Convention 135 (2013)
18. Martinez-Ramírez, M.A., Benetos, E., Reiss, J.D.: Automatic music mixing with deep learning and out-of-domain data. In: 23rd International Society for Music Information Retrieval Conf. (ISMIR), December (2022). https://doi.org/10.3390/app10020638
19. Celma, O., Herrera, P., Serra, X.: Bridging the music semantic gap, In: Bouquet P, Brunelli R, Chanod JP, Niederée C, Stoermer H, editors. In: Workshop on Mastering the Gap, From Information Extraction to Semantic Representation, With the European Semantic Web Conference; Budva, Montenegro Jun 11–14 (2006)
20. Kostek, B.: Towards searching the holy grail in automatic music and speech processing - examples of the correlation between human expertise and automated classification. In: Signal Processing: Algorithms, Architectures, Arrangements, and Applications (SPA), p. 16 (2022). https://doi.org/10.23919/SPA53010.2022.9927877
21. Slakh | Demo site for the Synthesized Lakh Dataset (Slakh). http://www.slakh.com/. Accessed 29 Dec 2022
22. Blaszke, M., Kostek, B.: Musical instrument identification using deep learning approach. Sensors **22**(8), 3033 (2022). https://doi.org/10.3390/s22083033
23. Koszewski, D., Görne, T., Korvel, G., Kostek B.: Automatic music signal mixing system based on one-dimensional Wave-U-Net autoencoders. EURASIP, 1 (2023). https://doi.org/10.1186/s13636-022-00266-3
24. Khalil, R.A., Jones, E., Babar, M.I., Jan, T., Zafar, M.H., Alhussain, T.: Speech emotion recognition using deep learning techniques: a review. IEEE Access **7**, 117327–117345 (2019). https://doi.org/10.1109/ACCESS.2019.2936124
25. Konangi, U.M. Y., Katreddy, V. R., Rasula, S. K., Marisa, G., Thakur, T.: Emotion recognition through speech: a review. In: 2022 International Conference on Applied Artificial Intelligence and Computing (ICAAIC), pp. 1150–1153 (2022). https://doi.org/10.1109/ICAAIC53929.2022.9792710

Multimedia Industrial and Medical Applications Supported by Machine Learning

Andrzej Czyżewski(✉)

ETI Faculty, Multimedia Systems Department, Gdańsk University of Technology, 80-233 Gdańsk, Poland
andcz@multimed.org

Abstract. This article outlines a keynote paper presented at the Intelligent Decision Technologies conference providing a part of the KES Multi-theme Conference "Smart Digital Futures" organized in Rome on June 14–16, 2023. It briefly discusses projects related to traffic control using developed intelligent traffic signs and diagnosing the health of wind turbine mechanisms and multimodal biometric authentication for banking branches to provide selected examples of industrial applications of intelligent decision technologies. In addition, the developed medical applications for communicating with the surroundings by unconscious people, advanced analyzing disordered speech, and an advanced non-contact respiratory-circulatory radar are presented, using intelligent data analysis and machine learning.

Keywords: variable speed limits road signs · wind turbine monitoring · multimodal biometric authentication · applications of human-computer interfaces · intelligent monitoring of vital signs

1 Introduction

Scientific and technological solutions intended for industrial and medical purposes have been the focus of the Multimedia Systems Department team at the Gdańsk University of Technology for many years. The results of undertakings of this kind in the past few years have been numerous studies and implementations, for example, intelligent road signs, automatic damage detection in wind turbines, and speech-impaired or unconscious patient communication systems.

A common feature of the projects selected to be outlined later in this paper is the use of intelligent decision technologies to enable the autonomous operation of the developed solutions and systems. Therefore, machine learning based on artificial neural networks, especially recurrent networks and autoencoders, was used.

Since there is not enough space in a short paper to present in detail many projects, a of cited literature was compiled for the Bibliography section. Using the provided list of literature references, interested readers may learn more about the discussed solutions and algorithms, particularly those based on intelligent data processing.

I. Czarnowski et al. (Eds.): KESIDT 2023, SIST 352, pp. 13–22, 2023.
https://doi.org/10.1007/978-981-99-2969-6_2

2 Intelligent Autonomous Road Signs for Adaptive Traffic Control

The developed intelligent road signs communicate the speed calculated based on measuring vehicle traffic and information received from a sequence of similar signs with variable content located along a motorway section connected via a wireless network. A unique feature of road signs is that they can operate autonomously, as the speed limit communicated by the road signs results from the traffic measurements they make. The recommended speed is expressed on an innovative modular electronic display and is transmitted wirelessly to vehicles equipped with the V2X interface (interface to the electronic communication system: vehicle infrastructure).

The development of active road sign designs required the solution of many research and technical problems, such as:

- efficient and weather-independent estimation of traffic parameters using innovative sensors constructed and data processing algorithms developed;
- a method of calculating recommended speeds for various traffic situations taking into account road topology;
- construction of a reliable wireless network;
- testing of innovative modular displays and autonomous power supply devices constructed;
- testing of prototypes in real traffic conditions.

The main objective of developing intelligent autonomous traffic signs is to prevent stacking and resulting vehicle collisions in highway and motorway traffic.

The application of the multi-modular structure of the Intelligent Road Sign allows for efficient traffic analysis, independent of weather conditions, performed based on simultaneous analysis of several types of data representation (acoustic, video, and microwave). In the course of research and experimental work, the Doppler microwave radar has been significantly improved by introducing a signal processing algorithm that eliminates noise and parasitic echoes, making it possible to measure vehicle speed more precisely [1, 5].

In addition, an original acoustic sensor was invented and developed that allows both the measurement of vehicle speed and the tracking of vehicle movement and facilitates their generic classification. The essence of the sensor's solution is its calibration method, the performance of which is essential to obtain accurate sound intensity indications [1, 2, 1–2].

Considering that a video camera is also used, several intelligent image processing methods have been developed and tested, which help determine safe speed under current road conditions. For example, an artificial neural network has been trained to detect road surface conditions from image analysis, providing an automatic classification efficiency of over 97% [6]. In addition, a novel experimental endeavor applies a deeply trained artificial neural network to transform an image recorded in visible light (using a regular RGB camera) into images similar to those obtained from thermal imaging cameras [3]. These experiments were conducted in parallel with work on applying a new type of neural network for traffic image analysis, the methodology of which is explained in a paper published in 2020 in the journal Applied Science [7].

3 Intelligent Monitoring of Wind Turbine Health

The project's subject is developing a measuring station for vibroacoustic monitoring of the operation of electrical machinery, particularly wind turbines.

The developed measuring stations use a multimodal set of sensors containing typically used accelerometers, plus an intensity acoustic probe [2, 4] and an inclinometer, the two last mentioned not earlier applied in this context. The measuring stations were included in a system that uses artificial neural network learning for mechano-electric machine vibroacoustic monitoring, especially for early detection of possible faults.

The acoustic probe through the signal transducer and an inclinometer are connected to the signal multiplexer. In turn, the central unit is connected further via a neural network module, i.e., a machine learning decision-making module, for subsequent transmitting acquired warnings through a communication module.

The measurements derived from environmental sensors are converted into electronic digital signals in analog-to-digital converters and transferred to the analytical system. The obtained parameters in the central unit are then classified by selected methods using neural networks advantageously spliced to process spectrograms, which are two-dimensional time-frequency signals [8].

In the decision module, a binary classifier determines the existence of an atypical condition in the monitored device. Classifiers based on intelligent calculation methods result in an algorithm for classifying a given period of turbine operation as typical or atypical, as well as estimating the degree of deviation from the nominal mode of operation of the mechanism. The use of the system makes it possible to perform multimodal analysis of the operation of electrical machinery based on the obtained physical quantities from environmental sensors, which affects the achievement of increased technical safety and environmental safety, especially in the environment of large-scale wind turbines.

For example, excessive bearing or bushing wear usually results in noise that reaches the microphones of an intensity acoustic probe, which is capable, according to its operating principle, of recording this noise along with the ability to determine the direction from which the noise is coming precisely. Thus, despite the complex machine emitting many sounds, it is possible to collect useful acoustic material for training the neural network, including the acoustic signal and data on the direction of its arrival [5]. The use of multiple accelerometers attached at different points of a running machine also provides vibroacoustic data along with information determining where it is emitted. After extracting distinctive features, feature vectors can be downloaded to train the artificial neural network on the cloud servers. After training, the synaptic weights of the network obtained as a result of the training are transferred to the measuring station's built-in hardware neural network module. Which will thus gain the ability to detect machine states that deviate from the long-term vibroacoustic and tilt norm [8]. Based on examining the output state of this built-in neural network, the decision-making module of the measuring station can distinguish between the accepted normative state and the pre-failure or failure states of the electric machine.

4 Multi-biometric Authentication of Banking Clients

The Multimedia Systems Department of Gdańsk University of Technology team has worked with the largest Polish bank in biometrics for many years. As a result, an experimental multimodal biometric workstation was recently constructed and replicated in 10 copies installed in bank branches. This workstation allowed us to investigate different banking biometric modalities' usability extensively. The biometric system includes a biometric pen constructed to verify handwritten signatures, 2D and 3D cameras, proprietary facial recognition software for processing 2D and 3D images, an audio chain that works with an artificial neural network to verify the speaker, and a hand-vein sensor. Moreover, a scanner of ID documents and a gaze tracker enable contactless control of the biometric authentication process. It also provides yet another biometric modality based on the analysis of saccadic movements. We named this testbed "Biometric Island" (the photo of it is included in Fig. 1). Embedding the software using intelligent algorithms for biometric fusion in the computing cloud is also an essential project outcome [9].

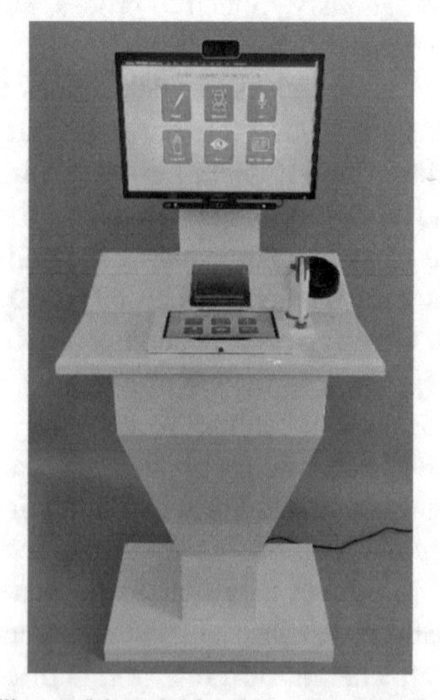

Fig. 1. The "Biometric Island" – multimodal banking stand engineered for convenient access to facial biometrics, voice, and hand vein modality, also employing ID scanning and gaze tracking-based control functions.

The experimental work at the first stage employed 7166 banking clients allowing validation of developed methods [10]. More recently, an automated analysis method for the dynamic electronic representation of handwritten signature authentication was researched. The developed algorithm is based on the dynamic analysis of handwritten signatures, employing a neural network to extract meaningful features from signatures acquired with an engineered electronic pen [11].

The original results of the project also include testing the resistance of the speaker authentication system to voice cloning attacks [12].

5 HCIBRAIN - Assessment of Impaired Cognitive Functions

The main research goal implemented under the project was to develop the concept and methodology for processing data obtained from Human-Computer Interfaces (HCI), enabling the assessment of cognitive functions of patients after brain injuries, using eye-tracking interfaces, electroencephalographic (EEG) helmets, and in some separate cases, electrodes implanted to the brain. The tool software developed as part of the project relates to several groups of diagnostic and therapeutic tasks.

A the beginning it was proved that the results of the subjective assessment of patient's condition used in neurology (GCS) correlate with the results obtained using the developed human-computer interfaces (EGT and EEG), thanks to which it is possible to objectify assessments of patients with cerebral palsy using modern information technologies [13].

The progressing study aimed to assess the level of awareness using objective eye-tracking measures during sessions involving test subjects in performing computer exercises. The results obtained based on the method developed during the creation of the system were compared with the results of standard neurological examinations of patients [14]. According to the first and subsequent diagnoses made during a standard neurological examination, one of these patients should have been in a vegetative state and the other in a coma. In contrast to the currently used subjective neurological scales, the developed method allowed mapping the patient's state of consciousness on the scale of real numbers, thus in an objectified way.

The research also aimed to study reading comprehension skills in patients awake from a coma who remain in a state of reduced consciousness. Eye-tracking technology has also been used to achieve this goal. The obtained results showed that people awakened from a coma, remaining in a state of reduced consciousness, could read with comprehension but had difficulty recognizing errors in the written text [15]. Furthermore, the obtained results made it possible to formulate recommendations regarding developing human-computer interfaces based on eye-tracking intended for people with awareness deficits [16].

Current global trends in consciousness research are concerned with detecting and analyzing the activity in many parts of the brain in the high-frequency range of electrical waveforms, i.e., gamma bands, high gamma bands, ripples, and in fast ripples. Since there is no possibility of transcranial recording of these frequencies, intracranial recording using electrodes placed directly in brain structures is used. Leading research in this area has been carried out by a member of our team, Dr. Michal Kucewicz, a specialist employed by the Multimedia Systems Department working with the Mayo Clinic (Rochester, USA). The results indicate that conscious activities: perception, memory, and recall of images from memory, are associated with increased high-frequency oscillations in visual, limbic, and higher cortical regions. Research in this area is currently being conducted in the aspect of mechanisms of recalling stimuli and information: visual, auditory, and verbal (verbal), based on cooperation with patients with deeply implanted electrodes in the structures of the hippocampus, amygdala body, and near the visual, auditory and motor cortex - on average up to 100 electrodes per person. Implantable electrodes and eye-tracking are also used in these studies as an interface to communicate with implanted patients [17].

6 Methodology of Polymodal Allophonic Speech Transcription

The project aimed to carry out research forming the basis for developing methods for automatic speech phonetic transcription (in English) based on the use of a combination of information derived from the analysis of audio and video signals.

In particular, basic research was conducted on the relationship between allophonic differentiation in speech, i.e., differences of the same sounds resulting from the different arrangement of articulating organs depending on the phonetic environment (i.e., neighboring phones or prosodic features) and objective signal parameters. In the focus of the researchers' attention were also the parameters of the speech signal (acoustic and vision) characteristic of Poles learning English, including those acquired using the mouth movement system, which allows obtaining additional data to deepen the analysis relating to how to pronounce sounds. In addition, the research considered the application of the technology under the development of a solution to benefit people with speech impairments. In particular, the goal was to create a computer bot in the future that could replace a speech therapist in the context of conducting tedious exercises for correct speech articulation.

The assumption was to develop detailed analysis methods that would allow for the differentiation of minor allophonic and accent variations. As a result of the conducted research, it was shown that thanks to the combined analysis of video and audio signals, phonetic speech transcription can be carried out with greater accuracy than using only the acoustic modality, as described in previous works available in the literature. In-depth research on the diversity of sounds in the context of auditory and visual signal parameters contributed to the advancement of the state of the art in audio-visual speech recognition and, thus, in the field of human-computer interaction. In the course of the research, the following hypotheses were experimentally verified: 1. the combined analysis of acoustic and vision data improves the efficiency of phonetic transcription of speech at the allophonic level. 2. the method of speech signal analysis allows a deeper analysis of advanced phonetic aspects of speech compared to the previous state of knowledge. 3. allophonic aspects, such as nazalization, rounding of vowels, aspiration, and characteristic features of lateral allophones, can be effectively detected by analyzing video signals. 4. differences in the speech signal resulting from allophonic and accent variations can be modeled using appropriate mathematical tools and recognized using machine learning methods. The practical result of the project is also the developed database of recordings, available at the address: http://www.modality-corpus.org/. The MODALITY database consists of over 30 h of multimodal recordings, including stereoscopic video streams of high resolution and audio signals obtained from the matrix of the microphone and the internal microphone embedded in a notebook-class computer [18, 19]. The corpus can be used to develop the AVSR (audio-visual speech recognition) systems employing machine learning because each statement was manually annotated [20, 21]. The Modality corpus extension, Alofon Corpus, contains audio-visual and face motion capture data [22]. The audio folder contains the audio tracks recorded by both the directional microphone and the Lavalier microphone and files for each camera system recording and capturing facial muscle movements.

7 CyberRadar

Researchers at Gdańsk University of Technology have developed a prototype contactless device useful in the fight against Long Covid-19 and in diagnosing patients with various respiratory and circulatory problems. Furthermore, to enable the development of research into new methods of monitoring the vital signs of patients with various types of diseases, the device, named CyberRadar, has successfully undergone clinical trials at the Medical University of Gdańsk. In particular, research on the optimization of algorithms underlying CyberRadar technology was realized in 2022, the results of which were published in the Nature Scientific Reports journal [24], following a year earlier publication relating to applications of cognitive methods in the field of respiratory rate quantification and abnormal pattern prediction [23].

The contactless device for monitoring respiratory and circulatory functions is presented in Fig. 2.

Fig. 2. The "CyberRadar" – the developed device for non-contact monitoring of respiratory and circulatory functions. The microwave sensor is located at the front of the housing under a recess. The protruding cylinder is a camera to facilitate patient positioning. On the top panel is an LCD display that first shows the camera image or graphs showing respiratory and circulatory data.

The technology and methodology for analyzing respiratory and circulatory patterns related to CyberRadar, once it reaches implementation in the form of a medical product, will enable patient monitoring:

- in research: through assessment of respiratory rhythm and respiratory patterns
- those diagnosed in hospital emergency departments,
- hospitalized - automatic detection of patient deterioration,
- treated at home (remote monitoring of the patient's condition),
- treated through telehealth and telecare,
- diagnosed and monitored with heart failure, lung disease, and after a stroke. Supplementing the solution with artificial intelligence makes it possible to detect breathing abnormalities.

The developed solution passed the stage of clinical trials, data analysis, and algorithm optimization completed in 2022. Consequently, the innovative non-contact monitoring of a patient's vital signs is now ready for the medical market, and its use in telemedicine and telecare, with applied data processing algorithms, has been optimized during clinical trials [24].

8 Conclusions

Of the many applications of technology based on sound or image processing, in other words, multimedia technology, being developed at Gdańsk University of Technology, six have been selected for presentation, which has been the subject of recent projects.

They represent industrial applications, such as deployed smart road signs, a wind turbine monitoring system, and a "Biometric Island" for installation in bank branches.

The keynote paper prepared for the Intelligent Decision Technologies conference also presents three diagnostic and therapeutic applications. The topics of two of them are related to two developed concepts of HCI interfaces. The first is concerned with assessing the condition of people who show no signs of consciousness, and the second is helpful in prospective robots used in speech therapy. Finally, the newest project is briefly described, which has resulted in the development of a non-contact device for monitoring patients' breathing and circulation, which finds applications in diagnosing a range of diseases, including long-covid symptoms.

It was impossible to discuss the details of developed solutions in this short paper. However, interested readers can find them in the numerous publications compiled in the Bibliography section. The reader of the cited papers and articles will note that a common feature of all the solutions described in these publications is intelligent decision-making algorithms based on machine learning.

Acknowledgments. A part of the presented research was subsidized by the Polish National Centre for Research and Development (NCBR) from the European Regional Development Fund within project No. POIR.01.01.01-0092/19 entitled: "BIOPUAP - a biometric cloud authentication system."

References

1. Czyżewski, A., Kotus, J., Szwoch, G.: Estimating traffic intensity employing passive acoustic radar and enhanced microwave Doppler radar sensor. Remote Sens. **1**, 110 (2019). https://doi.org/10.3390/rs12010110
2. Czyżewski, A., Kotus, J., Szwoch, G.: Intensity probe with correction system. Polish patent No. 236718 (2021)
3. Cygert, S., Czyżewski, A.: Style transfer for detecting vehicles with thermal camera. In: 23rd International Conference on Signal Processing: Algorithms, Architectures, Arrangements, and Applications, SPA, Poznań (2019). https://doi.org/10.23919/SPA.2019.8936707
4. Kotus, J., Szwoch, G.: Calibration of acoustic vector sensor based on MEMS microphones for DOA estimation. Appl. Acoust. **141**, 307–321 (2018). https://doi.org/10.1016/j.apacoust.2018.07.025
5. Czyżewski, A., et al.: Comparative study on the effectiveness of various types of road traffic intensity detectors. In: 6th International Conference on Models and Technologies for Intelligent Transportation Systems, pp. 1–7 (2019). https://doi.org/10.1109/MTITS.2019.8883354

6. Grabowski, D., Czyżewski, A.: System for monitoring road slippery based on CCTV cameras and convolutional neural networks. J. Intell. Inf. Syst. **55**(3), 521–534 (2020). https://doi.org/10.1007/s10844-020-00618-5
7. Cygert, S., Czyżewski, A.: Vehicle detection with self-training for adaptative video processing embedded platform. Appl. Sci. **10**(17), 1–16 (2020). https://doi.org/10.3390/app10175763
8. Czyżewski, A.: Remote health monitoring of wind turbines employing vibroacoustic transducers and autoencoders. Front. Energy Res. **10**, 858958 (2022). https://doi.org/10.3389/fenrg.2022.858958
9. Szczuko, P., Harasimiuk, A., Czyżewski, A.: Evaluation of decision fusion methods for multimodal biometrics in the banking application. Sensors **22**, 2356 (2022). https://doi.org/10.3390/s22062356
10. Hoffmann, P., Czyżewski, A., Szczuko, P., Kurowski, A., Lech, M., Szczodrak, M.: Analysis of results of large-scale multimodal biometric identity verification experiment. IET Biometrics **8**, 92–100 (2018). https://doi.org/10.1049/iet-bmt.2018.5030
11. Kurowski, M., Sroczyński, A., Bogdanis, G., Czyżewski, A.: An automated method for biometric handwritten signature authentication employing neural networks. Electronics **10**, 456 (2021). https://doi.org/10.3390/electronics10040456
12. Zaporowski, S., Czyzewski, A.: Investigating speaker authentication system vulnerability to the limited duration of speech excerpts and voice cloning. J. Acoust. Soc. Am. **148**(4), 2768–2768 (2020). https://doi.org/10.1121/1.5147706
13. Szczuko, P.: Real and imaginary motion classification based on rough set analysis of EEG signals for multimedia applications. Multimed. Tools Appl. **76**(24), 25697–25711 (2017). https://doi.org/10.1007/s11042-017-4458-7
14. Lech, M., Kucewicz, M.T., Czyżewski, A.: Human computer interface for tracking eye movements improves assessment and diagnosis of patients with acquired brain injuries. Front. Neurol. **10**(6), 1–9 (2019). https://doi.org/10.3389/fneur.2019.00006
15. Kwiatkowska, A., Lech, M., Odya, P., Czyżewski, A.: Post-comatose patients with minimal consciousness tend to preserve reading comprehension skills but neglect syntax and spelling. Sci. Rep. **9**(19929) 1–12 (2019). https://doi.org/10.1038/s41598-019-56443-6
16. Lech, M., Czyżewski, A., Kucewicz, M.T.: CyberEye: new eye-tracking interfaces for assessment and modulation of cognitive functions beyond the brain. Sensors **21**(22), 1–7 (2021). https://doi.org/10.3390/s21227605
17. Worrell, G.A., Kucewicz, M.T.: Direct electrical stimulation of the human brain has inverse effects on the theta and gamma neural activities. IEEE Trans. Biomed. Eng. **68**(12), 3701–3712 (2021). https://doi.org/10.1109/TBME.2021.3082320
18. Czyzewski, A., Kostek, B., Bratoszewski, P., Kotus, J., Szykulski, M.: An audio-visual corpus for multimodal automatic speech recognition. J. Intell. Inf. Syst. **49**(2), 167–192 (2017). https://doi.org/10.1007/s10844-016-0438-z
19. Korvel, G., Kurowski, A., Kostek, B., Czyzewski, A.: Speech analytics based on machine learning. In: Tsihrintzis, G.A., Sotiropoulos, D.N., Jain, L.C. (eds.) Machine Learning Paradigms. ISRL, vol. 149, pp. 129–157. Springer, Cham (2019). https://doi.org/10.1007/978-3-319-94030-4_6
20. Piotrowska, M., Korvel, G., Kostek, B., Ciszewski, T., Czyżewski, A.: Machine learning-based analysis of English lateral allophones. Int. J. Appl. Math. Comput. Sci. **29**(2), 393–405 (2019). https://doi.org/10.2478/amcs-2019-0029
21. Piotrowska, M., Czyżewski, A., Ciszewski, T., Korvel, G., Kurowski, A., Kostek, B.: Evaluation of aspiration problems in L2 English pronunciation employing machine learning. J. Acoust. Soc. Am. **150**(1), 120–132 (2021). https://doi.org/10.1121/10.0005480
22. Kawaler, M., Czyżewski, A.: Database of speech and facial expressions recorded with optimized face motion capture settings. J. Intell. Inf. Syst. **53**(2), 381–404 (2019). https://doi.org/10.1007/s10844-019-00547-y

23. Szczuko, P., et al.: Mining knowledge of respiratory rate quantification and abnormal pattern prediction. Cogn. Comput. **14**(6), 2120–2140 (2021). https://doi.org/10.1007/s12559-021-09908-8
24. Czyżewski, A., et al.: Algorithmically improved microwave radar monitors breathing more accurate than sensorized belt. Sci. Rep. **12**, 14412 (2022). https://doi.org/10.1038/s41598-022-18808-2

Main Track

Accessibility Measures and Indicators: A Basis for Dynamic Simulations to Improve Regional Planning

Victoria Kazieva[1(✉)], Christine Große[1], and Aron Larsson[1,2]

[1] Department of Communication, Quality Management, and Information Systems, Mid Sweden University, 851 70 Sundsvall, Sweden
{victoria.kazieva,Christine.Grosse,Aron.Larsson}@miun.se
[2] Department of Computer and System Sciences, Stockholm University, 164 07 Kista, Sweden

Abstract. The purpose of this paper is to define indicators and measures of accessibility that can support decision-makers in designing policies for sustainable regional development. The study identifies aspects of accessibility that may influence the attractiveness of certain areas for population and infrastructure investments, notably sparsely populated regions that lay outside urban environments. The paper's findings identify which accessibility indicators are relevant to the peculiarities of sparsely populated areas thereby defining the specific system requirements for possible simulation models. Such models can generate outcomes of policy scenarios that may be used to evaluate the consequences of policy-making. To support this objective, the study identifies the theoretical basis of accessibility indicators and their modeling potentials that address the complexity of the investigated domain. This paper is an important step toward developing a feasible modeling tool capable of generating policy-based scenarios that support effective decision-making in the context of regional planning.

Keywords: Accessibility Measures · Accessibility Indicators · Simulation Modeling · Decision-making · Regional Development

1 Introduction

Within Europe, sparsely populated regions have long been suffering from population decline, and efforts for improving population growth have been on the agenda for several decades [1]. Given the restricted opportunities available for such remote areas, investment and infrastructural resources are often limited. Therefore, it is important to identify high-priority needs in these areas to avoid the consequences of poor planning, which can have long-term effects that are hard to reverse [2]. Accessibility plays an important role in regional planning as it reflects the efficiency and effectiveness of a transport system. A deeper understanding of correlations between infrastructure, accessibility, and sustainable development can enable decision-makers to make the best use of the resources available [3].

I. Czarnowski et al. (Eds.): KESIDT 2023, SIST 352, pp. 25–34, 2023.
https://doi.org/10.1007/978-981-99-2969-6_3

This study focuses on accessibility and its role in regional development. The complex multidimensional concept of accessibility is known to comprise 4 components; transport (e.g., transport mode), land-use (e.g., infrastructure), temporal (e.g., opening hours), and individual (e.g., personal needs) [4, 5]. Attempts to redefine accessibility often lead to crude simplification or neglect certain important components of accessibility and their impact. As a result, the concept has been ignored to varying degrees by many decision-makers [5].

With respect to the field of systems science, the paper also considers the opportunities that simulation tools can provide to improve the knowledge about hidden correlations within the concept of the accessibility itself, between its components and processes. Simulation of a system modeling accessibility can be used to visualize the consequences of infrastructure developments as well as disclose dependencies critical for regional planning [6]. Standard analytical measures that disregard these factors of accessibility may generate intuitively understandable choices, but they risk neglecting the complex dynamics of the real-life system that the policy decisions target [7].

Despite technological progress and new possibilities for processing large volumes of data, there is still little practical application of advanced analytical tools that decision-makers use to evaluate different aspects of accessibility and their potential role to support planning [8]. Techniques that deploy complicated calculations and models are typically neither intuitive nor obvious, which leads to that they are often neglected [9]. Consequently, while the concept of accessibility is gaining more attention and the need to include it in regional planning is becoming more compelling, the gap between theory and real-life application persists. This lack of practical use often stems from the absence of models that can include aspects vital for accessibility and present them to decision-makers in a simple, easily interpretable way.

In this respect, simulation modeling can bridge the gap. As a visual medium that is responsive to complex process integration, it can be used as a complementary tool for planning to generate possible outcomes of a particular implemented policy. It can also improve understanding of the complex behaviors within a system and over time [10]. This is specifically useful when dealing with imprecise concepts such as accessibility when a link between cause and effect cannot be clearly defined for real-life dependencies for system processes, or when the correlations between input components and the actual impact of policy decisions are hard to ascertain.

To redress this situation, this paper focuses on the capability of existing modeling methods to support an efficient application of the accessibility concept and how it can be deployed in simulation tools to improve decision-making related to strategic regional planning. Through a review of the current state-of-the-art, this paper contributes to a better understanding of the gaps in this research field.

2 A System Theory Perspective

A deep understanding of accessibility complexity and its interdependencies is important for integrating the notion in the planning of a real-life system. Since the measures, indicators, and even evaluation criteria of accessibility are often explored from a certain perspective (e.g., economic benefits) and are generally not clearly defined [5], the key is to investigate the dynamics of the system that accessibility is embedded in and to explore how the behavioral patterns of a real-world environment change over time. This knowledge allows for a conscious choice that is crucial for generating feasible decisions.

When dealing with problem-solving in a complex environment, it is essential to include all known characteristics to enable proper decision-making about the system as a whole [7]. Problems arise if some of the system variables or behaviors are not clearly defined or are liable to change according to circumstance. This is true for accessibility which can have varying application and utility points (e.g., improved distribution of job opportunities), which suggests that the dynamic of the system can vary depending on the application goal and the focused characteristics of accessibility.

However, complex systems rarely behave in a linear, intuitive, or obvious cause-and-effect manner, generating easily interpretable results that decision-makers could rely on [11]. Complex systems typically create outcomes that provide further information about the problem which, if considered, can influence further actions and their outcomes, and so on. This generates a learning loop that feeds adjusted data back to the system and runs iteratively, both causing and reacting to dynamic changes.

The adjustment caused by data update from the learning loop can give rise to rather counterintuitive, unpredicted patterns of system behavior. For instance, the least favored policy may generate the most successful accessibility simulation scenario.

Nevertheless, studies show that such complexity is often ignored as policy-makers are prone to draw a decision based on a single cause event; in real life, in contrast, the consequences of a taken decision relate not only to the known system components and processes but to multiple side effects that are often neglected due to their unpredictable nature and a lack of knowledge about them [7]. This complexity implies a potentially large number of nonlinear dependencies that cannot be easily captured with known analytical methods but are nevertheless important contributors to behavioral patterns and should therefore be not excluded from system analysis [7, 11].

It is therefore important to examine the current state of the art in the field of accessibility and measures of evaluating it to support decision-makers with knowledge about the underlying behaviors of the complex system.

3 Method

The knowledge base on accessibility is constantly expanding as the concept receives more attention. Although the concept has evolved considerably from a simple measure of mobility in the past 45 years or more [12], there is still little consensus on what "accessibility" denotes and how it is to be measured. Thus, two main criteria have been used to identify research that is relevant to this study, namely:

- Studies that elaborate on the notion of accessibility and how it is evaluated.
- The most recent studies that include either a comprehensive review of the concept or practical tools for its application.

Based on these criteria, a literature search was performed on Scopus and Google Scholar resulting in 16 scientific papers in the field of accessibility evaluation. The papers mainly comprise peer-reviewed journal articles, but also conference proceedings, discussion papers, and a research report.

A further distinction has also been made between two commonly used notions:

- *Accessibility indicators* provide an important descriptive measure of spatial structure and performance [12].
- *Accessibility measures* define these indicators' impact on accessibility and how this impact can be evaluated (or measured).

The analysis focuses on representations of accessibility and their practical application. Of special interest is their suitability for simulation modeling approaches. Accessibility can be viewed as a tool for public decision-making in terms of attracting people and enterprises and identifying feasible infrastructure investments in sparsely populated remote areas. Indicators of accessibility should thus relate to the effectiveness of the transport system, the reachability of goods, services, and opportunities, and contributions to regional development.

4 Results

4.1 Requirements for Simulation Modeling

The definition of accessibility depends on its intended application [12]. Therefore, possible simulation scenarios of various decisions taken to improve accessibility call for the selection and evaluation of accessibility indicators that reflect the objectives of every policy-making process of planning for accessibility.

Simulating the system that accessibility is embedded in and including factors such as the influence of stakeholders and system process correlation (e.g., the influence of the number of job opportunities on the availability of healthcare) requires a certain level of system abstraction for simulation modeling [13] to handle real-world complexity. A simulation must represent aspects of the system that are crucial for the current modeling purpose while omitting details that do not reflect substantial behavioral characteristics of the system.

The level of abstraction appropriate for the scope of the accessibility application must take account of system components (here accessibility indicators), their dependencies (e.g., trends, behavioral patterns of the system processes, etc.), and the system actors (e.g., the population of a chosen area).

The simulated system must therefore reflect three major points:

- An appropriate level of abstraction.
- Easily interpretable results (e.g., visualization) for the model to be compelling enough for the end user (decision-makers).
- Support for representing the individual behaviors of system objects [10].

Two existing simulation modeling approaches for non-physical dynamic systems can meet these needs to varying degrees: System Dynamics (SD) and Agent-Based Modeling (ABM). The main advantage of these methods is that they capture a global dynamic of a system even if there is no known process trend at the system's top level. The ABM approach incorporates data at an individual level of active objects (i.e., agents), capturing their separate traits if at least some specific behavior is known [10], such as population mobility data. Adding SD features allows for more efficient model design whenever such agents communicate via environments, dynamically enabling certain patterns of iterations to show up over time at the simulation runtime. While ABM allows for building basic dependencies between stakeholders, SD enables the detection of correlations within the system at a higher level where notions of imprecise concepts, such as accessibility indicators, are not always measurable. However, even if empirical or statistical data is not available, these variables still must be a part of a model whenever their participation in the real-life system is predicted by observations or other implications [7].

4.2 Representations of Accessibility

Within its main field of application – transportation and sustainable development – accessibility is often referred to as "the physical access to goods, services, and destinations" [14]. However, as recent studies have emphasized, improved accessibility positively influences multiple aspects of human lives, such as the environment, public health, employment, and social inclusion. Thus, the conceptual shift from mobility toward "the potential for interaction" [4], which includes the efficacy of transportation and land-use to provide such potential, becomes central in accessibility research. It implies a need for relevant new tools that can capture dependencies arising from a complex approach. However, even well-known analytical methods, such as accessibility maps, still face challenges of which accessibility indicators to choose and how to evaluate them [4]. The issue is that no unified approach for choosing or measuring these indicators is available yet.

One of the reviewed papers provides an overview of 18 studies on sustainable transport system indicators [15]. Although the paper is restricted to sustainability in urban areas, the authors present 555 accessibility indicators. These variables to some extent either overlap or are unrelated to each other, which demonstrates an urgent need for a systematic approach. Geurs & Wee [5] suggest such an approach and classify the measures for evaluating accessibility. A summary of these measures and their characteristics is presented in Table 1.

These measures are mainly based on case-specific assumptions tied to urban areas. However, accessibility patterns not only differ in rural areas, but their impact is even greater because less densely populated regions can be more sensitive to infrastructural changes [1], which implies that poorly chosen investments in shrinking rural municipalities might incur greater expenses instead of aiding area development.

Table 1. The four major accessibility measures and their features.

Measures	Role	Evaluation Example	Advantage	Disadvantage
1. Infrastructure-based	Analyses the functioning and efficiency of the transport system	Travel times, level of congestion, etc	Data availability, ease of interpretation, and evaluation	Ignores the impact of land-use and changes in spatially distributed activities
	Is often used in transport planning and project evaluation			
2. Location-based: 2.1 Distance measures	Analyses the level of availability of spatially distributed activities – for more than 1 opportunity cumulative measures apply	The number of opportunities such as available jobs within a 30 min travel time; the distance between two points, etc	Data availability, ease of interpretation, and evaluation	Disregards competition effect (e.g., available jobs, schools); does not differentiate between the attractiveness of opportunities – all are weighted equally
	Is often used in geographical studies and urban planning			
2.2 Potential accessibility (gravity measures)	Analysis of the level of the potential availability of distributed activities – the closer the opportunity, the greater the influence	Apply an impedance function to evaluate opportunities in another zone vs the given one	Unlike the distance measure, it includes the effects of land-use, transport, and personal perceptions	Disregards competition effect (available jobs, schools, etc.) and temporal constraints (availability of opportunities at different times); not easily interpretable
	Is often used in geographical studies, urban planning, and evaluation of transport projects			
3. Person-based	Analysis of the level of accessibility of opportunities from an individual's perspective	Mapping individual/household travel patterns of personal choices regarding the perception of the availability of opportunities	Fills in behavioral gaps of other measures for better land-use evaluation	Complex for evaluation. Disregards long-time perspectives, competition effect, and restrictions such as activity duration
	Is mainly applicable to social-economic assessment			
4. Utility-based	Analysis of benefits from accessing spatially distributed activities	Evaluate the probability of choosing one opportunity over another, e.g., the logit model	Evaluation of transport and economic systems	Complex for evaluation and interpretation
	Is often used for social and economic evaluations and includes personal choice patterns			

This peculiarity calls for a certain shift in focus to identify which aspects of accessibility are most relevant for regional development and infrastructural investments outside of urban areas. In addition, individual perceptions of accessibility can play a role as a measure of region attractiveness both for the population and enterprises. However, this pillar of accessibility remains understudied as personal influences are subject to change and thus can only be ambiguously defined [4, 16]. Moreover, the clustering of enterprises to share the potential or specialized labor within a particular area has been proven effective for urban settings, which implies that even urbanized rural regions could benefit from such concentrations [17]. We now proceed with identifying accessibility indicators that are beneficial from this perspective, concerning the special characteristics and needs of a sparsely populated area and its inhabitants.

4.3 Accessibility Measures and Indicators

As the literature indicates, most accessibility indicators are not compliant with the criteria set for their comprehensive evaluation [5, 12] unless accompanied by complimentary ones. Interpretability and data availability requirements seem to ignore some measures due to their complexity. However, certain modeling methods, such as SD, allow for integrating them as variables.

To support decision-making in sustainable regional planning, it should be considered that accessibility is often measured by its actual state rather than the potential one [12]. Yet comparing both states can be used in simulation modeling to inform decision-makers about the consequences of a taken decision, and to signal which policies could generate a more effective outcome.

Based on these considerations with sustainability playing an essential role, the following measures (including environmental impact) are characterized in Fig. 1:

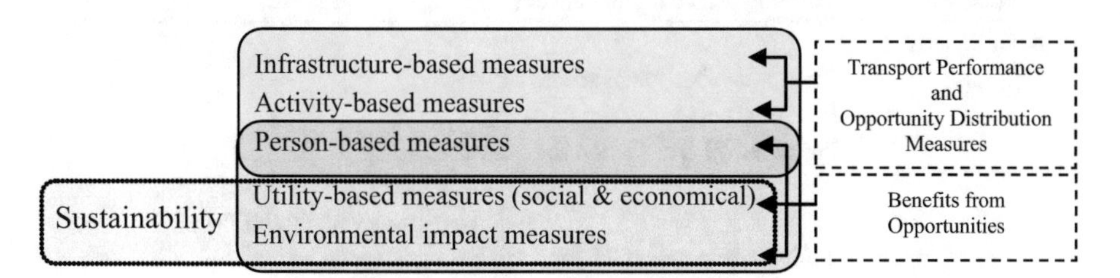

Fig. 1. Structure of accessibility measures

Despite the established pattern of measures and the numerous indicators available, only a few studies take advantage of such simulation modeling, mainly exploring distinctive networks (e.g., railways and airports) or indicators that focus specifically on accessibility in an urban context. These studies include evaluations based on the transportation disadvantage impedance index [18] for urban application; research into transportation network directness [3]; and even partial incorporation of SD loops for accessing the impact of a land-use and transportation interaction [19].

This review specifically considers accessibility evaluation criteria [5, 18, 20, 21]. The most relevant ones, according to the requirements discussed, are sensitivity to infrastructural or transportation system changes within the context of regional development. This means that the chosen indicators must reflect the region's attractiveness for enterprises and people and signal its critical infrastructure needs.

An important part of the review process was to assess which indicators could be aligned with the defined measures and thus to identify those indicators that would best reflect transport performance efficiency or demonstrate the benefits of opportunities or their spatial distribution. The key aim was to ensure that each indicator demonstrates the advantage of the group of measures (see Table 1) to which it belongs and therefore contributes to measuring accessibility when selected for the final set of indicators.

The environmental indicators are placed separately. This is done to measure the direct impact of the environmental factors on the system and to ensure their influence is considered properly in simulations.

The person-based indicator is not a single measurable unit as none of the existing indicators demonstrate the actual travel preferences, so mapping of personal choices must be done for accurate evaluation.

Based on the review, Table 2 presents the final selection of accessibility indicators.

Table 2. Regional accessibility indicators per measure category

Measure	Accessibility indicators	
1. Infrastructure-based	Distance to nearest supply point or PT (Public Transport) stop	
	Congestion level	
	Travel speed	
2. Activity-based	Availability of opportunities within a certain time threshold	
	Availability of opportunities in another zone vs the defined time threshold	
3. Person-based	Defining personal preferences (mapping travel patterns)	
4. Utility-based		
Social	Transport-related fatal accidents	
	Ease of access to nearest PT stop	
	Perceived PT satisfaction for	car-users
		non-car-users
		non-car-users with disabilities
	Taking advantage of PT for participating in social activities	
	Medical and childcare	
	Public safety	
	Education possibilities	
	Work opportunities	
Economical	GDP	
	Housing affordability	
	Travel costs	
	Local prices	
	Satisfaction with current income vs expenditure	
	Satisfaction with rates of wage growth vs expected spending	
5. Environmental impact	Air pollution	
	Non-renewable energy consumption	
	Renewable energy availability	

5 Concluding Remarks

This paper introduces accessibility measures and indicators as a basis for supporting decision-makers with context-relevant information at the policy design stage. It focuses on reviewing the potential of existing methods to ensure an efficient application of the accessibility concept and how it might be deployed with simulation and modeling tools to improve strategic regional planning. The findings presented above, lay a theoretic basis for an actual simulation of accessibility. They introduce a set of indicators that evaluate accessibility based on a comprehensive approach rather than by cherry-picking indicators due to their measurability and/or linearity, as typical of much of the literature reviewed.

Despite the absence of a straightforward tool for evaluating the effectiveness of a policy before its implementation [18], simulation modeling suggests ways to fill gaps in a low-data setting. By combining principles of ABM and SD, the suggested approach allows researchers to model "living" real-life systems that reflect hidden interdependencies and include non-linear, counter-intuitive features that are otherwise difficult to capture and model. The suggestion is to introduce available data about the objects and their behavior and let the system behavior emerge through multiple simulations, showing the advantages or drawbacks of policy decisions regarding these indicators of accessibility. For instance, high levels of congestion indicate low accessibility, while a short distance to the nearest stop signals a high level of accessibility. The challenge is to choose the appropriate abstraction level from the beginning due to a lack of concrete empirical data.

Further research can use the knowledge provided by this study to generate scenarios and evaluate the usability, usefulness, and useworthiness of the accessibility indicators, defining the outcomes of simulation modeling. Such simulation constitutes a substantial support tool [22] to enable planners to make sustainable and meaningful choices.

References

1. Grundel, I., Magnusson, D.: Planning to grow, planning to rock on – infrastructure management and development in shrinking municipalities. Eur. Plan. Stud. (2022). https://doi.org/10.1080/09654313.2022.2108311
2. López, E., Gutiérrez, J., Gómez, G.: Measuring regional cohesion effects of large-scale transport infrastructure investments: an accessibility approach. Eur. Plan. Stud. **16**(2), 277–301 (2008). https://doi.org/10.1080/09654310701814629
3. Ertugay, K.: A simulation-based accessibility modeling approach to evaluate performance of transportation networks by using directness concept and GIS. ICONARP Int. J. Archit. Plan. **7**(2), 460–486 (2019). https://doi.org/10.15320/ICONARP.2019.93
4. Lättman, K., Olsson, L.E., Friman, M.: A new approach to accessibility – examining perceived accessibility in contrast to objectively measured accessibility in daily travel. J. Transp. Econ. **69**, 501–511 (2018)
5. Geurs, K.T., Van Wee, B.: Accessibility evaluation of land-use and transport strategies: review and research directions. J. Transp. Geogr. **12**, 127–140 (2004). https://doi.org/10.1016/j.jtrangeo.2003.10.005

6. Kumagai, S., Gokan, T., Isono, I., Keola S.: Predicting Long-Term Effects of Infrastructure Development Projects in Continental South East Asia: IDE Geographical Simulation Model, ERIA Discussion Paper Series, Economic Research Institute for ASEAN and East Asia, Jakarta (2008)
7. Sterman, J.: Business Dynamics: Systems Thinking and Modeling for a Complex World. Irwin/McGraw-Hill, Boston (2000)
8. Miller, E.: Measuring Accessibility: Methods and Issues, International Transport Forum Discussion Papers, No. 2020/25, OECD Publishing, Paris (2020)
9. Straatemeier, T.: How to plan for regional accessibility? Transp. Policy **15**(2), 127–137 (2008). https://doi.org/10.1016/j.tranpol.2007.10.002
10. Borshchev, A., Filippov, A.: From system dynamics and discrete event to practical agent based modeling: reasons, techniques, tools. In: The Proceedings of the 22nd International Conference of the System Dynamics Society, Oxford, England (2004)
11. Forrester, J.: Some Basic Concepts in System Dynamics (2009). https://www.cc.gatech.edu/. Accessed 16 Dec 2022
12. Morris, J., Dumble, P., Wigan, M.: Accessibility indicators for transport planning. Transp. Res. Part A Gen. **13**(2), 91–109 (1979). https://doi.org/10.1016/0191-2607(79)90012-8
13. Crooks, A., Heppenstall, A.: Introduction to agent-based modelling. In: Heppenstall, A., Crooks, A., See, L., Batty, M. (eds.) Agent-Based Models of Geographical Systems. Springer, Dordrecht (2012). https://doi.org/10.1007/978-90-481-8927-4_5
14. Saif, M., Zefreh, M., Torok, A.: Public transport accessibility: a literature review. Periodica Polytech. Transp. Eng. **47**(1(Jan. 2019)), 36–43 (2019). https://doi.org/10.3311/PPtr.12072
15. Buzási, A., Csete, M.: Sustainability indicators in assessing urban transport systems. Periodica Polytech. Transp. Eng. **43**(3(Jan. 2015)), 138–145 (2015). https://doi.org/10.3311/PPtr.7825
16. Toth-Szabo, Z., Várhelyi, A.: Indicator framework for measuring sustainability of transport in the city. Procedia Soc. Behav. Sci. **48**, 2035–2047 (2012). https://doi.org/10.1016/j.sbspro.2012.06.1177
17. Naldi, L., Nilsson, P., Westlund, H., Wixe, S.: What is smart rural development? J. Rural. Stud. **40**, 90–101 (2015). https://doi.org/10.1016/j.jrurstud.2015.06.006
18. Duvarci, Y., Yigitcanlar, T., Mizokami, S.: Transportation disadvantage impedance indexing: a methodological approach to reduce policy shortcomings. J. Transp. Geogr. **48**(C), 61–75. Elsevier (2015). https://doi.org/10.1016/j.jtrangeo.2015.08.014
19. Wang, Y., Monzón, A., Di Ciommo, F.: Assessing the accessibility impact of transport policy by a land-use and transport interaction model -the case of Madrid. Comput. Environ. Urban Syst. **49**, 126–135 (2015). https://doi.org/10.1016/j.compenvurbsys.2014.03.005
20. Litman, T.: Developing indicators for comprehensive and sustainable transport planning. Transp. Res. Rec. (2007).https://doi.org/10.3141/2017-02
21. Haghshenas, H., Vaziri, M.: Urban sustainable transportation indicators for global comparison. Ecol. Ind. **15**(1), 115–121 (2012). https://doi.org/10.1016/j.ecolind.2011.09.010
22. Larsson, A., Ibrahim, O.: Modeling for policy formulation: causal mapping, scenario generation, and decision evaluation. In: Tambouris, E., et al. (eds.) ePart 2015. LNCS, vol. 9249, pp. 135–146. Springer, Cham (2015). https://doi.org/10.1007/978-3-319-22500-5_11

Image-Multimodal Data Analysis for Defect Classification: Case Study of Industrial Printing

Hiroki Itou[1], Kyo Watanabe[1], and Sumika Arima[2](✉)

[1] Degree Program in Systems and Information Engineering, University of Tsukuba, Tsukuba, Japan
{s2120409,s2120467}@u.tsukuba.ac.jp

[2] Institute of Systems and Information Engineering, University of Tsukuba, Tsukuba, Japan
arima@sk.tsukuba.ac.jp

Abstract. The final goal of this study is an integrated approach of walk-through from the classification of defect products to the feature estimation for preventing the cause of the defects in real-time or proactively. Particularly, this paper introduced the multimodal data analyses to classify the defect images of final products more accurate with inspection logs and factory process data in integrated manner.

Particularly, this paper focused on the industrial printing case in which the tiny and faint defects are difficult to be classified accurately. Motivation of this study is to clarify the possibility of image-wise classification instead of pixel-wise semantic segmentation of high annotation cost as well as low accuracy in partial of the similar shape classes in our previous study. It was introduced and numerically evaluated that various data augmentation as pre-process for imbalanced and small samples issue, image-multimodal data analyses with inspection log and/or the factory data, and ensemble of the multimodal and non-multimodal networks. As the result, the maximum accuracy of multimodal analyses is 79.37% of the model with test log and process data with 10 times augmented data. In addition, a confidence-based ensemble model with conditional branch results more accurate, 81.22% in summary of all classes. It is better than segmentation approach of pixel-wise in the previous study. Moreover, importance of additional variables is visualized as cause after multimodal analysis as aimed.

Keywords: multimodal analysis · defect patterns · classification · ensemble

1 Introduction

One of important study issues of 21st century is designing and managing sustainable production systems of material-energy-greenhouse gas emission conscious for the best batton passed to the next generation. Mass defective products and scraps or reworks are the worst manners of the production because it is only for consuming the material and energy as well as increasing greenhouse gas.

In the example of industrial printing, there is a rule that one roll (eg standard 4000 m) contains more than a specified number of defects, it will be returned as defective. This is due to the fact that they are used for packaging of products related to food, housing (e.g.

I. Czarnowski et al. (Eds.): KESIDT 2023, SIST 352, pp. 35–47, 2023.
https://doi.org/10.1007/978-981-99-2969-6_4

wallpaper), advertisements, and so on, and the quality is demanded to give customers an image of high quality. However, this is not a knowledge or standard that has been confirmed based on actual consumer sensibilities and purchase data. In fact, there are different in rules between countries and regions, and there is a room for reviewing industry standards and practices. In addition, the production side criteria also has a big impact, such industrial printing can be re-produced at high speed of about 30 min per roll, and it is faster than rejecting defective parts from the roll. As the result, large amounts of waste are generated daily. Even after environmental measures progressing all over the world, there is still demand for minimum necessary packaging and wallpaper for safety and health regardless of the materials used. In any case, to collect and utilize detail information on defects and preventing the defect occurrence are common industrial issues [1].

The main goal of this study is an integrated approach of walk-through from the classification of defect products to the feature estimation for preventing the cause of the defects in real-time or proactively. Particularly, this paper will introduce the multimodal data analytics in manufacturing process to integrately analyze the classification of defect patterns of final product quality and the causal variables of the defects collected at each process steps. Many papers are piled in more feasible image classification topics [2–6], however, discussion should shift to the defect cause analyses. That is because the classification itself does not reduce the occurrence of the defects and the consumption of materials and energy.

There are three stages to summarize the possibilities and limitations of the multi-modal data analyses. The first stage is the accuracy improvement of the classification of the multi-class defect patterns and their superpositions at low cost (small samples) by using additional feature/data besides image data. The second stage is adaptive selections and estimations of important features including interactions within a practical computational time and accuracy. The third stage is an integrated approach of walk-through from the classification to the feature estimation for proactively preventing the causes of the defect. This paper will mainly discuss the first stage and the preparation of the third stage, as the first report. On the other hand, the second issue of high-dimensional interaction modelling is not presented in this paper because of the paper volume and on-going patent matter.

A combination of an image and texts is a typical data set of image-multimodal data and the research is progressing [7, 8]. On the other hand, about analyses of product defect and its causes, multimodal data across product inspection and manufacturing processes is generally difficult to obtain due to confidentiality and complexity. At the present, there are few studies of defect cause analysis using big data of all process steps, except for Nakata and Orihara [9] and their related works. Moreover, their research is also limited to specify (rank) manufacturing equipment and processes for each defect pattern class. It is still on-going issue about more detail variables to be detected and controlled. Each defect is produced in one or more process steps physically or chemically. Based on that nature, we will propose and evaluate the image-multimodal data analysis.

This paper consists of 5 sections. Section 2 will introduce the dataset and previous studies. Section 3 will introduce a proposal method of a multimodal analyses and its evaluations. Section 4 will present the second proposal method of ensemble models of multimodal analysis and image data augmentations and its evaluations. Section 5 concludes this paper with summary and future works.

2 Objectives and Previous Studies

This paper presents a case study to improve the accuracy of automatic classification by image-multimodal data analysis. Data of industrial printing, gravure printing in particular, will be mainly used in this study.

2.1 Data Set

The data of this study includes the defect images [1] and variables acquired in production steps as well as test log of a final inspection of a printing fab. The first data of printing defects (Fig. 1(a) Right) is images taken by line cameras of industrial use. Each pixel has RGB values as the same as a standard image data. Difficulty here is the faint and tiny defect on a rich and various printing background, as well as imbalanced samples (Fig. 1(a) Left). This study particularly focuses on more precise classifications of 3 types of similar shape defects (Fig. 1(b)) by multimodal data analyses instead of human domain knowledge and pixel-wise annotations. For example, it is particularly difficult to discriminate two classes of line shape defects only by the defect image (Low accuracy even in pixel-wise segmentation), however skilled workers can judge the difference by using information of production environment in addition to the defect images [1].

Therefore, beside the image data, a test log and process environment data (Fig. 3(c)) are applied for the automatic classification by the multimodal analyses in this study. The test log includes attributes of tested images in mechanically detection of the defects such as a position and size of a defect on the printing film roll (Fig. 1(c) Upper part). Moreover, the variables of process environment have been measured in the printing process (Fig. 1(c) Bottom part) such as printing speed, ink viscosity as well as the wear of blade which is used to scrap ink (Fig. 2, ink scraping blade). Next is more description about printing process which relates to the multimodal data. Gravure printing is one major kinds of intaglio printing processes, which uses an engraved plate as an image carrier (Fig. 2). Each printing unit by color is composed of an engraved cylinder, an ink pan of the color, an impression roll and an ink scraping blade, and the image on the cylinder is transferred as a film passes between the cylinder and the roll. The gravure printing can realize very high-speed production (a few hundreds of meters per minute), and its application has become wider for various printed matter for industrial use.

A defect inspection step of the printing film is usually placed at the end of the printing process since sometimes defects such as ink splashes and linear scratches can occur during printing. The mainstream of the current inspection process is a method in which a person in charge determines whether candidates detected by image subtraction processing is a defect or not through visual inspection. There were two main problems with this method. One is that the relationship with the cause and the defects is unknown since candidates of defects are evaluated only by those size and color differences. The other is that the judgment highly depends on knowledge and experience of each skilled worker and may not be unified.

Some approaches including Auto Encoder and Deep Learning has studied to overcome these problems [10, 11]. However, those tend to require a template image to train model before each printing process, and thus not suited for flexible manufacturing system. Therefore, firstly we proposed a user-friendly annotation tool and a classification approach based on a pixel-wise semantic segmentation in our previous study [1]. However, as the result, two issues are clarified on heavy workload and variance of pixel-wise annotation (drawing) besides still no information of defect causes for prevention. Therefore, as the next step, we reconstruct the classification model of image-wise instead of the pixel-wise, as well as progress the acquisition and utilization of multimodal data which include candidates of the various defect cause.

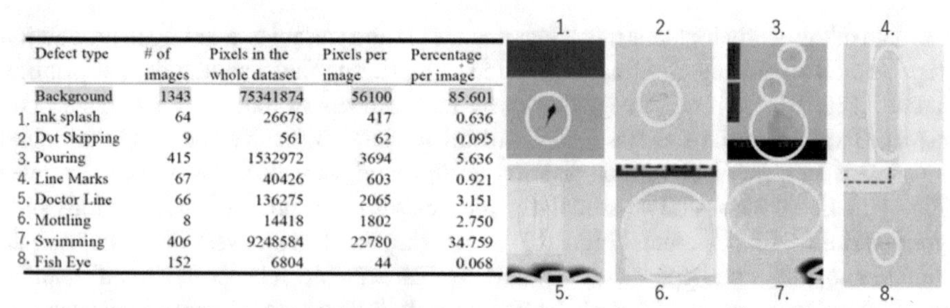

Defect type	# of images	Pixels in the whole dataset	Pixels per image	Percentage per image
Background	1343	75341874	56100	85.601
1. Ink splash	64	26678	417	0.636
2. Dot Skipping	9	561	62	0.095
3. Pouring	415	1532972	3694	5.636
4. Line Marks	67	40426	603	0.921
5. Doctor Line	66	136275	2065	3.151
6. Mottling	8	14418	1802	2.750
7. Swimming	406	9248584	22780	34.759
8. Fish Eye	152	6804	44	0.068

(a)8 typical defect classes, unbalance samples (Tsuji, et.al., 2020)

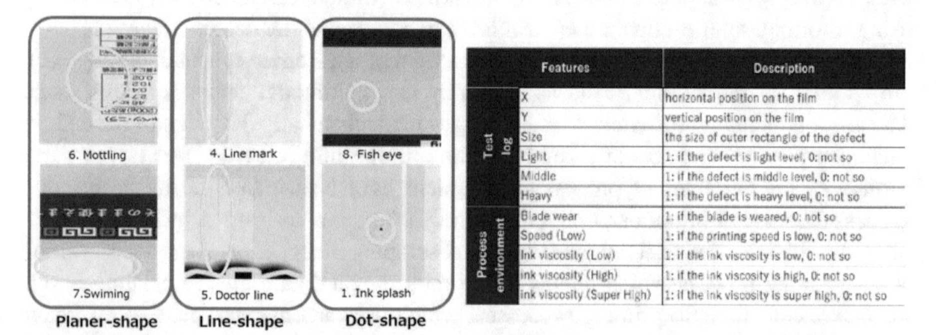

	Features	Description
Test log	X	horizontal position on the film
	Y	vertical position on the film
	Size	the size of outer rectangle of the defect
	Light	1: if the defect is light level, 0: not so
	Middle	1: if the defect is middle level, 0: not so
	Heavy	1: if the defect is heavy level, 0: not so
Process environment	Blade wear	1: if the blade is weared, 0: not so
	Speed (Low)	1: if the printing speed is low, 0: not so
	Ink viscosity (Low)	1: if the ink viscosity is low, 0: not so
	Ink viscosity (High)	1: if the ink viscosity is high, 0: not so
	Ink viscosity (Super High)	1: if the ink viscosity is super high, 0: not so

(b) 3 types shape of similar classes (c) Outer variables of the multimodal

Fig. 1. Data set.

2.2 Previous Studies

There are three specific issues to summarize the possibilities and limitations of the multimodal data for the analyses of defects and its causes. The first is the improvement of the classification accuracy of the defect patterns of multi-class and superpositions at lower annotation cost. The second issue is appropriate selections and estimations of important features including interactions and variety in a practical computational time and accuracy. Third issue is an integrated approach of walk-through from the

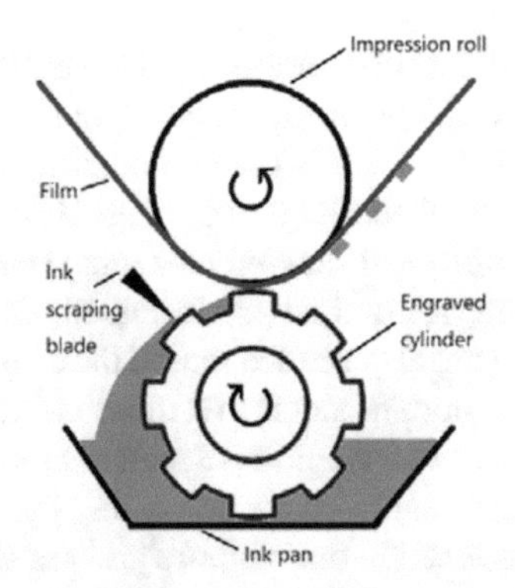

Fig. 2. Conceptual diagram of printing unit

classification to the feature estimation for proactively preventing the causes of the defect. This paper is focused on the first and the third stage a bit as the first report.

Now, for the first issue of the classification accuracy improvement, model buildings and pre-processes are mainly discussed in general [12, 13]. As the model building, for the industrial applications recent years, various CNN (Convolutional Neural Network) models are applied. However, Typical methods of the object detection and classifications cannot perform well such as R-CNN [14] and its derivation as well as SSD (Single Shot Multibox Detection), Auto-Encoder (AE), and so on because of the tiny and faint defect on various and rich background of print films [1].

Moreover, there are few studies about industrial printing. Some approaches including AE and Deep Learning has studied to over-come these problems [10, 11]. However, those tend to require a template image to train model before each printing process, and thus not suited for flexible manufacturing system. Tsuji et al. proposed and evaluated the pixel-wise semantic segmentation for the classification of faint and tiny defects on rich background [1]. As main process, simple segmentation models together with transfer learning of VGG16 is used. In concrete, based on SegNet [15], U-Net [16], and PSPNet [17], three networks SegNet-VGG16 (96 layers), UNet-VGG16 (42 layers), PSPNet-VGG16 (52 layers) are compared. The pixel-wise segmentation was selected in our previous study other than object detection and classification.

Pre-process are additional input channels and simple data augmentation of 30 times at least by random cropping of 384 * 384 px images from the original images (256 * 256 px) at least 30 images per class. In addition, the study also adopted Focal loss [18] as the loss function and Dice coefficient [19] modified as an evaluation function for multiclass segmentation of tiny defects. As the post-process, active learning of the Margin sampling, Least Confident, and Entropy Sampling are compared, and improve the learning efficiency by 75% at the number of images when the best combination of the Margin sampling and Bayesian SegNet with dropout. In this case 90 original images per class augmented by 30 times.

The approach of the previous research is fine under their motivation [1], but there are two problems. One is the high cost and variance of drawing annotation for the pixel-wise segmentation. There are also some problems in a classification by image. The common and unique factors between defect class are still hidden though model network becomes more complex and thus more efforts of model building is needed. That is more serious in resent superposition classifications [6]. The second problem is the lower accuracy and confidence of the classification of similar shapes (Fig. 1(c)). Particularly, it is known by veterans that those classification needs not only the image but also the information about the condition of process, material, and fab environment, that is the domain knowledge and sensing. Moreover, those are also needed to find the causes of defects in real-time or proactively. Here, the most important fact is that the classification itself does not reduce the occurrence of defects and the consumption of materials and energy. Discussion should shift to include the defect cause analyses for preventing the defects. That is because the nature that each defect is produced in one or more process steps physically, chemically, or mechanically.

As summery of this section, the classification not by the pixel but by image still has a room to study under a context of the multimodal data analyses. A combination of an image and texts is a typical data set of image-multimodal data and the research is progressing [7, 8]. On the other hand, about analyses of the product defect and its causes, multimodal data across product inspection and production processes are generally difficult to obtain due to confidentiality and complexity. In this study, we use the data collected in a company-university collaboration research for model evaluations.

3 Proposal Method #1: Multimodal Analyses

For the reasons mentioned in Sect. 2.2, this study aims accurate classification on classes by image without drawing pixel-wise annotation. Instead of the pixel-wise annotation, two kinds of data besides the image are used for higher accuracy and domain knowledge alternation. Note that the specification of machines used in this study is as follows: CPU: Intel(R) Core ™ i9 10900X (3.7/tb4.5 GHz), Memory: 256 GB RAM, GPU NVIDIA TITAN, RTX 4608cores 24 GB, OS: Ubuntu18.04LTS.

3.1 Pre-process

Deep convolutional neural networks have performed remarkably well on many Computer Vision tasks. However, these networks are heavily reliant on big data to avoid overfitting. Unfortunately, many application domains do not have access to big data, such as medical image analysis. The pre-process of data augmentation (DA) can improve the performance of their models and expand limited datasets to take advantage of the capabilities of big data [12]. DA encompasses a suite of techniques that enhance the size and quality of training datasets such that better Deep Learning models can be built using them.

Various and rare is the same in production defects as the medical data problem. This study firstly applies DA as a solution to the problem of limited data. 8 types of DA techniques are evaluated: flip and rotate of geometric transformations, crop, random zoom, colour shift, random erasing, mixing images (mixed-up), auto-contrast, and their all combination (Table 1(a)). The effect to the actual industrial printing data is evaluated after general effects of DA are evaluated by using image open data set of CIFAR-10 [20].

As the result of CIFAR-10 evaluations, each DA techniques are effective to improve the classification accuracy (Table 1(b)). However, for the actual case of Case1, a half of the techniques lead positive or no effects, but the rest half results negative effects (Table 1(c)). That is mainly because the features of defect image of very faint (low colour difference) and/or tiny (a few dozen pixels) is disappeared by DA. Note that the 10-layer CNN, batch size 128, epoch 300 is used for those basic DA evaluations.

In summary of the basic evaluation of 8 DA techniques, only three DA techniques of flip, crop, and rotate are effective for the printing defect case, and only a combination of those are applied the evaluations in the following sections. Concretely, random sampling and data augmentation of 5 or 10 times (DAx5, DAx10) are used.

Table 1. Eight DA techniques and effects (Case 1).

(a)Basic 8 DA techniques (Shorten, 2019)

No.	DA technique	Description
1	flip	image is fliped in horizontal/vertical axis
2	crop	crop up/down,left/right 4px
3	rotate	rotete of plus or minus 10°
4	random zoom	zoom at range [75, 125]%
5	colour shift	shift the colour in range [-50, 50]
6	random erasing	erase a rectangle of size 4*12px at 50% probability
7	mixed-up	images are mixed according to Beta(0.5,0.5) distribution
8	auto-contrast	auto contrast is done at 50% probability
9	all combination	All combinations of 1-8 DA techniques

(b) DA effects to CIFAR-10 data

No.	DA technique	Accuracy	Difference	Note
	None (Original)	78.09	–	–
1	flip	82.42	+4.33	effective for patterns with direction
2	crop	82.63	+4.54	effective to imbalanced location of objects
3	rotate	81.47	+3.38	Approproate range in which the object does not disappear
4	random zoom	**82.82**	**+4.73**	Effective to clear the shape of small objects
5	colour shift	81.13	+3.04	Effective to similar colour objects to give various colours
6	random erasing	80	+1.91	Effective if bigger objects are similar between images
7	mixed-up	80.46	+2.37	Problem is differences from human perceptions
8	auto-contrast	78.14	+0.05	Effective to the low contrast objects/images
9	all combination	86.06	+7.97	Combination influences to the accuracy

(c) DA effects to actual data of printing defect

No.	DA technique	Accuracy	Difference
	None (Original)	76.67	
1	flip	77.00	+0.33
2	crop	76.80	+0.13
3	rotate	77.40	+0.73
4	random zoom	76.67	0
5	colour shift	63.30	-13.37
6	random erasing	74.00	-2.67
7	mixed-up	70.00	-6.67
8	auto-contrast	-	-
9	all combination	73.30	-3.37

3.2 Proposal and Evaluations: #1-Multimodal Analyses

As the main process, multimodal data analyses and ensemble models are discussed and evaluated.

There are two problems of the previous research [1]. One is high cost and variance of drawing annotation for the pixel-wise segmentation. The second is the lower the accuracy of discrimination of similar shapes (Fig. 1(c)). Therefore, this study aims accurate classification on classes by image without drawing. For accuracy, two kinds of data besides the image are used. One is test log to describe the defect, for example location, size, etc. on the printing film (Table 2(a)). The test log mainly describes based on the features of defect area itself. In addition, the variables of process environment are measured by sensor devices and PLC controller developed for this study and set. These express the condition of jigs, process steps, and materials for the printing.

A succeed multimodal model in previous research [8, 21] ([3, 20]) is applied and extended as the multimodal analyses in this study. Figure 3(a) shows a base model of non-multimodal. Figure 3(b) describes the multimodal "log model" to which 6 test log variables are attached in addition to image data. Figure 3(c) shows the multimodal "env model" with 5 process environment variables in addition to the log model. Those three models are compared. Note that Fig. 3(d) is the same network as Fig. 3(c) and it describes the layer to observe the features effects.

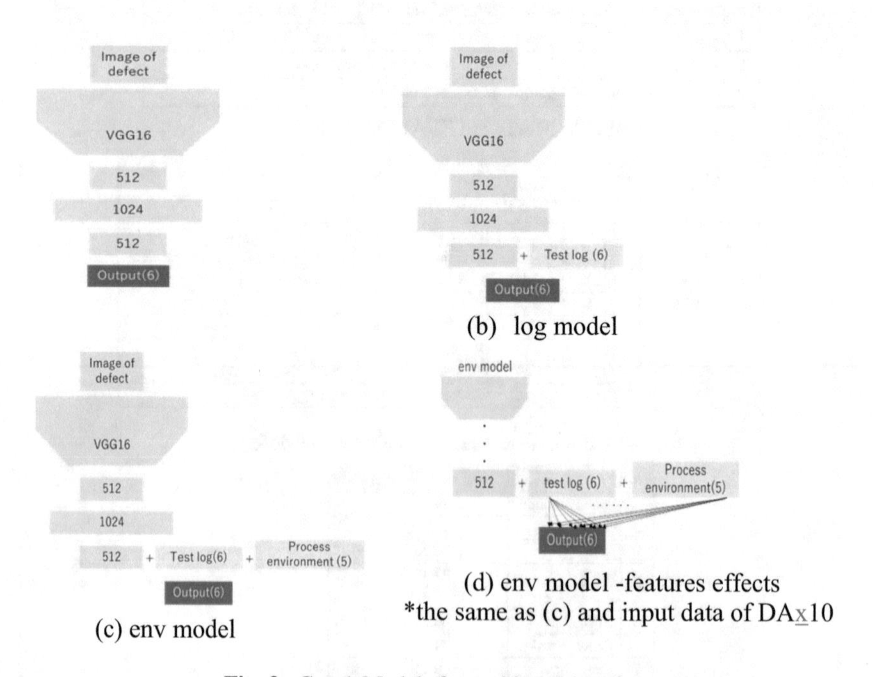

Fig. 3. Case1-Models for multimodal analyses.

For numerical evaluations, 63 defective images for each of 6 classes (Fig. 1(c)) are prepared by random sampling, divided into 75%:25%, and accuracy verification is performed by 4-division cross-validation (called ds1-ds4). Table 2 shows the result of the validation. Here, SGD is selected as the optimization algorithm because it is the best accuracy among SGD, Adam, RMSprop, Adadelta, Adadelta, Adamax, and Nadam.

As the result of Table 2(a)–(c), the maximum accuracy is 79.37% of the log model and env model with 10 times augmented data (DAx10). The accuracy improved by 4.77% from the base model accuracy of the original image. Regardless of DA, it is confirmed that the accuracy was improved by adding information other than images and learning in a multimodal manner. Particularly, the accuracy is significantly improved by about 6.4% (69.8%->76.2%) in "fish eye" class which is the tiniest and so difficult to precisely classify even by the pixel-wise segmentation. In addition to the accuracy, the importance of features the important features can be estimated which relate to the defect occurrence (Table 2(e)). Almost of the features of larger effect are reasonable by the domain knowledge (red box). However, note that important interactions, which is confirmed by the analysis of variance and the feature selection methods, are not yet

Table 2. Comparison of multimodal accuracy.

(a)Outer variables of the multimodal analyses

		Features	Description
Test	log	X	horizontal position on the film
		Y	vertical position on the film
		Size	the size of outer rectangle of the defect
		Light	1: if the defect is light level, 0: not so
		Middle	1: if the defect is middle level, 0: not so
		Heavy	1: if the defect is heavy level, 0: not so
Process	environment	Blade wear	1: if the blade is weared, 0: not so
		Speed (Low)	1: if the printing speed is low, 0: not so
		Ink viscosity (Low)	1: if the ink viscosity is low, 0: not so
		ink viscosity (High)	1: if the ink viscosity is high, 0: not so
		ink viscosity (Super High)	1: if the ink viscosity is super high, 0: not so

(b)Original images only (w/o data augmentation)

ds / mocel	base	log	env
ds1	69.79	69.79	70.83
ds2	70.83	71.88	72.92
ds3	77.08	78.13	81.25
ds4	78.89	80	81.11
average	74.6	74.87	**75.4**

(c) Case with data augmentation (5 times, i.e. DAx5)

ds / model	base	log	env
ds1	70.83	72.92	71.88
ds2	75	76.04	77.08
ds3	76.04	82.29	84.38
ds4	77.78	84.44	81.11
average	74.87	**78.84**	78.57

(d) Case with data augmentation (10 times, i.e. DAx10)

ds / model	base	log	env
ds1	72.92	73.96	76.04
ds2	78.13	78.13	77.08
ds3	80.21	82.29	82.29
ds4	80	83.33	82.22
average	77.78	**79.37**	**79.37**

(e) Important features (env model (Fig.3(d)))

		Features	5	8	1	6	7	4
			Doctor line	Fish eye	Ink splash	Mottling	Swimming	Line mark
Test	log	X	0.34530111	-0.0418744	0.24186233	-0.1346195	-0.0946818	-0.1156791
		Y	-0.2647776	0.30672762	-0.0962601	-0.3203445	0.34470902	0.13413539
		Size (outer rectangle)	-0.0076804	-0.3028154	-0.0380848	0.20273649	0.11385551	-0.015407
		Light	0.04848151	0.22166192	-0.2426282	-0.2224595	0.05229897	0.12562111
		Middle	-0.0306146	-0.0749583	0.05422942	0.00685698	-0.0023277	-0.0115973
		Heavy	-0.0040892	-0.0597156	0.09112348	0.09345266	-0.0254134	-0.0003095
Process	environment	Blade wear	-0.0715181	0.19648102	-0.1537671	0.11988847	-0.0537703	-0.1061544
		Speed (Low)	-0.0100722	0.14297145	0.01723352	-0.0280094	-0.0483365	-0.0496137
		Ink viscosity (Low)	-0.0564801	0.05497759	0.06007322	0.10810236	-0.120326	0.07738396
		ink viscosity (High)	0.09748301	-0.0958344	-0.0161259	-0.01035	-0.0234645	0.03818068
		ink viscosity (Super High)	0.01951562	-0.0563725	0.04256288	-0.0344534	0.0646426	-0.0119871

applied here though there are significant interactions between ink viscosity and the blade wear or the printing speed, etc.

4 Proposal and Evaluations: #2: Multimodal Ensemble Model

In addition, ensemble models are applied for higher accuracy and stability. Figure 4 shows two types of ensemble models (EM1, EM2). EM1 is a simple ensemble as a baseline (Fig. 4(a)), and EM2 is a confidence-based model with conditional branch (Fig. 4(b)). Concretely, if the confidence of the classification is lower than 50% in the best multimodal model (Fig. 4(b) Right part), then non-multimodal models are also used to be ensembled (Fig. 4(b) Left part) for more accuracy and stability. That is sometimes needed because the data augmentation cannot be simply applied the data except images.

As the summary of the results shown in Table 3, EM1 and EM2 improve the accuracy by 0.52% and 1.85% compared with the best performance of non-ensemble model, respectively.

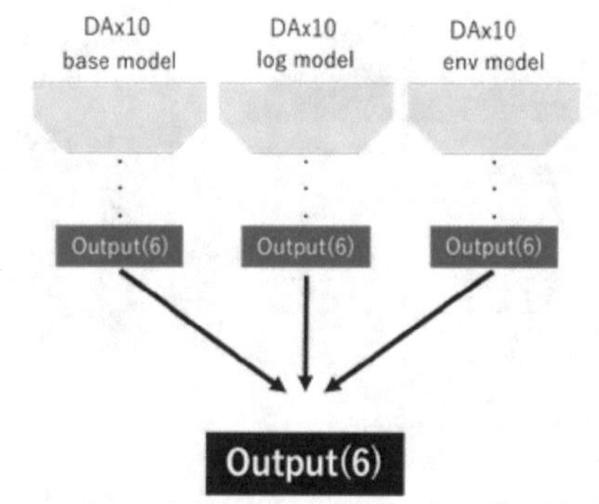

(a)Ensemble Model 1(EM1): simple ensemble

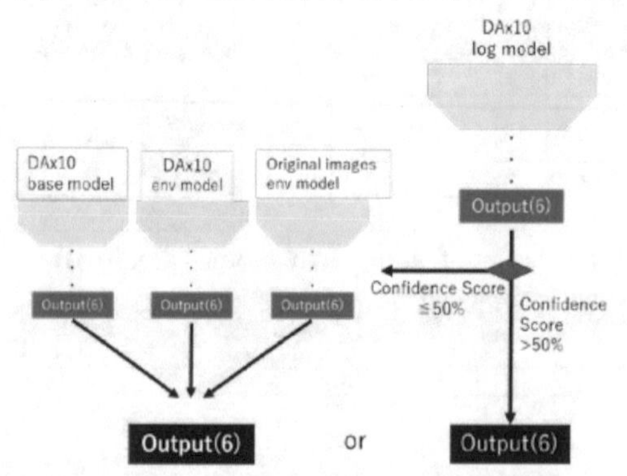

(b) Ensemble Model 2 (EM2): confidence-based model

Fig. 4. Ensemble models

Table 3. Accuracy of the ensemble models.

Ensemble model	Accuracy [%]
EM1	79.89
EM2	81.22

5 Conclusion

This paper presented case study of the accuracy improvement by image-multimodal data analysis, particularly the industrial film printing case. Instead of pixel-wise annotation and semantic segmentation of high cost, the image-wise multimodal data analysis is introduced and evaluated. The image-multimodal data includes defect images with class labels, test logs, and process environment variables measured in an actual fab. Data augmentation is also used because of small samples matter. As the result of full set of the image-multimodal with data augmentation performs the best. Concretely, the accuracy improved by 4.77%. Moreover, ensemble models are applied for higher accuracy and confidence. Two types of ensemble models, EM1 (simple ensemble) and EM2 (confidence-based ensemble) models are evaluated. EM1 and EM2 improve the accuracy by 0.52% and 1.85% compared with the best performance of non-ensemble model. In summary, the enough high accuracy (81.22%) is achieved by multimodal model even by image-wise classification instead of pixel-wise.

This study is still at a start line and will be concluded next few years. Future work about the film printing case is to accomplish an integration model with the interaction modelling of all variables of the fab environmental data such as temperature, humidity, air pressure, vibration, noise, CO_2 emissions, and VOT (volatile gas) emissions in addition to the variables presented in this paper. Each defect is produced in one or more process steps (or between steps) physically, chemically, or mechanically. Based on that nature, we will propose and evaluate the multimodal data analysis essentially for multi-class and superposition [6, 22] in the near future. Contents of this paper is for its preparation. Beyond sensing and knowledge of skilled workers, it will contribute to the automated integrated approach of walk-through from the classification of defect products to the feature estimation for preventing the cause of the defects in real-time or proactively.

Acknowledgements. We thank for all members of the company-university collaboration research. For data collection phase, this research is supported by Cross-ministerial Strategic Innovation Promotion Pro-gram (SIP), "Big-data and AI-enabled Cyberspace Technologies" (Funding Agency: NEDO). In addition, this research has been partly executed in response to support to KIOXIA Corporation. The authors appreciate all the supports.

References

1. Tsuji, T., Arima, S.: Automatic multi-class classification of tiny and faint printing defects based on semantic segmentation. In: Zimmermann, A., Howlett, R.J., Jain, L.C. (eds.) Human Centred Intelligent Systems. SIST, vol. 189, pp. 101–113. Springer, Singapore (2021). https://doi.org/10.1007/978-981-15-5784-2_9

2. Wu, M.-J., Jang, J.-S.R., Chen, J.-L.: Wafer map failure pattern recognition and similarity ranking for large-scale data sets. IEEE Trans. Semicond. Manuf. **28**(1), 1–12 (2015). https://doi.org/10.1109/TSM.2014.2364237
3. Nakazawa, T., Kulkarni, D.V.: Wafer map defect pattern classification and image retrieval using convolutional neural network. IEEE Trans. Semicond. Manuf. **31**(2), 309–314 (2018)
4. Kyeong, K., Kim, H.: Classification of mixed-type defect patterns in wafer bin maps using convolutional neural networks. IEEE Trans. Semicond. Manuf. **31**(3), 395–401 (2018)
5. Jin, C.H., Na, H.J., Piao, M., Pok, G., Ryu, K.H.: A novel DBSCAN-based defect pattern detection and classification framework for wafer bin map. IEEE Trans. Semicond. Manuf. **32**(3), 286–292 (2019)
6. Wang, R., Chen, N.: Detection and recognition of mixed-type defect patterns in wafer bin maps via tensor voting. IEEE Trans. Semicond. Manuf. **35**(3), 485–493 (2022)
7. Duong, C.T., Lebret, R., Aberer, K.: Multimodal Classification for Analysing Social Media (2017). https://arxiv.org/abs/1708.02099. Accessed 28 Dec 2022
8. Woo, L.J., Yoon, Y.C.: Fine-grained plant identification using wide and deep learning model. In: 2019 International Conference on Platform Technology and Service (PlatCon). IEEE (2019)
9. Nakata, K., Orihara, R.: A comprehensive big-data based monitoring system for yield enhancement in semiconductor manufacturing. IEEE Trans. Semicond. Manuf. **30**(4), 339–344 (2017)
10. Hirono, S., Uchibe, T., Murata, T., Ito, N.: Image recognition AI to promote the automation of visual inspections. Fujitsu **69**(4), 42–48 (2018)
11. Tamaki, T.: POODL–Image recognition cloud plat form for printing factory. https://www.slideshare.net/TeppeiTamaki/poodl-a-image-recognition-cloud-platform-for-every-printing-factory. Accessed 22 Jan 2019
12. Shorten, C., Khoshgoftaar, T.M.: A survey on image data augmentation for deep learning. J. Big Data **6**(1), 1–48 (2019). https://doi.org/10.1186/s40537-019-0197-0
13. Johnson, J.M., Khoshgoftaar, T.M.: Survey on deep learning with class imbalance. J. Big Data **6**(1), 1–54 (2019). https://doi.org/10.1186/s40537-019-0192-5
14. Girshick, R., Donahue, J., Darrell, T., Malik, J.: Rich feature hierarchies for accurate object detection and semantic segmentation. In: Proceedings of the IEEE Conference on Computer Vision and Pattern Recognition, pp. 580–587. IEEE, Ohio (2014)
15. Badrinarayanan, V., Kendall, A., Cipolla, R.: SegNet: a deep convolutional encoder-decoder architecture for image segmentation. IEEE Trans. Pattern Anal. Mach. Intell. **39**(12), 2481–2495 (2017)
16. Ronneberger, O., Fischer, P., Brox, T.: U-Net: convolutional networks for biomedical image segmentation. In: Navab, N., Hornegger, J., Wells, W.M., Frangi, A.F. (eds.) MICCAI 2015. LNCS, vol. 9351, pp. 234–241. Springer, Cham (2015). https://doi.org/10.1007/978-3-319-24574-4_28
17. Zhao, H., Shi, J., Qi, X., Wang, X., Jia, J.: Pyramid scene parsing network. In: Proceedings of the IEEE Conference on Computer Vision and Pattern Recognition, pp. 2881–2890. IEEE (2017)
18. Lin, T.Y., Goyal, P., Girshick, R., He, K., Dollár, P.: Focal loss for dense object detection. In: Proceedings of the IEEE International Conference on Computer Vision, pp. 2980–2988. IEEE (2017)
19. Lee, R.: Dice: measures of the amount of ecologic association between species. Ecology **26**(3), 297–302 (1945)
20. Krizhevsky, A.: The CIFAR-10 dataset (2009). https://www.cs.toronto.edu/~kriz/cifar.html. Accessed 12 Feb 2021

21. Cheng, H.-T., et al.: Wide & deep learning for recommender systems. In: Proceedings of the 1st Workshop on Deep Learning for Recommender Systems, pp. 7–10 (2016)
22. Wei, Y., Wang, H.: Mixed-type wafer defect recognition with multi-scale information fusion transformer. IEEE Trans. Semicond. Manuf. **35**(3), 341–352 (2022)

Image-Multimodal Data Analysis for Defect Classification: Case Study of Semiconductor Defect Patterns

Daisuke Takada[1], Hiroki Itou[1], Ryo Ohta[1], Takumi Maeda[1], Kyo Watanabe[1], and Sumika Arima[2(✉)]

[1] Degree Program in Systems and Information Engineering, University of Tsukuba, Tsukuba, Japan
s1911237@u.tsukuba.ac.jp

[2] Institute of Systems and Information Engineering, University of Tsukuba, Tsukuba, Japan
arima@sk.tsukuba.ac.jp, arimalab298@gmail.com

Abstract. The final goal of this study is an integrated approach of walk-through from the classification of defect products to the feature estimation for preventing the cause of the defects in real-time or proactively. Particularly, this paper will introduce the multimodal data analytics in production process integrations to analyze the classification of defect patterns of final product quality and the defect cause impact in integrated manner. There are three steps to summarize the possibilities and limitations of the multimodal data analyses. The first step is the accuracy improvement of the classification of the multi-class defect patterns of images at lower cost by using additional feature/data substituting human-being domain knowledge. That is also to prepare more accurate and effective classifications of superposition pattern of small samples data. The second step is adaptive selections and estimations of important features including interactions and variety within a practical computational time and accuracy. The third step is an integrated approach of walk-through from the classification to the feature estimation for proactively preventing the causes of the defect. As the first report, this paper introduces the first and the third issues through a case study. Particularly, this paper presents the semiconductor defect patterns case.

Keywords: Multimodal analysis · Defect patterns · Classification · Causal variables

1 Introduction

One of important study issues of 21st century is designing and managing sustainable production systems of material-energy-greenhouse gas emission concious for the best batton passed to the next generation. Mass defective products and scraps or reworks are the worst manners of the production because it is only for comsuming the material and energy as well as increasing greenhouse gas. In fact, it is reported that power consumption of cutting-edge semiconductor of the biggest foundry exceeds that of a developing country [1], on the other hand, much defective product and deadstock are produced [2].

I. Czarnowski et al. (Eds.): KESIDT 2023, SIST 352, pp. 48–61, 2023.
https://doi.org/10.1007/978-981-99-2969-6_5

The main goal of this study is an integrated approach of walk-through from the classification of defect products to the feature estimation for preventiting the cause of the defects in real-time or proactively. Particularly, this paper will introduce the multimodal data analytics in manufacturing process to integrately analyze the classification of defect patterns of final product quality and the causal variables of the defects collected at each process steps. Many papers are piled in more feasible image classification topics [3–7], however, discussion should shift to the defect cause analyses. That is because the classification itself does not reduce the occurance of the defects and the consumption of materials and energy.

There are three stages to summarize the possibilities and limitations of the multi-modal data analyses. The first stage is the accuracy improvement of the classification of the multi-class defect patterns and their superpositions at low cost (small samples) by using additional feature/data besides image data. The second stage is adaptive selections and estimations of important features including interactions within a practical computational time and accuracy. The third stage is an integrated approach of walk-through from the classification to the feature estimation for proactively preventing the causes of the defect. This paper will mainly discuss the first stage and the preperation of the third stage, as the first report. On the other hand, the second issue of high-dimensional interaction modeling is not presented in this paper because of the paper volume and on-going patent matter.

A combination of an image and texts is a typical data set of image-multimodal data and the research is progressing [8, 9]. On the other hand, about analyses of product defect and its causes, multimodal data across product inspection and manufacturing processes is generally difficult to obtain due to confidentiality and complexity. At the present, there are few studies of defect cause analysis using big data of all process steps, except for Nakata and Orihara [10] and their related works. Moreover, their research is also limited to specify (rank) manufacturing equipment and processes for each defect pattern class. It is still on-going issue about more detail variables to be detected and controlled. Each defect is produced in one or more process steps pysically or chemically. Based on that nature, we will propose and evaluate the image-multimodal data analysis.

This paper consists of 5 sections. Next Sect. 2 will introduce the dataset and previous studies. Section 3 will present and evaluate proposal method #1 of 2-step classification with resize, data augmentation, and CNN network tuning besides feature implementation. Section 4 will introduce multimodal analysis as proposal method #2 and its evaluations. Section 5 concludes this paper with summary and future works.

2 Objectives and Previous Studies

This paper presents semiconductor case study of the accuracy improvement by image-multimodal data analytics. As the first, target open dataset WM-811K to be extended is explained in Sect. 2.1, and previous studies will be introduced in Sect. 2.2.

2.1 Dataset

For the semiconductor case study, we use the open data set WM-811K of defect patterns (Fig. 1 (a)) of semiconductor wafer bin map (2D matrix) which is recorded in a final product test in an actual mass production [3]. About 1–50 thousand chips of small square-shape (called dies) on one wafer, and each die has an original label of 1 (Non-defective, green), 2 (Defective, yellow), or 0 (external of product, purple) (Fig. 1 (a)). In addition, the class label has been attached to 172950 wafers (=samples) in total by engineers with domain knowledge. There are difficulties on much different resolution of product-mix or technology-mix products, data imbalance between classes (Fig. 1 (b) table), random-cause defects appearance on all samples even the wafers of specific defect class (Fig. 1 (b) figure). Moreover, there are superpositions of some classes responding to different or common causes. Beside the data of 2D matrix with the class label, there is also a test log includes wafer attributes such as the number of columns and rows, or die sizes, defect rate, and so on. The 2D matrix is used in a multimodal data analysis as well as a pattern recognition and classification. Also, we will use it as data generation for superpositions near future because there is no exact superposition class label attached to WM-811K, and all superpositions are mixed in "Near-full" class.

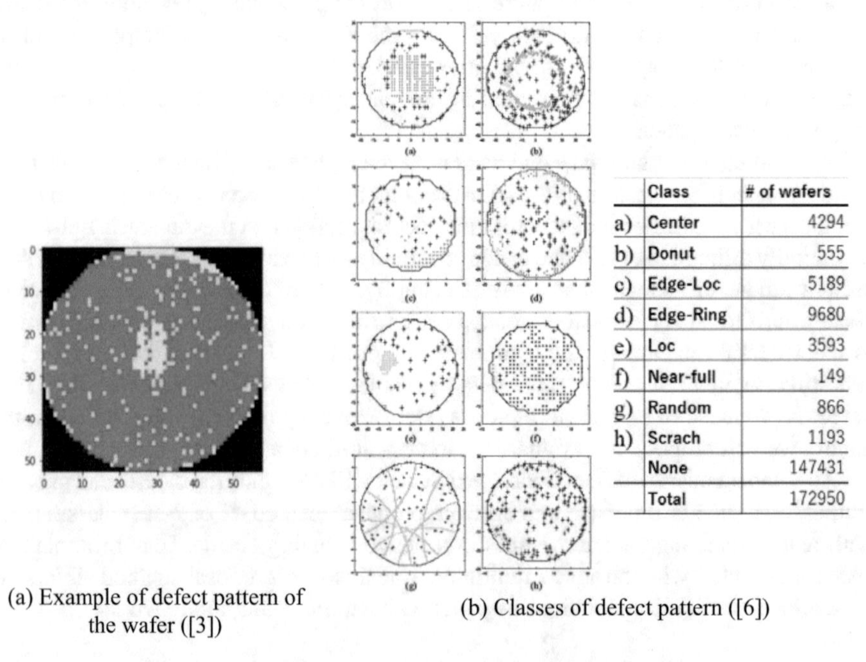

	Class	# of wafers
a)	Center	4294
b)	Donut	555
c)	Edge-Loc	5189
d)	Edge-Ring	9680
e)	Loc	3593
f)	Near-full	149
g)	Random	866
h)	Scrach	1193
	None	147431
	Total	172950

(a) Example of defect pattern of the wafer ([3])

(b) Classes of defect pattern ([6])

Fig. 1. Data of Semiconductor Defect Patterns - WM-811K.

2.2 Previous Studies

For the classification accuracy improvement which is the first issue in this study, pre-processes and model buildings are mainly discussed in general [11, 12]. As pre-process, data augmentation and feature constractions are mainly discussed and evaluated. The data augmentation is for train samples volume and variety, and the feature constraction is mainly for recognizing and utilizing the feature of the 2D defect pattern itself. For data augmentation, there is insufficient validation of basic pre-process validation of resolution and data augmentations of images, though there is a study in which data generation is used based on theoritical defect probability model [4]. For the feature construction, it is found that many basic geometry-based features are not so effective, on the other hand, just a few are succeed to improve accuracy such a Radon transformation [3], Tensor Voting [7], and so on. As the model building, for the industrial applications recent years, various CNN (Convolutional Neural Network) models are applied and extended for various single or multi-step classifications [4, 7], as well as simple Support vector machines (SVM) classification [3]. Moreover, Transformer is to attract attentions for more recent [13].

Now, the discussion topic is shift to the classifications accuracy of superposition defect patterns around 2022, however, the analysis is still limitted only by images. It is known that the accuracy deteriorates rapidly when overlaid more than 3–4 types of defect classes [7]. It becomes difficult in principle to achieve the high accuracy using only the DP data because the defects on the wafer become to spread to every location and also the defect dencity is high (like near-full) as the types of DP classes mixed in superposition. Discussion should again shift to the defect cause analyses. At the present, there are few studies of defect cause analysis using big data of all processes, except for Nakata and Orihara [10] and their related works. That is mainly because the data accumuration problem of huge (X00-X0000 process steps) and complex semiconductor manufacturing process. Moreover, their research is also limited to specifying (ranking) manufacturing equipment and processes for each defect pattern classification. Each defect is produced in one or more process steps (or between steps) pysically, chemically, or mechanically. Based on that nature, this paper is also for preparing the multimodal data analysis of the superpositions.

Now, for the integrated approach from the classification to the cause estimation for proactively preventing as the third issue, the analysis of multimodal data including images is one efficient choice. A combination of an image and texts is a typical data set of image-multimodal data and the research is progressing [8, 9]. On the other hand, about analyses of product defect and its causes, multimodal data across product inspection and manufacturing processes is generally difficult to obtain due to confidentiality and complexity. In this study, we extend open data, for evaluating possibility of multimodal analysis to motivate the analysis of actual company big data of confidential.

3 Proposed Methods and Evaluations

In this paper, we mainly introduce the multimodal analyses with semiconductor production cases based on actual data (WM-811K) for the classification of defect patterns of semiconductor wafer bin map (Fig. 1) determined in the product final test. We introduce 4 approaches as pre-process, main process, and post-process of analyses for the semiconductor case such as WM-811K. Resize resolution and data augmentation as the pre-process, CNN tuning as the post-process, and 2-step classification strategy and additional features activations as the main process. For the additional features, 2 proposal methods are presented and evaluated. One is to concatenate Radon transformation features (Sect. 3), and the second is multimodal analyses of images with additional outside variables with type-mix (categorical and numerical), various dimensions, and high missing rate (Sect. 4).

Note that the specification of machines used in this case study is as follows: Intel(R) Core ™ i7–8700 CPU @ 3.20 GHz, Memory: RAM32.0 GB, GPU: NVIDIA GeForce RTX 2080 VRAM: 8 GB, OS: Windows 10 Professional (64bit).

3.1 Pre-process: Resize and Data Augmentation

Pre-process is important part of accuracy improvement of actual problems. Particularly, the resize and data augmentation are applied for responding to various product-mix resolutions and imbalanced samples between classes in this study.

As resize, 5 kinds of kernels and 4 different resolutions are compared. As the kernel function, NEAREST, LINEAR, AREA, CUBIC, and LANCZOS4 are compared. As the resolution, size of mesh (row x column) is settled to 25 × 25, 37 × 37, 169 × 169, or 220 × 220 for comparison. Here, 25, 37, and 169 are the minimum, average, and maximum size of mesh of defect patterns of WM-811K, and 220 is similar size of the previous research [4] under our computational machine spec.

Data augmentation (DA) used in this study is flip and rotate. Flip type is "up/down" and/or "left/right". Rotate degrees are 45 and 60 often used in literatures.

Here, one of major networks of Nakazawa's CNN (Fig. 3(a)) is applied as a base model. As the numerical result is shown in Table 1, and the area kernel (AREA) is the best of five kernels in the view point of the level and stability of the accuracy (Table 1). About the resolution, the maximum value of 169 × 169 (max resolution of the data set) is the best of all of four evaluated sizes of "min", "average", "max" of the data set, and "limit". Here, "limit" is similar to the resolution of the previous research [4] under the machine spec limitation this study. Computational time of the combination of AREA and 169 × 169 is 11783.1 s, within acceptable time. Therefore, that combination is used in all following evaluation as preprocess.

Table 1. Comparison of accuracy beyond sizes and kernels.

Test_Acc	NEAREST	LINEAR	AREA	CUBIC	LANCZOS4	average	standard deviation
25×25 (min)	92.5%	90.2%	90.9%	90.8%	89.9%	90.8%	0.9%
37×37 (average)	90.5%	85.8%	93.9%	91.2%	87.8%	89.8%	2.8%
169×169 (max)	94.8%	95.5%	95.0%	95.1%	95.4%	**95.1%**	**0.3%**
220×220 (limit)	94.6%	94.3%	94.2%	92.0%	94.3%	93.9%	1.0%
average	93.1%	91.4%	**93.5%**	92.3%	91.8%		
standard deviation	1.8%	3.8%	**1.6%**	1.7%	3.1%		

Table 2 shows the result of evaluations of DA to WM-811K data. Table 2(a) shows the accuracy and computational time of 5 scenarios. The accuracy is expressed by (TP + TN)/(TP + TN + FP + FN), where TP, TN, FP, FN are True-Positive, True-Negative, False-Positive, and False-Negative of a confusion matrix, respectively. In Table 2 (b) and (c), the row is the true class, and the column is the forecasted class. The diagonal line represents Recall = (TP/(TP + FN)). Note that the test images (250 samples) are the same. For each DA cases (Table 2(a)), 5 data set in which 250 original images are randomly sampled independently are used for the trainings.

As the result of DA, the performance is the best when all combination of the flips and rotate operations. The rotation of 45° is selected because it is better than 60° though both are sometimes applied in literatures. Note that the class of "near-full" is excepted because it includes various superpositions but is not labelled correctly.

Table 2. Evaluation of Data augmentation

(a) Comparison of data augmentation

Scenario No.	Number of images used in training for each class	Accuracy of all class in average of 5 data sets	Computational time per trainning [sec.]
1	250 original images randomly sampled	71.50%	167.1
2	1000 images =250 (randomly sampled by the original image set) +750 (flips of up/down, left/right, abd up/down & left/right)	79.20%	669.8
3	1500 images =250 (randomly sampled by the original image set) +1250 (rotation by 60 degrees) training images	83.00%	1085.8
4	2000 images =250 (randomly sampled by the original image set) +1750 (rotation by 45 degrees) training images	83.50%	1412.9
5	2750 images =250(randomly sampled by the original image set) +750 (up/down, left/right, up/down/left/right flips) + 1750 (rotation by 45 degrees) training images for each class	84.70%	1930.5

(b) Confusion Matrix (Scenario No=1)

	Loc	Edge-Loc	Center	Edge-Ring	Scratch	Random	Donut	Accuracy	TrainTime
Loc	39.2%	14.2%	9.1%	0.6%	30.5%	1.6%	4.8%	71.5%	167.1
Edge-Loc	14.6%	51.7%	2.9%	6.2%	18.7%	5.8%	0.2%		
Center	4.9%	2.0%	86.6%	0.2%	5.0%	0.4%	0.9%		
Edge-Ring	0.3%	5.4%	0.2%	93.2%	0.9%	0.1%	0.0%		
Scratch	26.3%	16.2%	3.8%	1.6%	51.4%	0.6%	0.2%		
Random	1.2%	0.9%	1.0%	0.1%	0.1%	96.5%	0.2%		
Donut	8.9%	1.1%	1.6%	0.3%	1.4%	5.0%	81.7%		

(c) Confusion Matrix (Scenario No=5)

	Loc	Edge-Loc	Center	Edge-Ring	Scratch	Random	Donut	Accuracy	TrainTime
Loc	**56.3%**	9.4%	5.0%	0.2%	22.0%	1.3%	5.7%	**84.7%**	1930.5
Edge-Loc	5.4%	**82.3%**	1.5%	3.2%	5.2%	2.2%	0.2%		
Center	1.5%	0.5%	**93.7%**	0.1%	1.9%	0.8%	1.5%		
Edge-Ring	0.0%	3.1%	0.1%	**96.6%**	0.2%	0.0%	0.0%		
Scratch	18.1%	7.1%	1.2%	0.6%	**72.1%**	0.5%	0.5%		
Random	1.0%	0.6%	0.6%	0.2%	0.2%	**96.5%**	1.0%		
Donut	1.3%	0.5%	0.2%	0.0%	0.7%	1.9%	**95.4%**		

3.2 Proposal Method #1: 2-step Classification with Radon Features

Main Process (1): Two-step Classification. As main process of the WM-811K case, we introduce 3 approaches. The first approach is 2-step classification (Fig. 2). The first step is for classifying "None" and others before the second step classifying the multi-class defects for higher accuracy baseline of the classification to much imbalanced samples data (Fig. 1 (b)). The 2-step classification also reduce the computational time by 80% or so.

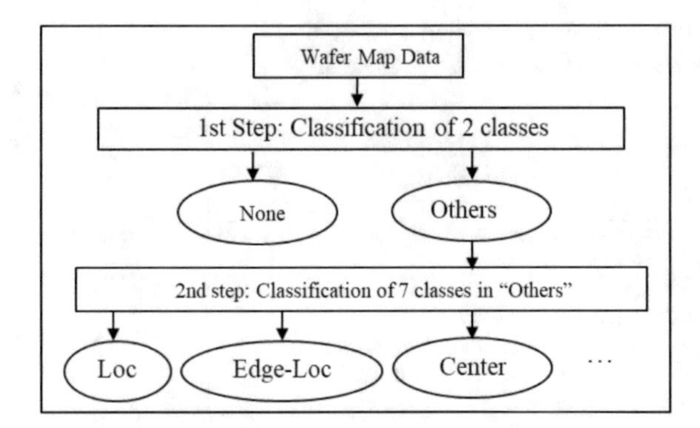

Fig. 2. Case2- Model of 2-step classification.

Main Process (2): Concatinate of Radon Transformation Feature. The second approach is multimodal analyses to introduce additional features. As the first step for WM-811K, the feature of defect area itself is evaluated. As mentioned in Sect. 3, a few are succeeded to improve accuracy such a Radon transformation [3], Tensor Voting [7], and so on. Here, we extend one of major networks of Nakazawa's CNN (Fig. 3(a)) for our multimodal analyses by implementing the Radon transformation feature (Fig. 3(b)). Particularly, it is expected that the Radon transformation feature improves the accuracy of classes of Loc, Scratch, and so on of low-level accuracy. Note that various combinations of the features and networks still have a room for detail comparisons. At least, Nakazawa's CNN is better than Kyeong's CNN [5] as the result of our evaluation.

Post-Process: CNN Network Tuning. From our additional benchmarking, hyperparameters are optimized for more accuracy than Table 2 results. A classification with the dropout rate of 0.875 improved the average accuracy by 1.6% for all. Moreover, epoch of 25 of lowest computational time improved the average Recall by 0.6% in the range that the loss is enough saturated and the loss of validation in training phase does not increase among 100 epochs. As the result of those improvement, Recall becomes 86.9% (Table 3(a)).

The model with radon transformation feature achieves the average Recall of 87.2% (Table 3(b)). This indicates that the additional effect of radon transformation features is small (+0.3%) for this model.

Now, As the post-process, the number of fully connected layer (FCL) is evaluated for further improvement. As summary, one additional FCLs (Fig. 3(c)) is the best performance. This achieves the average Recall of 88.3% (Table 3(c)), on the other hand a model with 2 FCLs leads lower Recall of 87.5%. Compared to the base model, a model with Radon transformation features and 1additional FCL improves average Recall by 1.4%. Particularly, "Loc" and "Scratch" which are difficult to be classified precisely are improved as in 67.0% and 82.7% of Recall (Table 3(c).

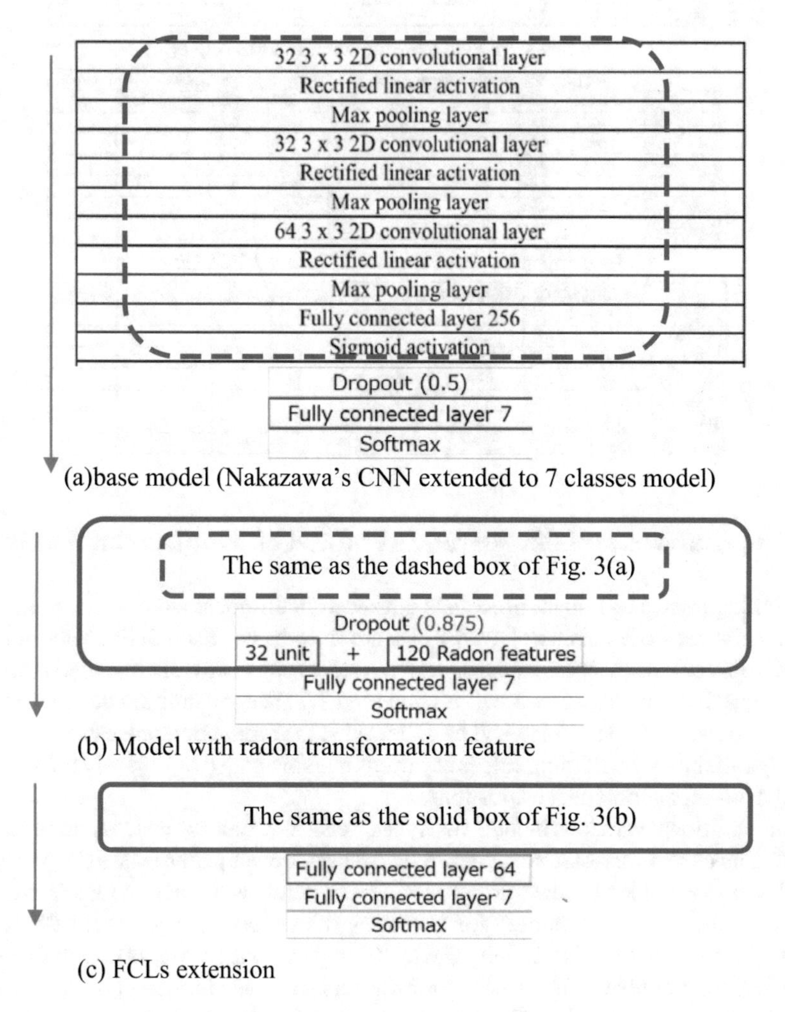

Fig. 3. Models proposed for multimodal analyses.

Table 3. Results of Proposal Method #1

(a) Only images (Fig.6(a))

	Loc	Edge-Loc	Center	Edge-Ring	Scratch	Random	Donut	Total average	TrainTime
Loc	63.4%	9.7%	4.5%	0.0%	15.8%	1.4%	5.1%	86.9%	**3146.0**
Edge-Loc	5.1%	85.6%	1.0%	1.7%	4.2%	2.1%	0.2%		
Center	3.0%	0.3%	93.1%	0.1%	1.0%	0.7%	1.8%		
Edge-Ring	0.0%	3.0%	0.0%	96.6%	0.3%	0.0%	0.0%		
Scratch	16.3%	4.3%	0.7%	0.5%	77.4%	0.2%	0.6%		
Random	1.2%	0.3%	0.5%	0.4%	0.0%	96.7%	0.9%		
Donut	2.3%	0.3%	0.2%	0.0%	0.3%	1.7%	95.1%		

(b) Image +Radon transformation features (Fig.6(b))

	Loc	Edge-Loc	Center	Edge-Ring	Scratch	Random	Donut	Total average	TrainTime
Loc	**65.3%**	6.8%	2.2%	0.1%	19.0%	1.8%	5.0%	**87.2%**	3260.2
Edge-Loc	5.9%	83.0%	1.0%	1.8%	5.4%	2.7%	0.2%		
Center	2.6%	0.4%	**93.8%**	0.0%	1.0%	0.6%	1.7%		
Edge-Ring	0.0%	3.0%	0.0%	96.4%	0.6%	0.0%	0.0%		
Scratch	15.0%	3.6%	0.5%	0.4%	**79.8%**	0.3%	0.3%		
Random	1.2%	0.4%	0.6%	0.1%	0.1%	**96.8%**	0.9%		
Donut	2.4%	0.2%	0.2%	0.1%	0.2%	1.4%	**95.4%**		

(c) Image +Radon transformation features + 1 FCL (Fig.6(c))

	Loc	Edge-Loc	Center	Edge-Ring	Scratch	Random	Donut	Total average	TrainTime
Loc	**67.0%**	8.9%	3.0%	0.0%	15.7%	1.7%	3.7%	**88.3%**	3153.1
Edge-Loc	3.7%	**87.4%**	1.1%	2.4%	3.5%	1.8%	0.1%		
Center	2.0%	0.3%	**95.5%**	0.0%	0.6%	0.5%	1.0%		
Edge-Ring	0.0%	2.6%	0.0%	**96.8%**	0.6%	0.0%	0.0%		
Scratch	12.6%	3.4%	0.4%	0.5%	**82.7%**	0.2%	0.2%		
Random	1.8%	1.1%	1.0%	0.2%	0.1%	94.4%	1.3%		
Donut	2.7%	0.2%	0.8%	0.0%	0.5%	1.3%	94.6%		

4 Proposal Method #2: Advanced Model of Multimodal Analysis

The third approach as a main process is advanced multimodal analyses in which additional variables from outside of WM-811K are introduced. Each defect pattern is produced in one or more process steps (or between steps) physically, chemically, or mechanically. Based on that nature, we will propose and evaluate the multimodal data analysis with additional variables which will be collected in the production process. In this study, dummy variables with missing values at various rates are added to WM-811K to motivate actual mass production data evaluations.

The additional variables include two types, "causal variables" and "randomized variables". The causal variables are attached by a pair of defect pattern class (DP class) and the cluster determined by the attributes of a wafer such as the defect rate. Particularly here, only the defect rate is used for clustering the wafers of an identical DP class to determine subdivisions. Concretely, GMM (Gaussian Mixture Model) clustering based on defect rate is applied (Fig. 4(b)) to form clusters well even for defect rates with multi-peak distributions (Fig. 4(a)). The GMM clustering is executed to 8 classes except two classes of "none" and "near-full". The exclusion is because the class "none" is to be classified in the first step of 2-step classification without additional variables, and the class "near-full" includes various superpositions without correct labels. The number of clusters are determined by BIC criteria for each class of defect patterns (DP) (Fig. 4 (c)).

After the cluster is determined under the identical DP class, "causal variables" is attached to each pair of a DP class and a cluster (Fig. 4(d)). The number of the causal variable is set to 5 for the cluster of the lowest defect rate in each DP class and to 3 for other clusters. In total, 117 causal variables are attached for each wafer (=sample). The same causal variables are attached to all wafers of the same pair of the DP class and the cluster. That means wafers can be classified exactly by the causal variables without image. Here, for the multimodal analysis, missing values at a given rate are set independently for each causal variable as a dropout. The missing value is often included in an actual data, and it is important to evaluate its influence on the classification accuracy.

In addition, we evaluate the influence of "randomized variables". The randomized variable that is randomly set to 1 with a predetermined probability is independently generated according to the Bernoulli distribution. Here, for instance, a total of 50 random variables are prepared, with 10 variables each having 5 different probabilities of 1 (0, 0.3, 0.5, 0.7, 1). The probability is set to 10 scenarios: (0, 0.1, 0.2, 0.3, 0.4, 0.5, 0.6, 0.7, 0.8, 1) in case 100/200 random variables of 0–1 form are used. Accuracy with and

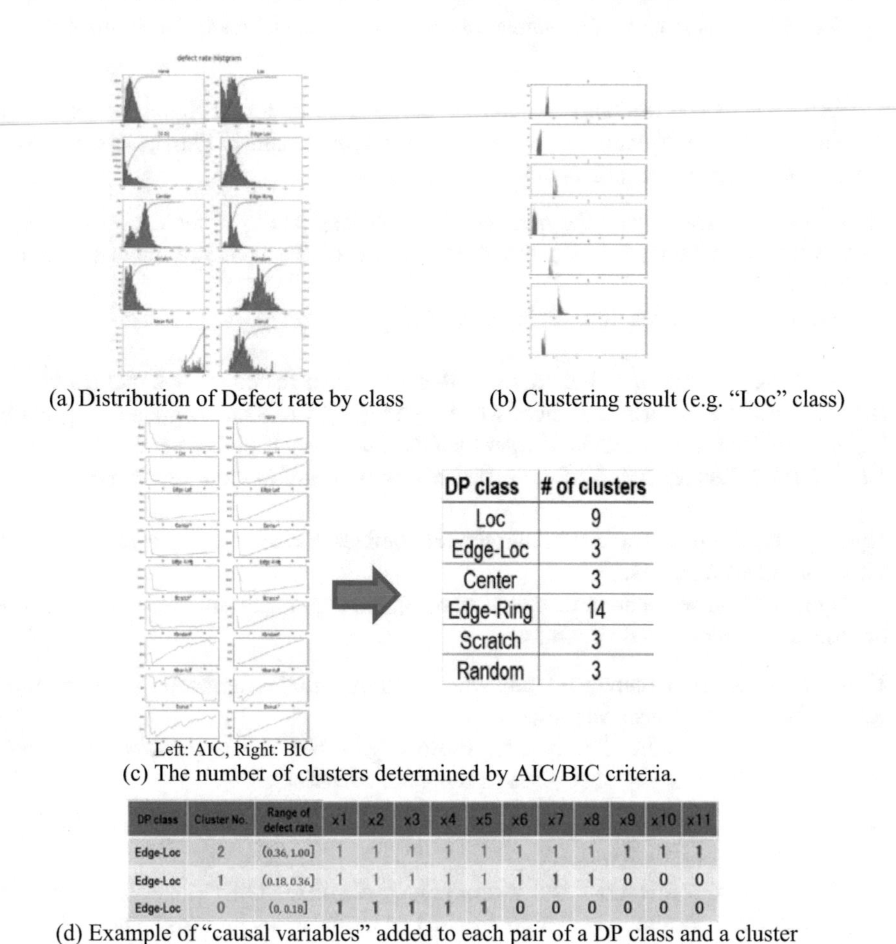

(a) Distribution of Defect rate by class

(b) Clustering result (e.g. "Loc" class)

DP class	# of clusters
Loc	9
Edge-Loc	3
Center	3
Edge-Ring	14
Scratch	3
Random	3

(c) The number of clusters determined by AIC/BIC criteria.

DP class	Cluster No.	Range of defect rate	x1	x2	x3	x4	x5	x6	x7	x8	x9	x10	x11
Edge-Loc	2	(0.36, 1.00]	1	1	1	1	1	1	1	1	1	1	1
Edge-Loc	1	(0.18, 0.36]	1	1	1	1	1	1	1	1	0	0	0
Edge-Loc	0	(0, 0.18]	1	1	1	1	1	0	0	0	0	0	0

(d) Example of "causal variables" added to each pair of a DP class and a cluster

Fig. 4. The clustering by DP class and "causal variables".

without the randomized variables is also compared. For generating random variable of numerical, standardized normal distribution with average 0 and variance 1 is used.

Figure 5 shows a model of the advanced multimodal data analysis which is an extended version of Fig. 3 (c).

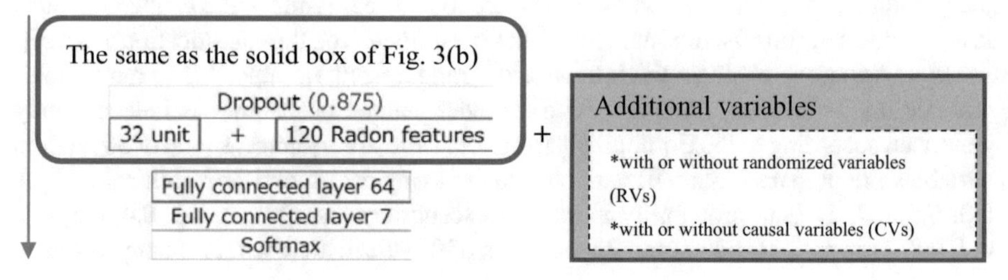

Fig. 5. Advanced model for multimodal analyses by introducing additional variables.

As the result of performance evaluation, additional variables are effective even though there are many missing values (e.g. 60, 80%). From the results in Fig. 6, we summarize considerations from four points of view as follows:

Hypothesis#1. The greater the number of the missing values of the causal variables, the lower the accuracy (A). Also, the boundary of accuracy deterioration is around 0.5 (50%) (B).

Hypothesis#1 verification results (Fig. 6(a)):

(A). Accuracy decreases with more missing values, as expected in the hypothesis.
(B). The accuracy deteriorates more when the missing value rate is greater, especially big deterioration slope when it is larger than 0.8. Accuracy is high if the missing values rate is 60% or less regardless of the presence or absence of randomized variables.

Hypothesis #2. the accuracy without the randomized variables is better than that with the randomized variables.

Hypothesis#2 verification results (Fig. 6(a)): No significant difference the presence or absence of randomized variables.

Hypothesis #3. The accuracy is better when the data includes the Radon transformation features besides the causal variables.

Hypothesis#3 verification results: From Fig. 6(b): The hypothesis is proved, regardless of the presence or absence of randomized variables.

Hypothesis #4. The type and dimension of the random variables (RVs) does not influence the accuracy significantly.

Hypothesis#4 verification results: From Fig. 6(d) and (e), the hypothesis is proved in the given domain. Particularly, it is confirmed that the accuracy improvement effect does not deteriorate even when type of RVs is mixed (with 01-RVs and N-RVs, Fig. 6(d)). Moreover, that is also confirmed even when the dimension of RVs is more than that of CVs (the case with RVs(200) > CVs (117))(Fig. 6(e)).

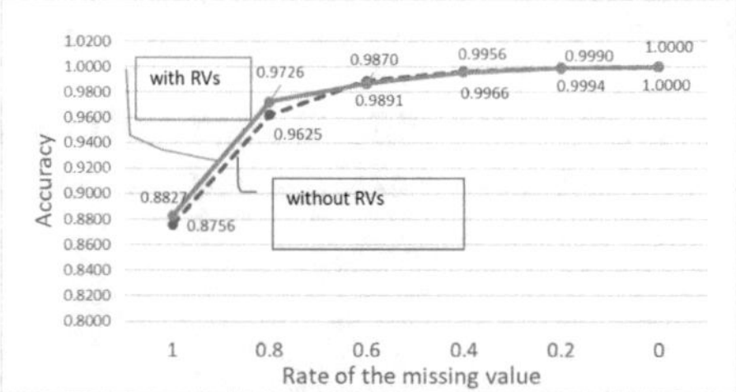

(a)Characteristics of Accuracy -Rate of missing values on the causal variables (with and without randomized variables (RVs))

		Random variables(RVs)	Causal variables (CVs) with	Causal variables (CVs) without
Radon transformation features (RFs)	with	with RVs	x1<a≦100%	x1=88.27%
		without RVs	x2<a≦100%	x2=87.56%
	without	with RVs	x3<a≦100%	x3=87.41%
		without RVs	x4<a≦100%	x4=86.9%

(b) Comparison summary: with and without CVs, RFs, and RVs.

		Random variables(RVs)	Rate of the missing value on the causal variables: 1	0.8	0.6	0.4	0.2	0
Radon transformation features (RFs)	with	with RVs	0.8827	0.9726	0.9870	0.9956	0.9990	1.0000
		without RVs	0.8756	0.9625	0.9891	0.9966	0.9994	1.0000
	without	with RVs	0.8756	0.9625	0.9891	0.9966	0.9994	1.0000
		without RVs	0.8741	0.9640	0.9930	0.9975	0.9987	1.0000

(c) Characteristics on the rate of missing value on causal variables

Accuracy (average)	Rate of the missing value on the causal variables					
	1	0.8	0.6	0.4	0.2	0
with 01-RVs(50)+N-RVs(50)	0.8829	0.9693	0.9886	0.9939	1.0000	1.0000
with 01-RVs(50)	0.8827	0.9726	0.9870	0.9956	0.9990	1.0000
with N-RVs(100)	0.8669	0.9657	0.9907	0.9943	0.9992	1.0000

(d) Influence of the type of RVs: 01-RVs, numerical (N-RVs), and those mixed variables

Accuracy (average)	Rate of the missing value on the causal variables					
	1	0.8	0.6	0.4	0.2	0
with 01-RVs (50)	0.8827	0.9726	0.9870	0.9956	0.9990	1.0000
with 01-RVs(100)	0.8861	0.9636	0.9848	0.9956	0.9981	1.0000
with 01-RVs(200)	0.8669	0.9657	0.9907	0.9943	0.9992	1.0000

(e) Influence of the dimension of RVs

Fig. 6. Numerical results of advanced multimodal analysis by introducing outside variables.

Note that the value of each evaluation is average value calculated by evaluation results of three data sets.

As the summary of the evaluations, those are all supportive to motivate actual mass production data evaluations of image-multimodal data analysis with causal variables.

5 Conclusion

This paper presented a case study of the accuracy improvement by image-multimodal data analysis. This paper focused on the defect patterns classification of semiconductor wafer bin map (2D matrix) of final product tested in actual foundry. We use the open data set WM-811K including about 1.7k images with class label. There are difficulties on much different resolution of product-mix products, data imbalance between classes, random-cause defects appearance on all samples even the wafers of specific defect class. Resize and data augmentation as the preprocess results effective as 13.2% improvement in Recall though the previous studies does not care so much because of the rich total samples even though the small sample class exists. As the main process of machine learning, two kinds of multimodal data analyses are proposed and evaluated. One is to introduce the feature of defect pattern itself, and the other is to add the defect causal variables. Concretely, for the former the Radon transformation feature is performed well to improve the classification accuracy. Combination of that features and one additional fully connected layer is considered as the best performance one in this case as the average Recall of 88.3%. Particularly, "Loc" and "Scratch" which are difficult to precisely classify are improved as in 67.0% and 82.7% from the Recall of the base model (only by using original images) 39.2% and 51.4%.

For the later, the causal variables (CVs) are attached to the pair of a defect pattern class and one of its subdivision clusters which is separated by GMM clustering and BIC criteria. Mainly the influence of the missing value rate of CVs and the randomized variables (RVs) are evaluated. The classification of accuracy stays very high when the missing value rate is less than 80% (that is, 20% rest) in evaluations in this study. That is positive for practical conditions, and the possibility of the multimodal data is appeared. In addition, the influence of combination with Radon transformation features are evaluated. The combination is compatible and positive because of higher accuracy regardless of RVs. In the end the influence of the type (numerical or categorical) and the dimension of RVs (particularly the condition of $|CVs| < |RVs|$) are evaluated. The proposed method is confirmed as robust in given domain and data sets without significant differences.

This study is still at a start line and will be concluded next few years. As future works about the semiconductor case, the next step is the data generation and multimodal analyses for superposition classifications and cause discoveries. In addition, various networks and features are needed to be evaluated, for example, the Tensor voting [7] and the Transformer [13]. Moreover, we will continue to study for our final goal, an integrated approach of walk-through from the classification of defect products to the feature estimation for preventing the cause of the defects in real-time or proactively.

Acknowledgements. This research has been partly executed in response to support to KIOXIA Corporation. We all thank for the support. We also appreciate the constructive comments of all reviewers.

References

1. Hou, B., Stapczynski, S.: Chipmaking's Next Big Thing Guzzles as Much Power as Entire Countries (2022). https://www.bloomberg.com/news/articles/2022-08-25/energy-efficient-computer-chips-need-lots-of-power-to-make#xj4y7vzkg. Accessed 16 Dec 2022
2. Arima, S., Kobayashi, A., Wang, Y.-F., et al.: Optimization of re-entrant hybrid flows with multiple queue time constraints in batch processes of semiconductor manufacturing. IEEE Trans. Semicond. Manuf. **28**(4), 528–544. IEEE (2015)
3. Wu, M.-J., Jang, J.-S.R., Chen, J.-L.: Wafer map failure pattern recognition and similarity ranking for large-scale data sets. IEEE Trans. Semicond. Manuf. **28**(1), 1–12. IEEE (2015). https://doi.org/10.1109/TSM.2014.2364237
4. Nakazawa, T., Kulkarni, D.V.: Wafer map defect pattern classification and image retrieval using convolutional neural network. IEEE Trans. Semicond. Manuf. **31**(2), 309–314. IEEE (2018)
5. Kyeong, K., Kim, H.: Classification of mixed-type defect patterns in wafer bin maps using convolutional neural networks. IEEE Trans. Semicond. Manuf. **31**(3), 395–401. IEEE (2018)
6. Jin, C.H., Na, H.J., Piao, M., Pok, G., Ryu, K.H.: A novel DBSCAN-based defect pattern detection and classification framework for wafer bin map. IEEE Trans. Semicond. Manuf. **32**(3), 286–292. IEEE (2019)
7. Wang, R., Chen, N.: Detection and recognition of mixed-type defect patterns in wafer bin maps via tensor voting. IEEE Trans. Semicond. Manuf. **35**(3), 485–493. IEEE (2022)
8. Duong, C.T., Lebret, R., Aberer, K.: Multimodal Classification for Analysing Social Media (2017). https://arxiv.org/abs/1708.02099. Accessed 28 Dec 2022
9. Woo, L.J., Yoon, Y.C.: Fine-Grained Plant Identification using wide and deep learning model. In: 2019 International Conference on Platform Technology and Service (PlatCon). IEEE (2019)
10. Nakata, K., Orihara, R.: A comprehensive big-data based monitoring system for yield enhancement in semiconductor manufacturing. IEEE Trans. Semicond. Manuf. **30**(4), 339–344. IEEE (2017)
11. Shorten, C., Khoshgoftaar, T.M.: A survey on image data augmentation for deep learning. J. Big Data **6**(60), 1–48. Springer (2019)
12. Johnson, J.M., Khoshgoftaar, T.M.: Survey on deep learning with class imbalance. J. Big Data **6**(27), 1–54. Springer (2019)
13. Wei, Y., Wang, H.: Mixed-type wafer defect recognition with multi-scale information fusion transformer. IEEE Trans. Semicond. Manuf. **35**(3), 341–352. IEEE (2022)

Decision-Making Model for Updating Geographical Information Systems for Polish Municipalities Using the Fuzzy TOPSIS Method

Oskar Sęk(✉) and Ireneusz Czarnowski

Gdynia Maritime University, Morska Str. 81-87, 81-225 Gdynia, Poland
o.sek@sd.umg.edu.pl, i.czarnowski@umg.edu.pl

Abstract. Geographic information systems (GISs) allow for the acquisition, processing, and sharing of spatial data and related descriptive information. Their primary function is to present spatial data in an interactive manner, and they are often used to solve complex planning and organisational issues. In Poland, GIS and land information systems are primarily used by administrative units such as municipalities and voivodships. This paper aims to define the problem, decision-making goals and criteria used when updating GISs in Polish municipalities and to conceptualise a descriptive model for decision making in a Polish municipality using the fuzzy TOPSIS method.

Keywords: Decision-making · Multi-criteria decision-making · TOPSIS · Polish municipalities · e-government · e-administration

1 Introduction

For nearly two decades now, geographic information systems (GISs) have been gaining popularity, as they allow for the acquisition, processing, and sharing of spatial data and related descriptive information about objects within the coverage area of the system. A GIS is a tool that allows for the collection, storage, analysis, retrieval, and visualisation of geographical data. Its primary function is to present spatial data in an interactive manner: with a GIS, users can select the specific spatial information they want, and view the related data associated with it [1].

Advanced technologies and applications enable instant access to data, selection, analysis and the production of reports, with the aim of facilitating decision making and allowing the user to select the best solution. Spatial information is obtained by interpreting spatial data, and includes objects, events, facts, processes, phenomena, ideas, etc. The backbone of a GIS is a database containing spatial and descriptive information about the objects in the real world that are represented in the system. The primary feature of a GIS is the ability to create a cartographic representation of data. GISs vary in terms of the information involved, their size and their function, depending on the needs of the user [2].

I. Czarnowski et al. (Eds.): KESIDT 2023, SIST 352, pp. 62–73, 2023.
https://doi.org/10.1007/978-981-99-2969-6_6

In Poland, the construction of GISs and land information systems (LISs) is led by administrative units, which are mainly municipalities and voivodships. A GIS is a crucial element in the development and functioning of local communities, as it provides valuable information for potential investors and is an invaluable resource for tourists. GIS store two types of data simultaneously, which are known as spatial and descriptive data. Spatial data contain information about the shape and location of objects in a specified reference system (geometric data) and the spatial relationships between objects (topological data), and are presented in the form of digital maps. Descriptive data, on the other hand, are any kind of information that does not have a spatial reference, such as the attributes of objects or phenomena, and are represented in tabular form [2].

A GIS is therefore a system that combines a database for storing spatial information with a set of tools for analysing those data. Its purpose is to help solve complex planning and organisational issues. GIS tools are essential for making strategic decisions that impact the development of a specific area, and the use of a GIS in the administrative decision-making process can play a significant role in management integration [3].

The primary aim of this paper is to define the problem, decision-making goals and criteria used to update GISs in Polish municipalities. A further aim is to conceptualise a descriptive model for decision making in Polish municipalities using the fuzzy TOPSIS method. This research was conducted using an integrated review method in order to generate a new conceptual framework for creating decision-making models for updating GISs in Polish municipalities. Thus, the contributions of this paper are related to the organisation and standardisation of current scientific knowledge regarding the decision-making process for updating GISs in Polish municipalities.

The remainder of the paper is organised as follows. Section 2 presents a general overview of decision-making methods. Section 3 describes the general objective of implementing a GIS, the conditions and criteria involved, and the potential alternatives in a decision to stay with the current GIS, to modernise it, or to implement a new one. Section 4 introduces the fuzzy TOPSIS method and the calculations carried out in the model. Conclusions and directions for future work are presented in the final section of the paper.

2 Decision-Making Models

A wide range of discrete multi-criteria decision methods (MCDMs) have been developed to address various types of issues. Each method has certain advantages and disadvantages, and can be classified into a specific group of decision support methods. The various types of multi-criteria decision support methods are outlined below, following Trzaskalik's typology [4]:

- Additive: These are based on developing a model of the decision maker's preferences using an additive linear function. Examples include SAW [5], F-SAW [6], SMART [7] and SMARTER [8];

- Analytical hierarchy (and related models): These are based on the development of a hierarchical model and the determination of preferences by, for example, pair-wise comparisons of criteria, sub-criteria and individual variants of a solution. Examples include AHP [9], REMBRANDT [10], F-AHP [11], ANP [12], F-ANP [6], MACBETH [13];
- Verbal: Here, the object of analysis is an unstructured problem consisting of qualitative parameters, for which there are no objective aggregation models. Examples include ZAPROS [14] and ZAPROS III [15];
- ELECTRE: This method is based on a critique of the multi-attribute utility theory approach, and builds on Roy's proposal to expand the basic preference situations to include cases of equivalence, weak or strong preference, and incomparability. Many variations of this method have been created [16–18];
- PROMETHEE: The focus of analysis in this case is to compare the ratings of different variants for all criteria, using preference functions as a measure of their importance, with a value ranging from zero to one. The larger the difference in the ratings, the stronger the preference for that variant, and the smaller the difference, the weaker the preference. Several versions of PROMETHEE methods have been developed [19–21] along with modifications under the name EXPROM [22–24];
- Reference points: These methods compare the selected alternatives with respect to assumed ideal and anti-ideal variants. Examples include TOPSIS [25], F-TOPSIS [26], VIKOR [27], DEMATEL + ANP + VIKOR [6] and methods named BIPOLAR [28–30];
- Interactive: This approach is based on the assumption that the decision maker can evaluate a single option or a small subset of options in a local environment. The multi-iterative approach is divided into two phases: the computational phase, and a dialogue with the decision maker. The dialogue phase involves expressing preferences and addressing the consequences of the choice of options in the form of probability distribution parameters. The process is repeated until a satisfactory solution is obtained. Examples include STEM-DPR [31], INSDECM [32] and ATO-DPR [33].

In addition, MCDM methods can be classified in terms of the type of data employed, as classical (crisp) and fuzzy MCDM methods. In crisp MCDM methods, the scores for the alternatives based on certain criteria and the weights of these criteria can be precisely evaluated and expressed as crisp numbers, whereas in fuzzy MCDM methods, the linguistic variables linked with fuzzy numbers are used to reflect the vagueness of the subjective expressions presented by the decision maker [34]. Fuzzy set theory, introduced by Zadeh, has been applied to make use of vague information in different contexts, especially in decision making [35].

3 Problem Formulation

3.1 Objective and Decision-Making Conditions

The main objective of a GIS is to "[satisfy] public needs for geographical space information". This includes fulfilling the needs of both public administrations and non-administrative users, such as businesses and individual citizens [36]. Specific objectives include streamlining collections of spatial information within the administration, improving data collection, maintenance and storage, enhancing accessibility to spatial data for both the administration and other data users, and creating new information products using existing data [36].

Access to reliable and up-to-date spatial information and its use in decision making are important issues not only in Poland, but also in other European Union member states. Solving these problems requires joint actions aimed at exchanging, using, accessing and utilising both spatial data and spatial data services at various levels of public authorities and in different sectors of the economy. This has given rise to the creation of national spatial information infrastructures that form part of the European spatial information infrastructure, known as INSPIRE. The INSPIRE Directive 2007/2/EC, established in 2007, created a spatial information infrastructure throughout Europe for the purpose of supporting environmental policies and activities that may affect the environment. The directive also had an overall impact on the improvement of spatial data in EU member states [37]. From the perspective of achieving the goals of the INSPIRE Directive, it is significant that the creators of databases containing spatial information are required to ensure the interoperability and consistency of these data [3]. This directive mandates that Member States develop and implement their own spatial information infrastructure as part of the larger spatial information infrastructure of the European Union [38]. Every GIS must therefore be compliant with current legislation, i.e. with the INSPIRE directive and other national laws, for example the Polish Spatial Information Infrastructure Act of 4th March 2010.

A GIS can be viewed as a type of information communication technology (ICT) that includes a combination of IT devices and software that work together to achieve specific goals. In this context, the GIS can handle, store, transmit and receive spatial and descriptive data through telecommunication networks by using the appropriate terminal equipment for each specific type of network [39]. As a form of ICT, a GIS is associated with e-government, which refers to the government's utilisation of technology (and specifically web portals) to increase access to and distribution of government information and services to citizens, business partners, employees, other government agencies and other agencies [40]. E-government requires a comprehensive approach that encompasses multiple aspects (organisational, economic, cultural, social, political and organisational) and stages of maturity (from information to personalisation), at both the strategic and technical levels [41].

A crucial challenge involves determining the factors that influence the success of e-government adoption. These success factors refer to the areas and operations that should be given priority in order to achieve the best results from e-government adoption, using the theory of critical success factors. This provides a solid foundation for identifying the key criteria that should be followed during e-government adoption or, in the case examined here, for making a decision on whether or not to update a GIS in a Polish municipality [41].

3.2 Decision-Making Criteria

Various standards have been used in the literature to evaluate the implementation of GISs. The application of GISs within e-government in one research work included a check on the availability and accessibility of the current GIS, and determining whether the data in the GIS were up-to-date and whether the GIS technology and interactivity were outdated [1]. In another study, the author created an economic model for the implementation of a GIS that included the scope and options for the GIS infrastructure, the phases and options in terms of the benefits of GIS implementation, the projected benefits and costs, the cost-effectiveness indicators determining the success or failure of GIS implementation, and the period of economic analysis over which the amortisation period of the GIS investment was defined [36]. Research into knowledge vacuums has considered factors impacting the success of e-government innovation and implementation, such as information and data, information technology, organisational and managerial, legal and regulatory, and institutional and environmental aspects [42]. Other researchers have used a critical success factor approach to the successful implementation of ICT in e-government [41, 43], as mentioned in Sect. 3.1.

The following criteria were adopted for our model:

- Economic aspects, such as the financial situation of the municipalities, potential economic risks and benefits of GIS modernisation, and an overview of public outlays;
- Socio-cultural aspects, such as the information culture, the potential exclusion of citizens due to numerous factors, and public demand for e-services;
- Technological aspects, such as licensing, standardisation, the interoperability and integration of systems, and the quality and maturity of e-services;
- Organisational aspects, such as compliance with e-government strategy, the ability to support top management, and the adaptation of new management models [41, 43].

3.3 Decision Alternatives

This paper examines three options for a local government that is considering updating its GIS. These options are maintaining the current system without any changes, upgrading the current system using existing technology within existing standards, and implementing a completely new system using new technology, which renders the previous system obsolete. Ceasing to use a GIS or shutting it down are not options, because this is widely used software, especially in decision-making processes for public government (such as municipalities).

4 Proposed TOPSIS Decision-Making Model

4.1 Fuzzy TOPSIS Method

The fuzzy TOPSIS method, which was first introduced by Hwang and Yoon in 1981, is a widely utilised technique for multi-criteria decision making in uncertain or vague situations. It is used to rank alternatives in a fuzzy environment [25].

The first step is to produce a decision matrix by determining the criteria, the alternatives and the fuzzy scale used in the model. A decision matrix is then created by evaluating the alternatives in terms of the selected criteria using the abovementioned fuzzy scale.

The second step is to create a normalised decision matrix. Based on the positive and negative ideal solutions, a normalised decision matrix can be calculated using the following relations for a positive ideal solution (1) and a negative ideal solution (2).

$$\tilde{r}_{ij} = \left(\frac{a_{ij}}{c_{j*}}, \frac{b_{ij}}{c_{j*}}, \frac{c_{ij}}{c_{j*}}\right), \; c_{j*} = \max_i c_{ij} \tag{1}$$

$$\tilde{r}_{ij} = \left(\frac{a_j^-}{c_{ij}}, \frac{a_j^-}{b_{ij}}, \frac{a_j^-}{a_{ij}}\right), \; a_j^- = \min_i a_{ij} \tag{2}$$

The next step is to create a weighted normalised decision matrix. This can be calculated by multiplying the weight of each criterion in the normalised fuzzy decision matrix, according to the following formula (3), where $\tilde{w}_{ij}$ represents the weight of criterion c_j.

$$\tilde{v}_{ij} = \tilde{r}_{ij} * \tilde{w}_{ij} \tag{3}$$

The fourth step is to determine the fuzzy positive ideal solution (FPIS, A^*) and the fuzzy negative ideal solution (FNIS, A^-). The FPIS (4) and FNIS (5) of the alternatives can be defined as follows, where $\tilde{v}_i^*$ is the max value of i for all the alternatives, and $\tilde{v}_i^-$ is the min value of i for all the alternatives. B and C represent the positive and negative ideal solutions, respectively.

$$A^* = \{\tilde{v}_1^*, \tilde{v}_2^*, \ldots, \tilde{v}_n^*\} = \left\{\left(\max_j v_{ij} | i \in B\right), \left(\min_j v_{ij} | i \in C\right)\right\} \tag{4}$$

$$A^- = \{\tilde{v}_1^-, \tilde{v}_2^-, \ldots, \tilde{v}_n^-\} = \left\{\left(\min_j v_{ij} | i \in B\right), \left(\max_j v_{ij} | i \in C\right)\right\} \tag{5}$$

The next step is to calculate the distance between each alternative and the FPIS A^* and the distance between each alternative and the FNIS A^-. The distance between each alternative and FPIS (6) and the distance between each alternative and FNIS (7) are calculated as follows. Note that $d(\tilde{v}_{ij}, \tilde{v}_j^*)$ and $d(\tilde{v}_{ij}, \tilde{v}_j^-)$ are crisp numbers.

$$S_i^* = \sum_{j=1}^{n} d\left(\tilde{v}_{ij}, \tilde{v}_i^*\right) \; i = 1, 2, \ldots, m \tag{6}$$

$$S_i^- = \sum_{j=1}^{n} d\left(\tilde{v}_{ij}, \tilde{v}_i^-\right) \; i = 1, 2, \ldots, m \tag{7}$$

The parameter d is the distance between two fuzzy numbers. Given two triangular fuzzy numbers (a_1, b_1, c_1) and (a_2, b_2, c_2), the distance between them can be calculated as follows (8):

$$d_v\left(\tilde{M}_1, \tilde{M}_2\right) = \sqrt{\frac{1}{3}\left[(a_1 - a_2)^2 + (b_1 - b_2)^2 + (c_1 - c_2)^2\right]} \tag{8}$$

The final step is to calculate the closeness coefficient and rank the alternatives. The closeness coefficient (9) of each alternative can be calculated as follows. The best alternative is closest to the FPIS and farthest from the FNIS.

$$CC_i = \frac{S_i^-}{S_i^+ + S_i^-} \tag{9}$$

4.2 Fuzzy TOPSIS Calculations

In this study, we consider four criteria and three alternatives that are ranked using the fuzzy TOPSIS method, which was implemented using an online tool called Online Output. The types and weights of the criteria were based on research in which a group of factors was related to a model of e-government adoption and certain factors were measured in terms of their influence on a scale of one to five, rated by experts [41]. In each of the four groups considered here (economic, socio-cultural, technological and organisational factors), the impact values for the individual assessments were summed, and the weighted mean and standard deviation (crisp numbers) were calculated based on these assessments (see Table 1).

Table 1. Characteristics of Criteria

Criterion	Fuzzy weight	Crisp weighted mean	Crisp standard deviation
Economic aspects	(0.200, 0.263, 0.325)	0,263	0,063
Socio-cultural aspects	(0.128, 0.216, 0.305)	0,216	0,089
Technological aspects	(0.217, 0.275, 0.333)	0,275	0,058
Organizational aspects	(0.182, 0.246, 0.310)	0,246	0,064

A case study was considered in which several alternatives were evaluated, involving a hypothetical situation where the currently implemented GIS may be outdated for a small municipality (see Table 2). To evaluate the alternatives, fuzzy scale codes were used (see Table 3).

Table 2. Assumptions made for the decision matrix

	Economic aspects	Socio-cultural aspects	Technological aspects	Organizational aspects
Maintain current GIS	4 – maintaining current GIS is the cheapest option in this point of time, but it might be more costly in the future	1 – users stop using current GIS because of lack of needed e-services functionality and high latency	1 – GIS is inoperable with other internal systems and poses data security risks	3 – officials are used to the processes, but functionality and data are outdated in current GIS
Upgrade current GIS	3 – upgrading current GIS would be costly, but not as implementation of a new one	2 – new functionality solves only part of the problem and GIS latency problem remains	2 – GIS will still lack most of its interoperability with internal systems	4 – officials will adjust to the new functionality while they maintain processes
Implement new GIS	1 – implementing new GIS is the most expensive alternative in comparison to the municipality budget	5 – latest technology used in GIS will be more appealing to the users in case of e-services available	5 – using latest technology will make GIS interoperable, while new data formats are available	2 – official will need to adjust to the new processes in the new GIS with new functionality

Table 3 shows the fuzzy scale used in the model. When evaluating the alternatives, the following question was posed: "What negative impact will a certain decision have on the different factors included in the criteria?".

Table 3. Fuzzy Scale

Code	Linguistic terms	L	M	U
1	Very high	1	1	3
2	High	1	3	5
3	Medium	3	5	7
4	Low	5	7	9
5	Very low	7	9	9

The alternatives in terms of various criteria were evaluated based on the decision matrix (see Table 4), the normalised decision matrix (see Table 5) and the weighted normalised decision matrix (see Table 6). Positive and negative ideal solutions were then calculated for these criteria (see Table 7) together with the distance of each alternative from these positive and negative ideals (see Table 8). Calculations of the closeness coefficient suggest that the best alternative for the example mentioned above is the implementation of new GIS (see Table 9). In this case, the benefits of the added functionality and capabilities of the new technology outweigh the financial expenses and the officials' recalcitrance to implement a new GIS. However, if there are funding issues, the municipality in this example may choose to upgrade its existing GIS instead.

Table 4. Decision Matrix

	Economic aspects	Socio-cultural aspects	Technological aspects	Organizational aspects
Maintain current GIS	(5, 7, 9)	(1, 1, 3)	(1, 1, 3)	(3, 5, 7)
Upgrade current GIS	(3, 5, 7)	(1, 3, 5)	(1, 3, 5)	(5, 7, 9)
Implement new GIS	(1, 1, 3)	(7, 9, 9)	(7, 9, 9)	(1, 3, 5)

Table 5. Normalized decision matrix

	Economic aspects	Socio-cultural aspects	Technological aspects	Organizational aspects
Maintain current GIS	(0.556, 0.778, 1.000)	(0.111, 0.111, 0.333)	(0.111, 0.111, 0.333)	(0.333, 0.556, 0.778)
Upgrade current GIS	(0.333, 0.556, 0.778)	(0.111, 0.333, 0.556)	(0.111, 0.333, 0.556)	(0.556, 0.778, 1.000)
Implement new GIS	(0.111, 0.111, 0.333)	(0.778, 1.000, 1.000)	(0.778, 1.000, 1.000)	(0.111, 0.333, 0.556)

Table 6. Weighted normalized decision matrix

	Economic aspects	Socio-cultural aspects	Technological aspects	Organizational aspects
Maintain current GIS	(0.111, 0.205, 0.325)	(0.014, 0.024, 0.102)	(0.024, 0.031, 0.111)	(0.061, 0.137, 0.241)
Upgrade current GIS	(0.067, 0.146, 0.253)	(0.014, 0.072, 0.169)	(0.024, 0.092, 0.185)	(0.101, 0.191, 0.310)
Implement new GIS	(0.022, 0.029, 0.108)	(0.100, 0.216, 0.305)	(0.169, 0.275, 0.333)	(0.020, 0.082, 0.172)

Table 7. Positive and negative ideal solutions

Criterion	Positive ideal	Negative ideal
Economic aspects	(0.111, 0.205, 0.325)	(0.022, 0.029, 0.108)
Socio-cultural aspects	(0.100, 0.216, 0.305)	(0.014, 0.024, 0.102)
Technological aspects	(0.169, 0.275, 0.333)	(0.024, 0.031, 0.111)
Organizational aspects	(0.101, 0.191, 0.310)	(0.020, 0.082, 0.172)

Table 8. Distance from positive and negative ideal solutions

Alternative	Distance from positive ideal	Distance from negative ideal
Maintain current GIS	0.433	0.225
Upgrade current GIS	0.343	0.325
Implement new GIS	0.281	0.377

Table 9. Closeness coefficient

Alternative	Ci	Rank
Maintain current GIS	0.342	3
Upgrade current GIS	0.487	2
Implement new GIS	0.573	1

5 Conclusions

The model presented in this paper aims to offer guidance to municipalities in determining the usefulness of implementing a GIS before establishing the specifics of implementation. The application of this type of decision-making model in municipalities would therefore allow for a more conscious and rational use of the budget for the public outlays of a local government unit. The solution proposed here is based on the critical success factor approach for selecting criteria and the fuzzy TOPSIS method. This approach should be verified using case studies to see if government units can avoid incurring unnecessary costs and wasting time on unnecessary new GIS implementations.

A different approach could involve identifying specific criteria that have a significant impact on the decision-making process. The fuzzy TOPSIS model presented in this paper is a preliminary and provisional approach rather than a final version. Before implementing it in municipalities, it needs to be improved through the involvement of the municipalities that are interested in using it to enhance their GISs.

References

1. Ramadhan, A., Sensuse, D., Arymurthy, A.: Assessment of GIS implementation in Indonesian e-Government system. In: Proceedings of the 2011 International Conference on Electrical Engineering and Informatics, Bandung, Indonesia (2011)
2. Monarcha-Matlak, A.: Wykorzystanie systemów informacji przestrzennej w administracji publicznej. In: Szpor, G., i Czaplicki, K. (eds.) Internet: informacja przestrzenna, pp. 3–12, C.H. Beck, Warsaw (2018)
3. Ganczar, M.: Implementation of the INSPIRE directive into national legal order in the field of infrastructure for spatial information. Studia Prawnicze KUL **3**(83), 91–95 (2020)
4. Trzaskalik, T.: Wielokryterialne wspomaganie decyzji. Przegląd metod i zastosowań. Zeszyty Naukowe Politechniki Śląskiej, Seria: Organizacja i Zarządzanie **74**, 239–263 (2014)
5. Churchman, C., Ackoff, R.: An approximate measure of value. J. Oper. **2**(1) (1954)
6. Tzeng, G., Huang, J.: Multiple Attribute Decision Making. Methods and Applications. CRC Press, London (2011)
7. Edwards, W.: Social Utilities, Engineering Economist, Summer Symposium, Series 6 (1971)
8. Edwards, W., Barron, F.: SMARTS and SMARTER: improved simple methods for multiattribute measurement. Organ. Behav. Hum. Decis. Process **60**, 306–325 (1994)
9. Saaty, T.: The Analytic Hierarchy Process. McGraw Hill, New York (1980)
10. Lootsma, F.: The REMBRANDT system for multi-criteria decision analysis via pairwise. Faculty of Technical Mathematics and Informatics, Delft University of Technology, Delft (1992)
11. Mikhailov, L., Tzvetinov, P.: Evaluation of services using a fuzzy analytic hierarchy. Appl. Soft Comput. J. **5**, 23–33 (2004)
12. Saaty, T.: Decision Making with Dependence and Feedback. The Analytic Network Process. RWS Publications, Pittsburgh (1996)
13. Bana e Costa, C., Vansnick, J.: Sur la quantification des jugements de valeur: L'approche MACBETH. Cahiers du LAMSADE **117** (1999)
14. Larichev, O., Moskovich, H.: ZAPROS-LM – a method and system for ordering. Eur. J. Oper. Res. **82**(3), 503–521 (1995)
15. Larichev, O.: Ranking multicriteria alternatives: the method ZAPROS III. European **131**, 550–558 (2001)
16. Roy, B., Bouyssou, D.: Aide Multicritere a la Decision: Methodes at Cas. Economica, Paris (1993)
17. Zaraś, K., Martel, J.: Multiattribute analysis based on stochastic dominance. In Munier, B., Machina, M. (eds.) Models and Experiments in Risk and Rationality. Kluwer Academic Publishers, Dordrecht (1994)
18. Nowak, M.: Preference and veto thresholds in multicriteria analysis based on stochastic. Eur. J. Oper. Res. **158**, 339–350 (2004)
19. Brans, J.: L'ingénièrie de la décision; Elaboration d'instruments d'aide à la décision. La méthode PROMETHEE. In: Nadeau, R., Landry, R. (eds.) L'aide à la décision: Nature, Instruments et Perspectives d'Avenir, pp. 183–213. Presses de l'Université Laval, Québec (1982)
20. Górecka, D., Muszyńska, J.: Spatial analysis of the innovation of Polish regions. Acta Universitatis Lodziensis Folia Oeconomica **253**, 55–70 (2011)
21. Nowak, M.: Investment project evaluation by simulation and multiple criteria decision. J. Civ. Eng. Manag. **11**, 193–202 (2005)
22. Diakoulaki, D., Koumoutsos, N.: Cardinal ranking of alternative actions: extension of the PROMETHEE method. Eur. J. Oper. Res. **53** (1991)
23. Górecka, D., Szałucka, M.: Country market selection in international expansion using multicriteria decision aiding methods. Multiple Criteria Decis. Making **8** (2013)

24. Górecka, D.: Zastosowanie metod wielokryterialnych opartych na relacji przewyższania do oceny europejskich projektów inwestycyjnych. In: Nowak, M. (ed.) Metody i zastosowania badań operacyjnych'10, pp. 100–125. Wydawnictwo Uniwersytetu Ekonomicznego w Katowicach, Katowice (2010)
25. Hwang, C., Yoon, K.: Multiple Attribute Decision Making Methods and Applications: A State of the Art Survey. Springer, New York (1981)
26. Jahanshahloo, G., Hosseinzadeh, F., Izadikhah, M.: Extension of the TOPSIS method for decision-making problems with fuzzy data. Appl. Math. Comput. **181**(2) (2006)
27. Opricovic, S.: Multicriteria optimization of civil engineering systems. Technical report. Faculty of Civil Engineering, Belgrade (1998)
28. Konarzewska-Gubała, E.: Bipolar: multiple criteria decision aid using bipolar reference. LAMSADE "Cashier et Documents" **56** (2009)
29. Trzaskalik, T., Sitarz, S., Dominiak, C.: Unified procedure for Bipolar method. In: Zadnik, L., Żerovnik, J., Poch, J., Drobne, S., Lisec, A. (eds.) SOR 2013 Proceedings (2013)
30. Górecka, D.: Wielokryterialne wspomaganie wyboru projektów europejskich. TNOiK, Toruń (2009)
31. Nowak, M.: Interaktywne wielokryterialne wspomaganie decyzji w warunkach ryzyka. Wydawnictwo Akademii Ekonomicznej w Katowicach, Katowice (2008)
32. Nowak, M.: INSDECM – an interactive procedure for stochastic multicriteria decision. Eur. J. Oper. Res. **175**, 1413–1430 (2006)
33. Nowak, M.: Trade-off analysis in discrete decision making problems under risk. In: Jones, D., Tamiz, M., Ries, J. (eds.) New Developments in Multiple Objective and Goal Programming. LNE, vol. 638, pp. 103–115. Springer, Heidelberg (2010). https://doi.org/10.1007/978-3-642-10354-4_7
34. Sotoudeh-Anvari, A.: The applications of MCDM methods in COVID-19 pandemic: a state of the art review. Appl. Soft Comput. **126** (2022)
35. Zadeh, L.: Fuzzy sets. Inf. Control **8**(3), 338–353 (1965)
36. Sambura, A.: Założenia modelu ekonomicznego SIP w Polsce. Prace Instytutu Geodezji i Kartografii **XLVI**(99), 89–110 (1999)
37. Izdebski, W.: Spatial information in Poland – theory and practice. Roczniki Geomatyki XV **2**(77), 175–186 (2017)
38. Ogryzek, M., Tarantino, E., Rząsa, K.: Infrastructure of the spatial information in the European community (INSPIRE) based on examples of Italy and Poland. Int. J. Geo-Inf. **9**(12) (2020)
39. Pepłowska, K.: System EZD RP w polityce rządowej. In: Barciak, A., Drzewiecka, D., Pepłowska, K. (eds.) Rola archiwów w procesie wdrażania systemów elektronicznego zarządzania dokumentacją, pp. 119–125. Wydawnictwo Uniwersytetu Śląskiego, Katowice (2018)
40. Ziemba, E., Papaj, T.: Implementation of e-government in Poland with the example of the Silesian Voivodship. Bus. Inform. **3**(25), 207–221 (2012)
41. Ziemba, E., Papaj, T., Żelazny, R.: A model of success factors for e-government adoption - the case of Poland. Issues Inf. Syst. **14**(2), 87–100 (2013)
42. Choi, T., Chandler, S.: Knowledge vacuum: an organizational learning dynamic of how e-government innovations fail. Gov. Inf. Q. **37**(1) (2020)
43. Ziemba, E., Papaj, T., Żelazny, R., Jadamus-Hacura, M.: Factors influencing the success of e-government. J. Comput. Inf. Syst. **56**(2), 156–167 (2016)

An Ontology-Based Collaborative Assessment Analytics Framework to Predict Groups' Disengagement

Asma Hadyaoui(✉) and Lilia Cheniti-Belcadhi

ISITC, PRINCE Research Laboratory, Sousse University, Hammam Sousse, Tunisia
asmahadyaoui@gmail.com, lilia.cheniti@isitc.u-sousse.tn

Abstract. Project-Based Collaborative Learning (PBCL) enables a group to collaborate on a real-world, complex project while systematically completing learning tasks. However, maintaining everyone's attention and interest in the collaborative process is difficult, which may affect performance. The primary purpose of this paper is to predict disengagement among PBCL learner groups. To achieve this, we developed a Collaborative Assessment Analytics Framework (CAAF), which is built on ontologies and the collection of formative assessment data. The experiment involved 312 undergraduate students registered in the first year of the degree program in transportation technology and engineering. We divided the students into diverse groups based on their Python programming skills. We conducted a collaborative assessment analysis to determine the different assessment patterns that could indicate disengagement in this situation. The assessment of the group's behavior by our analytical layer is backed by an ontology inspired by xAPI. Based on this semantic layer, supervised machine-learning algorithms were used to predict group disengagement. Using a Decision Tree algorithm, we were able to predict group disengagement with a high degree of accuracy of 86.96%, but using Random Forest, we were able to attain an astounding degree of accuracy of 95.65%. In such instances, instructors will be able to provide early intervention and feedback based on these findings.

Keywords: PBCL · formative assessment · self-group assessment · peer-group assessment · disengagement · learning analytics · Ontology · xAPI · prediction · Decision Tree · Random Forest

1 Introduction

Student engagement is a buzzword that has garnered considerable attention in recent decades. Despite the popularity of student engagement, the significance of engagement and its opposite, disengagement, are frequently overlooked [1]. According to [2], disengagement is the simultaneous withdrawal and defense of an individual's preferred self through behaviors that foster a lack of connections, physical, cognitive, and emotional absence, and passive, incomplete role performances. It has been demonstrated that students frequently experience boredom and inattention. Despite teachers' best efforts,

I. Czarnowski et al. (Eds.): KESIDT 2023, SIST 352, pp. 74–84, 2023.
https://doi.org/10.1007/978-981-99-2969-6_7

it is challenging to maintain a high level of student interest and prevent learner disengagement. The authors in [3] found that student discontent fully mediates the relationship between perceived academic element quality and student desire to disengage. When students want to abandon learning assignments, they are likely to do so. Perceived disengagement behavior cost is a major predictor of student intention and actual disengagement behavior. Maintaining a high level of student attention and preventing disengagement from the learning process is a pedagogical difficulty, particularly in the context of Online Collaborative Learning (OCL). It is still challenging to measure and evaluate each student's engagement in learning groups. Currently, a variety of newly developed technologies expand the scope of collaboration and offer enhanced learning opportunities. In this instance, we cite the work of [4] which showed that students' cognitive engagement in AR-assisted activities improved significantly. They concluded that compared to receiving subject information created by experts, students engage in self-generated context-enabling learning activities more frequently. In addition, research on the use of FlipGrid and VoiceThread [5], discussion tools that combine dynamic media such as audio and video, showed how these collaborative media tools might encourage student involvement and collaboration. While all the preceding referenced research focuses on the individual learner, to our knowledge no research has examined the disengagement of a group of learners. In addition, we have not discovered any attempt to predict group disengagement, be able to interact with it, and prevent it from impacting the performance of learners. This research work's major objective was to investigate the factors that can lead to group disengagement in a collaborative online learning environment, specifically in the context of PBCL. To comprehend the group's behavior during project completion, we opted for a collaborative analysis of the formative assessment process. This analysis served two purposes. One objective was to examine the potential association between formative assessment outcomes and group disengagement within the context of a PBCL. Another objective was to predict group disengagement based on the formative assessment data, which included peer group assessment and self-group assessment. This effort is guided by three research queries: How to optimally gather, organize, and characterize evaluation data within the framework of PBCL? What is the connection between a group's formative assessment achievement and its potential disengagement? Can we predict group disengagement depending on formative assessment data? To answer the first issue, we utilized social analytics assessment since it provides a clear picture of a group's behavior by collecting usable data through formative assessments using applicable technologies. Nonetheless, a question does arise: What, exactly, is formative assessment? Although numerous definitions of formative assessment have been provided, there is no obvious logic for defining and delimiting it within broader pedagogical ideas. The distinction between the summative and formative roles was first proposed by [6] in the context of program evaluation. According to Scriven, summative assessment results are used to determine the overall value of an educational program (compared to an alternative), whereas formative assessment data are designed to support program improvement. The assessment data provide indicators for analyzing the social presence of learners, in addition to monitoring and measuring the academic achievement of groups. These collected facts prompted us to build formal structures for their optimal utilization. The requirement to manage and interpret assessment data is crucial in the field. Thus, our approach is based on Semantic web technologies which provide a valuable foundation for the semantic integration of multi-source e-learning data, allowing for systematic consolidation, linking, and sophisticated querying. Therefore, we developed an ontological model called OntoGrAss and influenced by the xAPI standard to represent assessment data collected and reasoning rules to make a set of group-level and assessment data-level inferences. Ontologies are essential components that can be defined as formal descriptions of knowledge, consisting of a set of concepts and their possible relationships [7]. Exploratory Data Analysis (EDA) was used to first discover disengaged groups and then to determine the criterion that distinguishes and identifies these groups. EDA is a fundamental method that comprises conducting preliminary research on data to detect patterns, outliers, and test and validate hypotheses using summary statistics and graphical representations. EDA collects information, provides a greater comprehension of the data, and eliminates odd or unnecessary numbers. This allows a machine learning model to predict our data set more precisely. In addition, it enables the selection of a supervised machine-learning model. For pattern detection and prediction of group disengagement, we employed the Decision Tree and Random Forest algorithms in our methodology. This article's remaining sections are organized as follows. Section 2 describes the proposed Collaborative Assessment Analytics Framework (CAAF) by outlining its many layers. In Sect. 3, we provide OntoGrAss, our ontology. Then, we describe our experimental framework, the experimental outcomes, and the prediction models. We conclude with a few observations.

2 The Collaborative Assessment Analytical Framework (CAAF) Description

PBCL is a strategy that focuses on students and enables them to cooperate on a complicated, real-world project that improves their knowledge and abilities as they carefully complete smaller learning activities that add up to a larger project. A study [8] provided evidence of the effectiveness of PBCL to engage and improve a student's learning within an introductory programming subject in CSCL. It was found that focusing programming courses on authentic problems made the course more interesting for students [9]. We based our collaborative assessment methodology on this premise. A collaborative project to measure the mastery of particular skills by groups of learners is offered. The final objective is attained by completing a sequence of sub-objectives, each of which results in an assessment activity that is allocated to each group member based on the abilities required. Teachers frequently characterize the evaluation of group work as difficult and challenging, with individual assessment and fairness presenting concerns [10]. In this regard, our suggested framework includes two forms of formative assessments in addition to summative assessments. In our suggested system, summative assessment is the ultimate grade assigned by the instructor, who decides whether or not to validate the project on a 0-to-20-point scale. Regarding formative assessment, we've explored two assessment approaches: self-group assessment and peer-group assessment. A meta-analysis conducted in [11] revealed the positive effects of self- and peer-assessment interventions on academic performance in various contexts. We can cite the study conducted in [12] that casts light on how self-assessment complements peer assessment by comparing the use of self-assessment and peer assessment for an academic writing assignment among undergraduate students in Hong Kong. Self-group assessment is defined in our work as any sort of feedback offered by one group member to another group member in response to a question posed or to comment on a finished and delivered task. Each student has his/her profile, objective, prior knowledge, and performance about an assessment activity, etc., and he/she is encouraged to share his/her work with the other members of his/her group utilizing the associated discussion forum. Following each activity submission, all feedback should be posted in the forum. All of the forum contributions are gathered, analyzed, and interpreted to offer relevant and reliable data about self-group feedback. As a metric for evaluating the self-group assessment, the rate of group contributions, the contributions of presentation of the outcomes of the identified activities, corrections provided, and ideas for improvement are all analyzed. During the duration of the project, we also considered the amount of time each group member spent connected to the framework. For the peer group assessment, chat rooms are used to coordinate peer group assessment sessions. Each group is required to show its work to its peers, and a debate is held thereafter to ask questions, request explanations or justifications, offer corrections, etc. The peer then determines whether to approve (and hence increase the project's final grade) or disapprove the presented effort (and subsequently lower the final grade). All of the aforementioned assessment data are collected, evaluated, and interpreted to provide relevant and dependable data for our ontological models, which serve as training data to predict group disengagement and allow the teacher to deliver appropriate feedback. In the subject of assessment analytics, the necessity to organize and evaluate assessment data is crucial. All these types of formatives should be taken into consideration when designing the architecture for the CAAF. To achieve this objective, we propose an architecture that supports the analytics of assessment data predicting the likelihood of group disengagement during the project's completion.

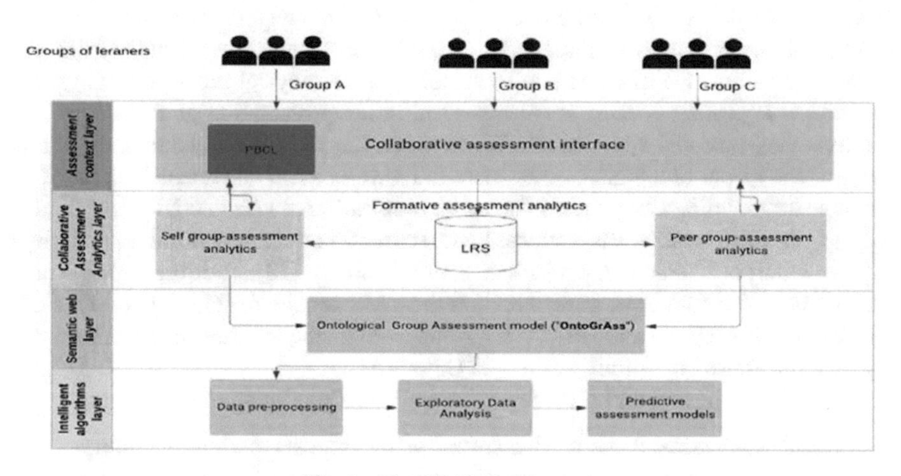

Fig. 1. The CAAF architecture

Figure 1 illustrates an overview of the CAAF architecture's main layers. The proposed architecture consists of four basic and ultimate layers: Learning context layer: it describes all interactions of groups of learners with the environment. We used Moodle to put our PBCL strategy into action because it is the most used Learning Management System (LMS) in Tunisian colleges. Collaborative groups of students utilize the LMS's discussion forum to communicate, assist one another, and express their opinions. Groups created multimedia presentations describing their project's progress and assessed the work of others using an assessment rubric; Assessment data analytics layer includes all assessment saved data in the Learning Record System (LRS). As previously indicated, these data include self-group assessment data and peer-group assessment data; Semantic layer: contains modeling ontology called OntoGrAss of the assessment data and reasoning rules to perform a set of inferences related to group feedback and formative assessment data. The semantic layer's primary function is to facilitate communication across the various parts of the architecture using a standardized representation of assessment data. This structure was conceived based on the xAPI data model. Consequently, it has a fundamental understanding of the assessment process, the assessment outcome, and the assessment setting; Intelligent layer: This layer enables the construction of prediction models of the potential disengagement of learner groups. To forecast this, our models are developed using supervised Machine Learning algorithms with OntoGrAss as input. Quality decisions must be based on quality data. Data preprocessing is important to get quality data.

3 Assessment Data Ontology Modeling

To optimize the benefits of sharing and automatically processing assessment data, all assessment data must be expressed using semantic web formalisms. Only by precisely defining the meaning of assessment data can computers and humans effectively collaborate. Our suggested collaborative assessment framework relies heavily on the modeling of assessment data derived from our group assessment analytics. For this reason, we focus primarily on data from three types of assessments: self-assessments, peer assessments, and summative assessments. Developing an ontological model for assessment analytics is appealing because it ensures a uniform representation of assessment data and has many potential applications. We relied on the xAPI specs standard during the construction of our ontological assessment framework. xAPI statements consist of the fundamental building components "actor verb object". An xAPI activity is a kind of object with which an actor has interacted. The action, when combined with a verb, might indicate a unit of instruction, performance, or experience [13]. To further complement our assessment method, we have modified the xAPI requirements to cover three distinct assessment types: summative, peer-group assessment, and self-group assessment. Figure 2 illustrates our ontological model OntoGrAss.

In the following, we will provide a more in-depth description of our ontology:

- ProjectAssessmentStatement: This describes the proposal made to the groups.
- ProjectAssessmentVerb: Six types of activities can be performed in the evaluation context, depending on the actor:

  – Spend: A group spends "x" total hours using the collaborative platform.
  – Deposit: A learner uploads his/her work to his/her group's discussion forum.

Fig. 2. OntoGrAss description of classes

- Grade: The instructor assigns a final grade ranging from 0 to 20.
- Comment: A student may participate in the discussion forum.
- Increment: A group can increment the final grade of its peer's project if it considers it validated.
- Decrement: A group can decrement the final grade of its peer's project if it does not validate it.

- ProjectAssessmentActor: An actor can represent the teacher in the context of summative assessment, the student in the scenario of self-group assessment, or a group in the context of peer group assessment.
- ProjectAssessmentObject: In each context, the following factors will be evaluated: the rate of participation in the discussion forum, the time spent online during the performance of the project, the final grade assigned to the project, and the grade assigned by each learner to his/her groupmate.
- ForumContribution: It describes the contributions made to the discussion forum by the groups. Students could post messages by indicating beforehand the type of message to be sent. We have divided the contributions to the learners' forum into three categories:

  - OtherContribution: Participants who are experiencing trouble completing their tasks as part of a larger collaborative endeavor can publish such messages. Also, they can provide answers to questions or problems raised by other group members. In addition, such contributions may be made by group members who desire to share a resource, or a tip, or plan a synchronous meeting with other group members.

  - ProvideFeedback: A learner can provide feedback in response to a submitted activity to propose adjustments, suggest improvements, or affirm a result.
  - DepositActivity: A learner can share his or her activity to solicit feedback from the group.

- Context: Assessment contexts include:

  - Summative assessment.
  - Formative assessment: peer group assessment or self-group assessment.

- Result: Since we aim to develop a comprehensive, accurate, and adaptable assessment analysis ontology model that successfully supports assessment analysis, the assessment outcome class should be specified with extensive metadata. A learner's result is determined by averaging the grades assigned to his or her submitted work by other group members. The result also refers to the grade assigned by the instructor for the group's project following peers' potential additions or deletions.

4 Methodology

This research work aims to predict the likelihood of group disengagement in a Project-Based Collaborative Learning (PBCL) environment. The ensuing research question was developed in light of the stated objective: Is it possible, based on formative assessment, to predict group disengagement? Our research is predicated on evidence indicating a strong link between disengagement and bad assessment results. We investigated 312 undergraduates. During the first semester of the academic year 2022–2023, they are enrolled in the first-degree program at our higher education Institute. Each group was heterogeneous before the experiment. The participants were divided into 84 groups of three to four pupils each (60 groups of 4 and 24 groups of 3). Each group was given a pre-test to determine how well they performed and to ensure that each group consisted of individuals with varying programming levels. In addition, the numerous learning resources accessible in Moodle were presented and shown to the attendees. To identify duties for each member and devise tactics for completing the task at hand, groups began talking with one another over message boards. Within the context of self-group assessment, the forum is our primary source of assessment data. As a result of encouraging greater online connection and collaboration, a discussion forum was developed on Moodle for each group of the sample to enhance intra-group engagement and resource exchange. Included are both the files from completed collaborative projects and Moodle activity logs that track the amount of time and contributions made by different forum groups. We analyzed the data and polled the forum participants to determine how the experiment went. The primary experiment was carried out for a total of six weeks during the first semester. First, EDA was conducted to identify the groups exhibiting disengagement during the project's completion. EDA is the crucial process of conducting preliminary research on data to identify patterns, identify outliers, test hypotheses, and confirm ideas using summary statistics and graphical representations. This enables a machine learning model to predict our dataset more precisely. In addition, it helps us choose the best Machine Learning model. We made use of Google Colab for the implementation.

5 Results

5.1 Knowing About Groups Disengaged

The boxplot was utilized to identify and effectively communicate the disengaged groups. According to [14], a boxplot is a descriptive statistics technique used to graphically portray numerical data groups by their quartiles. Boxplots are a common approach for illustrating the distribution of data using a five-number summary: minimum, first quartile, median, and maximum. This graph style is used to identify outliers quickly. In terms of our methodology, these outliers are very important. A statistical outlier is a point that dramatically deviates from the other observations. As demonstrated in Fig. 3, the outliers in our experiment provide the groups with very low final project grades. And while all the groups are diverse, the programming levels for all the groups are virtually the same. Since the same project is assigned to all groups for the same time and under the same conditions, we can only ascribe this disparity in the final appraisal of the project to the group's disengagement.

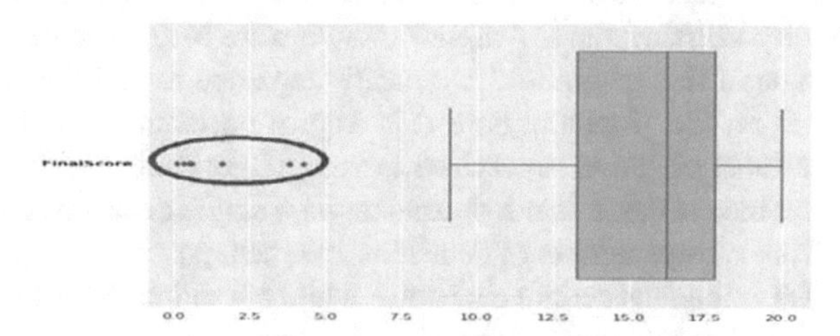

Fig. 3. Boxplot of final group project scores

Observing the boxplots revealed that our outliers are groups having a final project grade of no more than 5. Consequently, we consider a group to be disengaged if its final project score is below 5.

5.2 Predicting Groups' Disengagement Depending on Formative Assessment

We applied the Decision Tree (DT), which is a supervised learning technique, to answer our core question: Can we predict the disengagement of a group before the summative evaluation so that we can intervene and improve the outcomes?

Using attribute selection measures (ASM) to segment the data, the DT algorithm discovered the most predictive characteristic for group disengagement based on formative assessment data (The self-group assessment: forum contribution, times online, the peer group assessment: peerGroupApprovedOrNot). This is a significant indicator of the group's potential disengagement, as indicated by peerGroupApprovedOrNot. Figure 4 shows this criterion as a decision node that splits the data set into two large portions. Each algorithm iteration calculates multiple Gini indexes. The Gini Index, often known as the Gini impurity, measures the chance of misidentifying a randomly picked attribute: $\text{GiniIndex} = 1 - \sum_{i=1}^{n} P_i^2$. Where P_i is the probability of an element belonging to a

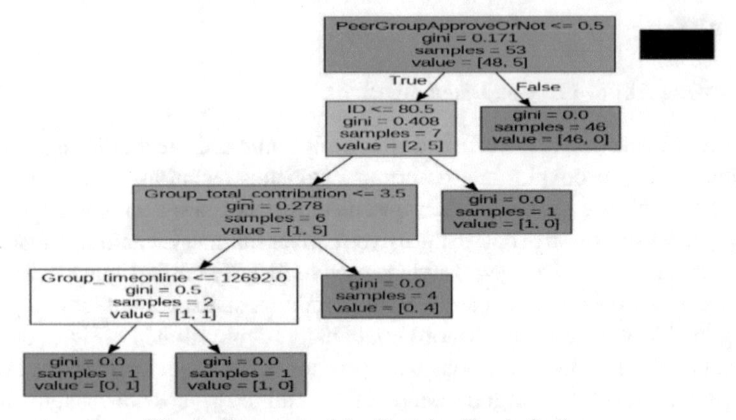

Fig. 4. Summary of the Decision Tree's finding

particular class. As with any prediction algorithm, it is necessary to calculate some metrics to determine its efficacy. Comparing actual test set values to expected test set values yields accuracy. By utilizing the DT algorithm, we were able to achieve an accuracy of 86.96%. A value of this magnitude is typically regarded as satisfactory. We moved to a second algorithm (i.e., Random Forest) to improve accuracy. The Random Forest ranks among the most effective supervised learning algorithms and is quite popular. The primary difference is that it is not dependent on a single decision. It compiles random decisions from a large number of decisions and then makes a conclusion based on the majority. A DT integrates certain decisions, while a Random Forest mixes multiple Decision Trees [15]. The precision attained 95.65% using the Random Forest algorithm. We utilized DT because it is simple to understand and interpret, which is especially essential when attempting to determine why a group of learners may have disengaged. Random Forest, on the other hand, offers a comparable level of interpretability with greater precision. Moreover, collaborative e-learning environments are intricate systems with multiple variables and interactions. DT and Random Forest are capable of capturing non-linear relationships between these variables, which linear models may overlook.

6 Discussion and Conclusion

We proposed a collaborative assessment analytics technique for the prediction of PBCL context group disengagement. Specifically, we were interested in learning groups rather than individual learners. We also focused on formative assessment data. For that, we have presented a formative assessment strategy based on two axes: the self-group assessment, which takes into account the group interactions, the activities posted in the discussion forum, the feedback from the group members, and the platform connection time during the project. In addition, the peer group assessment allows each group to present their project to their peers and receive verbal criticism in a chat room, as well as a decision to approve or reject the work. We have presented an architecture for our proposed CAAF based on semantic web technologies. Our architecture is guided by the xAPI specifications and is built on an ontological model suited for assessment analysis. OntoGrAss is the ally of our prediction model of probable disengagement. Our research is based on evidence demonstrating a high correlation between disengagement and very negative outcomes. As revealed by [16], there are substantial correlations between online collaborative tools, collaboration with peers, student involvement, OCL activities, and student learning achievement. Also, according to [17], disengagement from task content in educational apps could severely impair learning outcomes. To do predictions, we initially employed DT as a methodology, a supervised, simple classification instrument for predictive analytics. The prediction results for the training data, i.e., the data known by the algorithm, were favorable, while the accuracy for the test data was only 86.96%. To further improve our results, we employed Random Forest, which enabled us to achieve a prediction accuracy score of 95.65. Comparing our prediction accuracy found to those of the researchers [18] who have studied the feasibility of leveraging early dropout predictions of DNN models, 85% accuracy was validated. This indicates that our outcomes are remarkable, and as a result, it demonstrates the viability of the suggested method. It is possible that DT and Random Forest are not the only machine learning techniques that have the potential to provide accurate results for predicting group disengagement; we can also investigate Support Vector Machines (SVMs), for instance. Lastly, being able to predict the disengagement of a group from their behavior within the framework of the formative assessment of a collaborative project is a way to devise tools to prevent this disengagement from occurring again, maintain the motivation of the members, and thereby guarantee good results and a higher level of performance. The recommendations are intended for both the affected group and the instructor to prevent unfavorable results.

References

1. Balwant, P.T.: The meaning of student engagement and disengagement in the classroom context: lessons from organisational behaviour. J. Furth. High. Educ. **9486**, 1–13 (2017). https://doi.org/10.1080/0309877X.2017.1281887
2. Kahn, W.A.: Psychological conditions of personal engagement and disengagement at work. Acad. Manag. J. **33**(4), 692–724 (1990). https://doi.org/10.5465/256287
3. Pham, T.T.K., Vu, D.T., Dinh, V.H.: The impact of academic aspect quality on student disengagement in higher education. Educ. Sci. **12**(8), 507 (2022). https://doi.org/10.3390/educsci12080507
4. Wen, Y.: Augmented reality enhanced cognitive engagement: designing classroom-based collaborative learning activities for young language learners. Educ. Tech. Res. Dev. **69**(2), 843–860 (2020). https://doi.org/10.1007/s11423-020-09893-z
5. Saçak, B., Kavun, N.: Rethinking flipgrid and voicethread of in the context online collaborative learning theory. In: Research Anthology on Remote Teaching and Learning and the Future of Online Education, pp. 331–348 (2022). https://doi.org/10.4018/978-1-6684-7540-9.ch018
6. Scriven, M.: The methodology of evaluation. In: Program Evaluation, p. 16 (1967)
7. Gruber, T.R.: A translation approach to portable ontology specifications. Knowl. Acquis. **5**(2), 199–220 (1993)
8. Yeom, S., Herbert, N., Ryu, R.: Project-based collaborative learning enhances students' programming performance. In: Proceedings of the 27th ACM Conference on on Innovation and Technology in Computer Science Education, vol. 1, pp. 248–254 (2022)
9. Wilson, O.A., Essel, D.D.: Learning computer programming using project-based collaborative learning: students' experiences, challenges and outcomes. Int. J. Innov. Educ. Res. **9**(8), 191–207 (2021)

10. Forsell, J., Forslund Frykedal, K., Hammar Chiriac, E.: Group work assessment: assessing social skills at group level. Small Gr. Res. **51**(1), 87–124 (2020). https://doi.org/10.1177/1046496419878269
11. Yan, Z., Lao, H., Panadero, E., Fernández-Castilla, B., Yang, L., Yang, M.: Effects of self-assessment and peer-assessment interventions on academic performance: a meta-analysis. Educ. Res. Rev. 100484 (2022). https://doi.org/10.1016/j.edurev.2022.100484
12. Cheong, C.M., Luo, N., Zhu, X., Lu, Q., Wei, W.: Self-assessment complements peer assessment for undergraduate students in an academic writing task. Assess. Eval. High. Educ. **48**(1), 135–148 (2023). https://doi.org/10.1080/02602938.2022.2069225
13. Samuelsen, J., Chen, W., Wasson, B.: Enriching context descriptions for enhanced LA scalability: a case study. Res. Pract. Technol. Enhanc. Learn. **16**(1), 1–26 (2021). https://doi.org/10.1186/s41039-021-00150-2
14. Fitrianto, A., Muhamad, W.Z.A.W., Kriswan, S., Susetyo, B.: Comparing outlier detection methods using boxplot generalized extreme studentized deviate and sequential fences. Aceh Int. J. Sci. Technol. **11**(1), 38–45 (2022). https://doi.org/10.13170/aijst.11.1.23809
15. Quinlan, J.R.: Induction of decision trees. Mach. Learn. **1**(1), 81–106 (1986). https://doi.org/10.1007/bf00116251
16. Ng, P.M.L., Chan, J.K.Y., Lit, K.K.: Student learning performance in online collaborative learning. Educ. Inf. Technol. **27**(6), 8129–8145 (2022). https://doi.org/10.1007/s10639-022-10923-x
17. Kristensen, J.K., Andersson, B., Torkildsen, J.V.K.: Modeling disengaged guessing behavior in a vocabulary learning app using student, item, and session characteristics. In: 2022 International Conference on Advanced Learning Technologies (ICALT), pp. 414–416. IEEE (2022). https://doi.org/10.1109/ICALT55010.2022.00128
18. Park, H.S., Yoo, S.J.: Early dropout prediction in online learning of university using machine learning. Int. J. Inf. Vis. **5**(4), 347–353 (2021). https://doi.org/10.30630/JOIV.5.4.732

Artificial Intelligence Innovation in Daily Life

On Sensing Non-visual Symptoms of Northern Leaf Blight Inoculated Maize for Early Disease Detection Using IoT/AI

Theofrida Julius Maginga[1](✉), Deogracious Protas Massawe[2], Hellen Elias Kanyagha[2], Jackson Nahson[2], and Jimmy Nsenga[1]

[1] African Centre of Excellence in Internet of Things (ACEIoT), University of Rwanda, Kigali, Rwanda
tjmaginga@gmail.com

[2] Sokoine University of Agriculture, 3000 Morogoro, Tanzania

Abstract. Conventional plant disease detection approaches are time consuming and require high skills. Above all, it cannot be scaled down to smallholder farmers in most developing countries. Using low cost IoT sensor technologies that are gas, ultrasound and NPK sensors mounted next to maize varieties for profiling these parameters on a given period. Here we report an experiment performed under controlled environment to learn metabolic and pathologic behavioral patterns on healthy and NLB inoculated maize plants by generating time series dataset on profiled Volatile Organic Compounds (VOC), Ultrasound and Nitrogen, Phosphorus, Potassium (NPK). Dataset has been preprocessed with pandas and analyzed using machine learning models which are dickey fuller test and python additive statsmodel and visualized using matplotlib library to enable the inference of an occurrence of a disease a few days post inoculation without subjecting a plant to an invasive procedure. This enabled a deployment and implementation of noninvasive plant disease detection prior to visual symptoms that can be applied on other plants. With analyzed data, the IoT technology in this experiment has enabled the detection of NLB disease on maize disease within seven days post inoculation because of monitoring VOC and ultrasound emission.

Keywords: NLB · Maize · IoT · timeseries · VOC · ultrasound · NPK

1 Introduction

Plant diseases are caused by chronic or emerging pathogens that result in stagnant growth of about 10% in the plant system [1]. Meanwhile, [2] has reported maize loss due to plant disease by 40% in East Africa, and these diseases keep spreading to other areas. [3] realized that food security as a part of zero hunger sustainable development goal number two (SDG-2) is becoming almost unattainable given the increase in global human population and plant diseases causing diverse effects on livestock health and human demographic patterns.

I. Czarnowski et al. (Eds.): KESIDT 2023, SIST 352, pp. 87–96, 2023.
https://doi.org/10.1007/978-981-99-2969-6_8

Existing technology-based approaches for early disease detection vary from the wide range of biomolecular approaches like Polymerase Chain Reaction (PCR) and Enzyme-Linked Immunosorbent Assay (ELISA) [4]. Despite these approaches being accurate, the access to equipment and skills to perform such experiments is not affordable for smallholder farmers in developing countries. Furthermore, these experimentation approaches require a destructive procedure to a sample (plant) [5]. On the other hand, as an alternative to overcome those limitations [4] implemented a noninvasive approach that profiles volatile organic compounds (VOCs) to late blight-infected tomato using an imperceptible sensor patch integrated with graphene-based sensing materials, which captures the plant's Deoxyribonucleic Acid (DNA) properties for real time detection. However, development of such technology is highly expensive to make it viable for the mass market of smallholder farmers.

Moreover, the detection of crop diseases via VOCs has also been confirmed by [6–8] where plants emit VOCs in a peculiar pattern when infected by a disease and this can be profiled as a plant's mode of communication. Additionally, a study done by [9] confirmed the rapid growth interest in profiling plants' VOCs as a factor for identifying metabolic and pathologic processes in the plant system. In parallel to VOC, there are different modes of plant's communication, it has also been reported in [9] that when a plant is stressed/unhealthy it emits sounds. A few research works have been done on this area includes [10] who observed that tomato and tobacco when cut (stressed) emits a mean value of 65 dB for an airborne sound, compared with the ambiance sound level for the quiet urban daytime or suburban area which ranges from 45–50 dB [11]. Therefore, sound is an additional parameter that may be sensed to spot diseased crops and it has also been observed that the fertilizer consumption rate varies when a crop becomes unhealthy as per study done by [12], hence a special interest on these parameters.

In that regard, on leveraging the latest advances on IoT sensing technologies, this paper aims to present the data collection experimental approach for noninvasive disease detection by using affordable and low powered Internet of Things (IoT) technology. Moreover, given the lack of open dataset for characterizing diseased maize during the pre-visual symptom disease cycle, the paper describes the procedures for time series dataset generation for both healthy versus laboratory inoculated maize crop through data collection of VOCs, ultrasound and NPK consumption over a period starting from when the maize plant is cultivated to inoculation, up until when visual symptoms appear, in our experimentation this period was about 35 days. This paper is hereby categorized as follows; Sect. 2 is focused on the experimentation approach for data collection by providing a clear depiction of the methodology used for generating inoculums and introducing the spores on maize plants; Sect. 2.2. describes the IoT data collection devices used in our experimentation campaign. Section 3 provides the highlight of the analysis on the collected dataset and predictive machine learning model training. Section 4 concludes on the highlighted analysis of data versus the early detection of maize disease.

2 Materials and Methods

2.1 Experimentation Approach for Data Collection

Northern Leaf Blight on Maize. In this study, we identified Northern Leaf Blight (NLB) as the most prominent and almost neglected disease that affects maize plants in the region. NLB is caused by a fungus scientifically known as Exserohilum turcicum [13] and it can be identified by relatively large gray elliptical or cigar-shaped lesions that develop on leaves ranging from 1 to 6 inches long. It is favored by high relative humidity and cool to moderate temperature conditions. NLB occurrence results in yield loss of up to 30 to 50% when it develops early in the season, with the diseased plant undergoing a premature death due to inability to photosynthesize caused by leaf blighting [14, 15]. Under these circumstances, if the plant is left untreated or unattended, sporangia can spread to other plants since plant pathogens take up to 2 weeks [16] to spread from infected plants to uninfected plants; thus, causing a huge loss to the farmers.

Study Area and Field Management. Controlled environment experiments were laid out at Sokoine University of Agriculture in Morogoro, Tanzania. Four maize varieties that are either resistant or susceptible to NLB disease were selected for this experiment considering that they are commonly used by small holder farmers and highly recommended by seed suppliers in the region. These varieties are: DK8033, DK9089, SeedCo 719 (Tembo) and SeedCo 419 (Tumbili). Maize seeds were sown in four liters plastic buckets, four seeds per pot, at 7 cm distance from each other (Fig. 1). The experiments were made up into two sets where set 1 contained treatment One (T1) as control (healthy) and set two with treatment Two (T2) NLB inoculated plants. Each set of treatment had eight buckets randomized in such a way each maize variety had equal chance of receiving light, temperature, and humidity gradients. The plants were irrigated twice per week and fertilized with 10 mg of NPK two weeks post sowing and after 6 weeks from first application (Fig. 1).

Fungal Isolation, Inoculum Preparation & Application. We isolated E. turcicum from NLB diseased plants collected from nearby maize fields. The diseased tissue pieces of about 0.5 cm were prepared and sterilized by soaking into 70% alcohol for one minute before rinsing with sterile distilled water and blotting using sterile paper. Sterilized pieces of tissue were aseptically transferred into sterile potato dextrose agar (PDA) medium and incubated at 24 °C for 14 days at 12-h light/dark cycles (HK & JN) to induce sporulation (Fig. 1). The resulting fungal colonies were checked and confirmed microscopically for presence of E. turcicum spores and subcultured into fresh PDA medium for pure culture production.

Following 14 days of growth pure cultures were selected for inoculum production. Spores' suspension was prepared by aseptically flooding Petri plates containing pure culture with sterile distilled water and scrapping the colonies using sterilized toothbrush. Thoroughly mixed spore suspension was poured into a bottle containing a known amount of sterile distilled water. Spore concentration from the prepared suspension was quantified by counting the spore's using a hemocytometer. A known amount of suspension was pipetted into a hemocytometer and mounted on a microscope stage for counting. The quantified spore suspension was then adjusted to 106 spores/ml by adding more sterile water or spores and later sieved using sterilized cheesecloth into a sterile bottle. The resulting suspension were filled into prepared bottles ready for inoculation onto six leaved maize seedlings (8 weeks old).

Set of eight control plants (T1) was inoculated by pouring 5 ml of plain sterile distilled water and another set of eight plants (T2) were inoculated by pouring 5 ml of the prepared suspension between the leaf and maize stem during cool hours (around 1700 h in the evening) (Fig. 1). This was done to facilitate spore survival and maintain an adequate humidity condition. The two sets of the plants were placed in a separated table that are 5 m away from each other. The plants were raised for 90 days under screen house conditions. Each of the experimental pots was maintained by regular watering and checked for disease and pests twice per week. This has been done to ensure that plant on T1 stay healthy all over the experiment and for T2 only NLB is the only active disease.

Fig. 1. Maize Plant Experiment set with inoculum, spore on the middle and last is the application of inoculum spores on maize plants

2.2 IoT Based Data Collection Approach

IoT sensors in this study have been identified as a low-cost technology for implementing a noninvasive procedure for disease detection. IoT sensors were placed next to both plants on day 35 post inoculation because spores require a slightly matured plant to act as a host for reproduction and proper infection. During this experiment, we measured the following parameters to monitor non-visual NLB symptoms: (1) total volatile organic compounds (VOCs), (2) soil's nitrogen, phosphorus, potassium (NPK), (3) ultrasound. Furthermore, as part of monitoring and validating the experimentation conditions other environmental parameters have been measured such as temperature, relative humidity, and barometric pressure for both control and inoculated maize plant varieties.

Total Volatile Organic Compound. For collecting such data, a Bosch BME688 Development Kit was used. This tool is a new technology widely used for gas sensing in different use case scenarios such as detecting leakage of harmful or noxious gasses. In our study, the identified tool contains gas sensors that can measure the unique electronic fingerprints that enable it to identify the gas emission pattern of a particular object. A study done by [17] identifies that concentration of different gases can be converted to standard electrical signals using gas sensor technology. To guarantee 3 days of operation for this development kit, we connected it to a power bank of 5000 mAh. Additionally, this development kit uses ESP32 microcontroller, with CR1220 coin cell battery for real time tracking and a 32 GB microSD for data storage. Figure 2 displays how the device was placed next to a maize plant so that it can be able to smell the VOCs emitted by a plant over time.

Ultrasound. Data regarding the value of sound data emitted by a plant was collected by two sound sensors interchangeably. Sensors used were OSEPP Electronics Multiple Function Sensor Development Tools and DAOKI Sound Microphone Sensor both programmed on ESP8266 using the Arduino IDE. Sensors were sending data over the cloud-based tool known as ThingSpeak. ESP8266 comes with a WiFi module that facilitates smooth real time data transfer over the cloud to be exported later for data analysis. The IoT devices were powered by a 5000 mAh shared with another set for VOC data collection. Figure 2 shows how data was collected by placing a microphone next to a plant stem and leaves so that the device may be able to closely listen to a plant for the non-human audible sound values.

Nitrogen, Phosphorus & Potassium (NPK). Fertilizer consumption overtime was measured by collecting data using the Taidacent Soil NPK sensor and JXCT soil NPK sensor. All two sets of sensors were programmed on Arduino IDE using ELEGOO Nano Board CH 340/ATmega+328P as one of the components and RS-485 for handling serial port for enabling communication between signals received from the NPK sensor and ones sent back to the microcontroller. This sensor itself required a separate 12 V adapter and powering the microcontroller required about 3/5 V and the source of power was a direct current (DC) powerline. Data was captured directly from the USB port and saved using CoolTerm software that is used for capturing the inputs for the indicated USB serial port. Figure 2 shows how the NPK sensor was placed on the identified maize variety.

Fig. 2. Data collection using IoT sensors on Volatile Organic Compounds, microphone sensors and NPK fertilizer

3 Results and Discussion

3.1 Dataset and Data Preprocessing

In the above experimentation campaign, the data collected between healthy and inoculated maize crops are time-series that are organized in rows and columns format. VOC data for healthy were 34,812 and for inoculated were 38,621 rows. Data from the sound sensor was captured via Decibel measurement with a total of 16949 rows for control (T1) and 172595 rows for inoculated (T2). NPK fertilizer consumption data collected summed up to 37440 rows for control (T1) maize variety and 23955 rows for inoculated (T2) maize variety. Data cleaning was done to omit other parameters from VOC general dataset which included temperature, barometric pressure and relative humidity and as well be able to acquire the univariate feature for our dataset.

3.2 Checking to Stationarity (ADF)

For any timeseries data, it is recommended to check the data stationarity so that it may be able to relate towards the parameter values versus time. In this case we used the Dickey fuller Test (ADF) that is an approved statistical test to check the stationarity of our data. The idea behind is to check whether our data agrees or rejects the null hypothesis. According to [18] in statistics, if obtained values are greater than the threshold value (0.05) then the null hypothesis is accepted and vice versa. Results for the stationarity check for both of our dataset on VOC and ultrasound are shown on Table 1.

Table 1. Dickey Fuller Test for Control (T1) and Inoculated (T2) Maize Plants on VOCs and Ultrasound Time Series Data

	Volatile Organic Compounds	Ultrasound
Control Maize (T1) p-value	4.2538635031500405e−18	1.3281864632799207e−09
Inoculated Maize (T2) p-value	9.40807059183648e−13	8.689417848955201e−25

From the results shown on given tables can be inferred as follows: Total VOC emission, and ultrasound values have p-values greater than the threshold value, for this case it has failed to reject the null hypothesis (data is nonstationary).

In this case [19] argues that, when data is not stationary it can only mean that, there's and observed strong trend and seasonality in term of volatile organic compounds emission on both plants and ultrasound: therefore, confirming that it is possible to acquire predictable pattern results [20].

3.3 Identification of Volatiles Patterns

The profiled emission for the control maize plant seemed to decrease overtime while for the NLB inoculated maize plant showed a steady increase overtime. Figure 3 shows the total mean of a calculated sample of gas sensor values for the volatile emission over time, and the acquired results show a clear decrease and an increase for the control variety and inoculated variety consecutively.

To furthermore observe the meaning of our data, we implement the statsmodels library to decompose our time series data into trend and seasonality. For this case, in the trending aspect of our decomposed data we are looking at a pattern of VOC emission that spans across daily periods for both control and inoculated maize variety. The selection

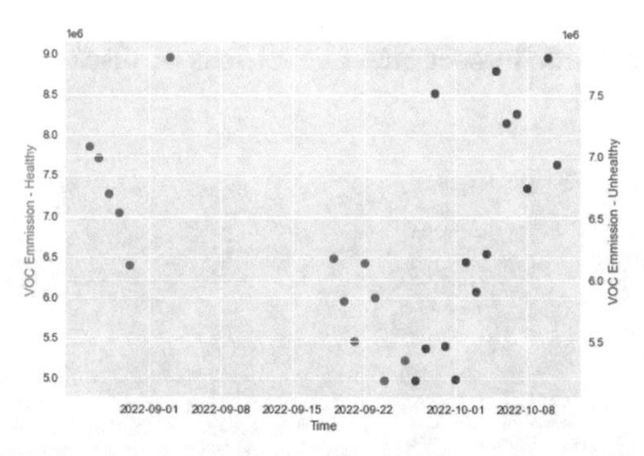

Fig. 3. Calculated mean sample of VOCs emission for healthy vs inoculated maize

of the statsmodel is because it is a powerful tool for generally understanding the time series data and as well provides an abstract of a better understanding of what our problem presents [21]. The general formula for the statsmodel additive formula is presented as shown on equation one.

$$Y[t] = T[t] + S[t] + e[t] \tag{1}$$

Figure 4 shows the trend of VOC pattern emission for the identified maize varieties. Control maize variety (healthy/T1) has shown that the trend is decreasing while for inoculated maize variety (unhealthy/T2) the trend is increasing.

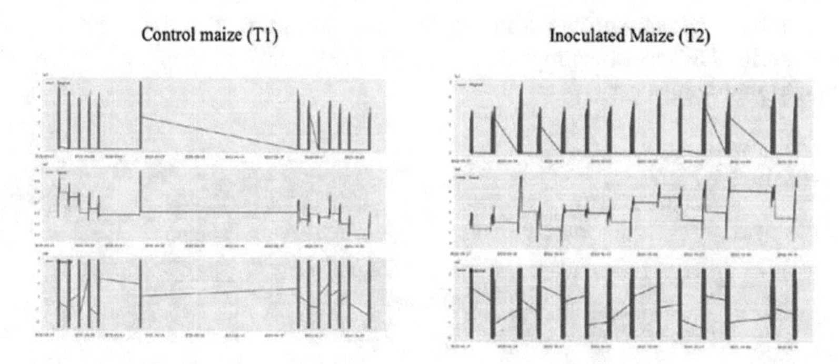

Fig. 4. Trend of the VOCs emission for healthy and inoculated maize crops

3.4 Ultrasound Emission Patterns

Figure 5 shows the general values for ultrasound values for healthy and unhealthy that differ in terms of values, where healthy maize has shown to emit ultrasound values of 50 dB that is close or equal to ambient level of emitted sound. While for the inoculated maize the captured values are greater than the ambient sound level. Moreover, the daily seasonal values for healthy maize averaged on 20 dB and for the inoculated maize averaged on 80 dB. This observation can only mean that, with plants or particularly for maize crop sound is emitted under stress conditions of disease inoculation above the ambient ultrasound level.

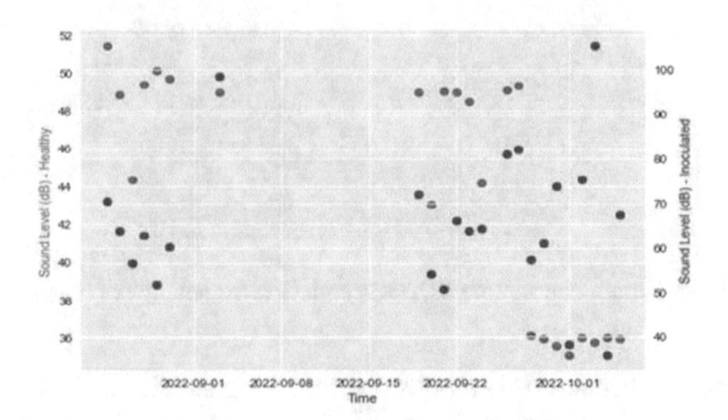

Fig. 5. Ultrasound emission for healthy and inoculated maize

3.5 NPK Consumption Pattern

Observation done on the total NPK data were not able to give a conclusive result as far as whether the fertilizer consumption had a particular trend on healthy versus inoculated maize variety. However, Fig. 6 shows a bit more systematic pattern of the NPK being lower on healthy maize, that could be translated as a proper consumption rate of NPK fertilizer to help the plant grow while a randomized pattern with higher NPK values that can be described as less consumption rate as the NPK rate on inoculated maize variety maintain higher values rate.

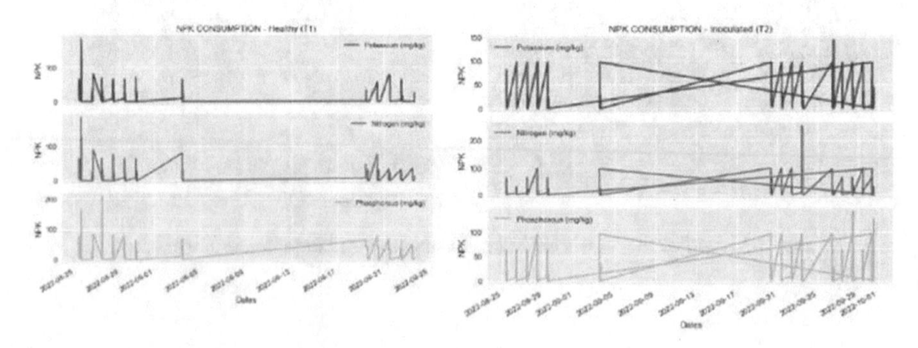

Fig. 6. Trend of the NPK consumption for healthy and inoculated maize variety

4 Conclusion

This paper presents a research experimentation leveraging the latest IoT sensing technologies to collect non-visual symptoms on diseased maize, this as a first step to build an early disease detection device. The experimentation particularly focused on Northern

Leaf Blight which is the main critical maize disease in east Africa. The study has generated time series datasets that can be reused under the sub-Saharan countries' context for different cases of machine learning model development. Dickey Fuller Test has been applied to check the stationarity of the data, statsmodel library was used to decompose the data into trend and seasonality. Data observation shows that given a maize plant has been inoculated with spores, VOC and sound level patterns increase over time as signaling processes from a plant to communicate its distress. Furthermore, from collected data analysis of non-visual symptoms, it has proven that NLB maize diseases can be detected within less than 7 days while visual symptoms appear after 14 days. Hence monitoring non-visual symptoms can help to detect diseases early and therefore reduce the overall yield loss.

Acknowledgements. This work is financially supported by The PASET Regional Scholarship and Innovation Funds as a part of PhD work scholarship and as well hosted at the African Centre of Excellence in Internet of Things Rwanda. Experimental works have been hosted by Sokoine University of Agriculture, Tanzania.

References

1. Strange, R.N., Scott, P.R.: Plant disease: a threat to global food security. Annu. Rev. Phytopathol. **43**, 83–116 (2005). https://doi.org/10.1146/annurev.phyto.43.113004.133839
2. Wangai, A.W., et al.: First report of maize chlorotic mottle virus and maize lethal necrosis in Kenya. Plant Dis. **96**(10), 1582 (2012). https://doi.org/10.1094/PDIS-06-12-0576-PDN
3. National Agricultural Research Organization (NARO). Pests and diseases management in maize (2011). https://teca.apps.fao.org/teca/fr/technologies/7019. Accessed 18 July 2022
4. Li, Z., et al.: Real-time monitoring of plant stresses via chemiresistive profiling of leaf volatiles by a wearable sensor. Matter **4**(7), 2553–2570 (2021). https://doi.org/10.1016/j.matt.2021.06.009
5. Hussain, S., Lees, A.K., Duncan, J.M., Cooke, D.E.L.: Development of a species-specific and sensitive detection assay for Phytophthora infestans and its application for monitoring of inoculum in tubers and soil. Plant Pathol. **54**(3), 373–382 (2005). https://doi.org/10.1111/j.1365-3059.2005.01175.x
6. Balodi, R., Bisht, S., Ghatak, A., Rao, K.H.: Plant disease diagnosis: technological advancements and challenges. Indian Phytopathol. **70**(3), 275–281 (2017). https://doi.org/10.24838/ip.2017.v70.i3.72487
7. Li, Z., et al.: Non-invasive plant disease diagnostics enabled by smartphone-based fingerprinting of leaf volatiles. Nat. Plants **5**(8), 856–866 (2019). https://doi.org/10.1038/s41477-019-0476-y
8. Skoczek, A., Piesik, D., Wenda-Piesik, A., Buszewski, B., Bocianowski, J., Wawrzyniak, M.: Volatile organic compounds released by maize following herbivory or insect extract application and communication between plants. J. Appl. Entomol. **141**(8), 630–643 (2017). https://doi.org/10.1111/jen.12367
9. Gagliano, M., Mancuso, S., Robert, D.: Towards understanding plant bioacoustics. Trends Plant Sci. **17**(6), 323–325 (2012). https://doi.org/10.1016/j.tplants.2012.03.002
10. Khait, I., et al.: Plants emit informative airborne sounds under stress. https://doi.org/10.1101/507590
11. PSU Noisequest. https://www.noisequest.psu.edu/noisebasics.html. Accessed 01 Nov 2022

12. Downer, J.: Effect of fertilizers on plant diseases - topics in subtropics - ANR blogs. Topics in Subtropics (2013) https://ucanr.edu/blogs/blogcore/postdetail.cfm?postnum=12364. Accessed 18 Oct 2022
13. Bucheyeki, T.L., Tongoona, P., Derera, J., Msolla, S.N.: Combining ability analysis for northern leaf blight disease resistance on Tanzania adapted inbred maize lines. In: Advances in Crop Science and Technology, vol. 05, no. 02 (2017). https://doi.org/10.4172/2329-8863.1000266
14. Jackson, T.: Northern corn leaf blight, Nebraska Extension (2015)
15. Onwunali, M.R.O., Mabagala, R.B.: Assessment of yield loss due to northern leaf blight in five maize varieties grown in Tanzania. J. Yeast Fungal Res. **11**(1), 37–44 (2020). https://doi.org/10.5897/jyfr2017.0181
16. Fry, W.E., et al.: The 2009 late blight pandemic in the eastern United States - causes and results. Plant Dis. **97**(3), 296–306 (2013). https://doi.org/10.1094/PDIS-08-12-0791-FE
17. Ge, L., Mu, X., Tian, G., Huang, Q., Ahmed, J., Hu, Z.: Current applications of gas sensor based on 2-D nanomaterial: a mini review. Front. Chem. **7** (2019). https://doi.org/10.3389/fchem.2019.00839
18. Aditya Satrio, C.B., Darmawan, W., Nadia, B.U., Hanafiah, N.: Time series analysis and forecasting of coronavirus disease in Indonesia using ARIMA model and PROPHET. Procedia Comput. Sci. **179**, 524–532 (2021). https://doi.org/10.1016/j.procs.2021.01.036
19. 8.1 Stationarity and differencing | Forecasting: Principles and Practice, 2nd edn. https://otexts.com/fpp2/stationarity.html. Accessed 18 Nov 2022
20. Moreno-Torres, J.G., Raeder, T., Alaiz-Rodríguez, R., Chawla, N.V., Herrera, F.: A unifying view on dataset shift in classification. Pattern Recognit. **45**(1), 521–530 (2012). https://doi.org/10.1016/j.patcog.2011.06.019
21. Introduction—statsmodels. https://www.statsmodels.org/stable/index.html. Accessed 21 Nov 21

Inference Analysis of Lightweight CNNs for Monocular Depth Estimation

Shadi Saleh(✉), Pooya Naserifar, and Wolfram Hardt

Department of Computer Engineering, Chemnitz University of Technology, Straße der Nationen 62, 09111 Chemnitz, Germany
{shadi.saleh,pooya.naserifar, wolfram.hardt}@informatik.tu-chemnitz.de
https://www.tu-chemnitz.de/informatik/ce/index.php

Abstract. Monocular Depth Estimation (MDE) from a single image is a challenging problem in computer vision, which has been intensively investigated in the last decade using deep learning approaches. It is essential in developing cutting-edge applications like self-driving cars and augmented reality. The ability to perceive depth is essential for various tasks, including navigation and perception. Monocular depth estimation has attracted much attention. Their popularity is driven by ease of use, lower cost, ubiquitous, and denser imaging compared to other methods such as LiDAR scanners. Traditional MDE approaches heavily rely on depth cues for depth estimation and are subject to strict constraints, such as shape-from-focus and defocus algorithms, which require a low depth of field of scenes and images. MDE without some particular environmental assumptions is an ill-posed problem due to the ambiguity of mapping between the depth and intensity of color measurements. Recently, Convolutional Neural Networks (CNN) approaches have demonstrated encouraging outcomes in addressing this challenge. CNN can learn an implicit relationship between color pixels and depth. However, the mechanism and process behind the depth inference of a CNN from a single image are relatively unknown. In many applications, interpretability is very important. To address this problem, this paper tries to visualize a lightweight CNN (Fast-depth) inference in monocular depth estimation. The proposed method is based on [1], with some modifications and more analyses of the results on outdoor scenes. This method detects the smallest number of image pixels (mask) critical to infer the depth from a single image through an optimization problem. This small subset of image pixels can be used to find patterns and features that can help us to better formulate the behavior of the CNN for any future monocular depth estimation tasks.

Keywords: Convolutional neural networks · Depth estimation · Monocular depth estimation · Inference analysis · Visualization

1 Introduction

Like some other fields of study, artificial neural networks have tried to fill the gap between the performance of machines and human capabilities. One of these regions is the ability to perceive the depth of the machine. In recent decades, many works have

I. Czarnowski et al. (Eds.): KESIDT 2023, SIST 352, pp. 97–108, 2023.
https://doi.org/10.1007/978-981-99-2969-6_9

been accomplished to enable the computer to understand the depth of information. We have a single photo and its pixels as input in monocular depth estimation. The output is a photo (tensor) containing each pixel's corresponding distance to the camera. In these recent years, the performance of deep convolutional neural networks has shown a significant breakthrough in this field [1]. Standard depth detectors such as LiDARs are usually cumbersome, heavy, and consume high energy. Consequently, there is a growing interest in monocular depth estimation cameras due to their low cost and size and high energy efficiency [2]. However, on the other hand, despite the increasing usage of deep convolutional neural networks for monocular depth estimation, it is widely unclear how and under what circumstances a CNN can accurately predict depth from a single RGB image [1]. In addition, with the growing usage of deep neural networks in embedded applications, there is a pressuring need to analyze the depth inference of lighter CNN used in these platforms and to know their behavior.

Based on some psychophysical studies, for monocular depth estimation, multiple cues can be considered to visually assist humans in determining depth. Figure 1 depicts some of these cues. Does CNN use other depth cues or the same cues for monocular depth estimation? How similar is the perception of the CNN of depth to human visual

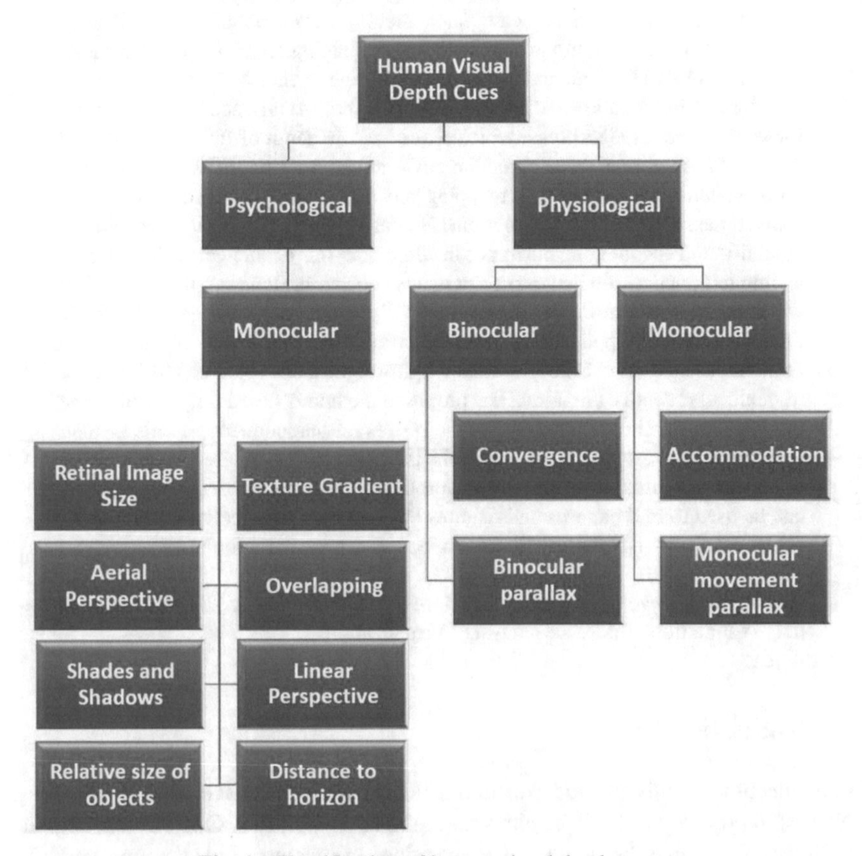

Fig. 1. Classification of human visual depth cues.

perceptions of depth? Responding to these questions is not straightforward since it is ambiguous, even for the human visual sense.

2 Related Works

The old methods of estimating depths for a scene heavily relied on depth cues and, as a result, needed some kinds of equipment, such as configured cameras for stereo depth estimation [3]. The new trend for depth estimation is mainly followed by the monocular paradigm, especially with the emergence of deep learning. Especially for the outdoor depth estimation, more efforts have been made to collect a more reliable dataset plus synthetical data to support a deep learning-based model. This will be integrated with a new learning approach and traditional depth estimation method to obtain the best results [3]. On the other hand, deep learning-based methods still have problems to be solved. Only a few studies and research have been conducted for outdoor depth estimation [3]. It is even less for models and methods following embedded application constraints. Besides, outdoor depth estimation or depth estimation in the wild needs more large datasets since they are mainly designed for indoor applications to support robots and 3D reconstruction (indoor activities). [2, 4–7] are all trying to give novel ideas to improve depth estimation, but the need for accurate training data is a big challenge. Complementary research about the inference analyses of CNN in monocular depth estimation can be practical to produce more useful data to train the model.

Many studies endeavor to decode the inference of CNNs, although most have focused on image classification tasks. In fact, only a few works among all these attempts are considered effective [1]. Methods such as LIME (Local Interpretable Model-agnostic Explanations [8]), CAM (class activation maps) [9], gradient-based methods [10], and prediction difference analysis [11].

The abovementioned methods were developed for classification inference analysis, not to analyze CNNs for depth estimates. CNN provides a two-dimensional map for applications involving depth estimation. On the other hand, classification is a score for a category. Under this logic, gradient-based methods, CAM, and their variants are all automatically disregarded.

The masking methods that utilize constant shape masks or super-pixels (Prediction Difference Analysis, for example), employing low-level image features, are no match for the purpose. In other words, their shapes should be interpretable and comparable with depth cues used by CNN to estimate depth which they may not match. Besides, they are based on complicated mathematics and with using high approximation.

3 Proposed Methodology

Identifying relevant pixels can be considered and formulated as a sparse optimization problem. The output of this method is an estimated mask that selects the smallest number of pixels from which the target CNN can provide the maximally similar depth map that it predicts from the original input [1]. In fact, instead of optimization the output of the CNN w.r.t. input, the additional CNN is responsible for estimating the mask from the input image. In other words, besides the depth estimator CNN, which is the target of the visualization, another CNN should be trained to estimate the mask, which is a representation of the inference of the CNN. The methodology is to train a convolutional network G to predict the mask M. The assumption is that the network N can correctly estimate depth from only a specified set of sparse pixels. The model's accuracy using only this subset of the input is similar to those they infer from the complete input [1].

Assume a network N that calculates the depth map of a scene from an RGB image as input:

$$Y = N(I) \tag{1}$$

In the above formula, I is the normalized version (z-score normalization) of the input RGB image, and Y is the estimated depth map. As previously mentioned, human vision can use some of the so-called depth cues to infer depth. These visual depth cues are only from a small subset of image pixels. Therefore, it is assumed that CNNs can infer depth maps equally reasonably from a small subset of pixels of I [1]. As mentioned, if we directly use the output of the CNN (N) with respect to input to perform the optimization, it often gives us unexpected results. Accordingly, instead of optimizing individual elements of M, to have more visually interpretable masks, an additional network G is used to generate the M mask [1]. During this method, two CNNs are used. Network G is a CNN network for predicting the set of relevant pixels or the mask M. G is trained, which is quite similar to Y (the original estimation from the whole image) [1].

$$G(I) = M \tag{2}$$

M is the estimation of the most critical pixels of I for N to predict its depth map Y. M will be maximally sparse. As mentioned, the network N is trained to estimate the depth map of an input image I (RBG image). In fact, N is the target of visualization. During the process of the training, N must be constant. The output M is elementwise multiplied by I and inputted to N, yielding an estimate of $\hat{Y}$ of the depth map.

$$\hat{Y} = N(I \otimes M) \tag{3}$$

If the network G is adequately trained, it is possible to generate a mask so that the $\hat{Y}$ can be as close as Y. In other words, the output of N, with/without using the mask, should be very close. In that case, mask M can represent the essential pixels involving depth estimation. According to the above explanations, the optimization problem is for mutated as follows:

$$\min l_{dif}(Y, N(I \otimes G(I))) + \lambda \frac{1}{n} \|G(I)\|_1 \tag{4}$$

where ldif explains how much Y and $\hat{Y}$ are similar, λ is a hyperparameter that controls the sparseness of M; n is the number of image pixels, and ||G(I)||1 is the norm 1(of a vectorized version) of M. Please note that the Sigmoid activation function was applied to the output layer of G, which constrain its output in the range between 0 and 1 [1]. The desired output for the mask is to remove unimportant pixels of I so that the results stay similar to the original estimation $Y = N(I)$. At the same time, the rate of getting sparse for mask M should be at a reasonable speed. The reason is to give enough time for the mask to capture complicated patterns inside the mask. This allows M to contain complicated depth cues before getting too sparse to be interpretable. Equation (5) raises the vector to the power of two before calculating its norm of it.

$$\min l_{dif}(Y, N(I \otimes G(I))) + \lambda \frac{1}{n} \left\| G^2(I) \right\|_1 \tag{5}$$

Fig. 2. The difference of generated mask. (a): solving Eq. 5, (b): solving Eq. 4. Other training parameters were constant for (a) and (b)

Figure 2 shows that solving (4) as the optimization problem tends to generate sparse masks with fewer details on the Kitti dataset. In fact, in order to address this optimization, the mask tends to get sparse very fast, revealing only a few details and with a lower final accuracy (the output of N with/without masks is significantly different). This method also results in less accuracy on the test set since the mask is too sparse. On the other hand, modification of the term related to the level of sparseness in (5) will generate a mask with more details. It helps the model detects more patterns before the produced mask gets sparse (since the sparseness loss acts stronger than other loss functions even with lower sparseness parameters).

3.1 Loss Function

Most of the existing losses for training depth prediction networks yield an output that suffers from loss of spatial resolution in the predicted depth maps. This loss of spatial resolution can have potential symptoms, such as the distorted and blurry reconstruction of object boundaries. In [12], a tripled loss has been introduced to focus on the depth maps with higher spatial resolution and for having more accurate estimation. This tripled loss causes finer resolution reconstruction and measures errors in-depth, gradients, and surface normal, contributing to improving accuracy in a complementary fashion [12] (Fig. 3).

$$l_{dif} = l_{depth} + l_{grad} + l_{normal} \tag{6}$$

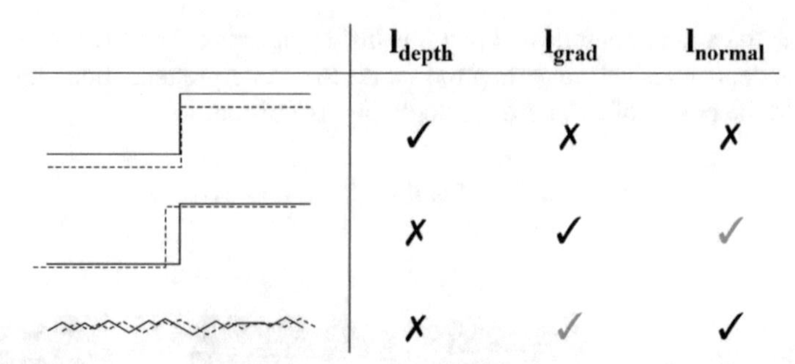

Fig. 3. The three loss functions retain orthogonal sensitivities to different types of errors in predicted depth maps. The solid and dotted lines shown in the first column denote two depth maps under comparison, where one-dimensional depth images represent them for the sake of depth, and the horizontal axis is, for example the x-axis of the images [12].

3.2 Mask Network

In this method, the mask network plays a significant role. As mentioned, the mask network aims to predict the essential pixels (mask) for depth estimation. The CNN model used for producing masks follows the Encoder-Decoder structure.

The model architecture is DRN-D-22, Dilate Residual Network (DRN) with 22 layers (pre-trained on ImageNet). This encoder architecture keeps local structures of the input image because it has a smaller number of down-samplings [1]. The last layer of DRN-D-22 (the last fully connected layer) was removed. It yields a feature map with 512 channels and 1 to 8 resolutions of the input image. Up-convolution block consists of unpooling plus Conv(5×5), which is followed by a 3×3 convolution after the up-convolution. A projection connection was also added from the lower-resolution feature map to the result. The unpooling operation only needs to be applied once for both branches [13]. Having successive up-projection blocks lets high-level information be more efficiently moved forward in the network and, at the same time, leads to increasing feature map sizes [13]. The output of the decoder is a feature map with 64 channels and the same size as the input image. This will be followed by a 3×3 convolutional layer to output M [1].

4 Depth Network

The depth network is the target of the visualisation. In fact, the Mask network aims to visualise the depth inference of this depth network. The lightweight, energy-efficient network architecture proposed in [2] is an efficient network architecture for embedded applications. This architecture addresses the problem of fast depth estimation on embedded systems. This architecture follows the same Encoder-Decoder architecture but with different encoder and decoder types. In particular, it is concentrated on the design of a low-latency decoder. It is feasible to attain equal accuracy as prior work on depth estimation but at inference speeds that are an order of magnitude faster [2].

In fact, State-of-the-art single-view depth estimation algorithms stand based on complicated deep neural networks [2]. Improving accuracy in these kinds of networks leads to computation-intensive algorithms and makes them too slow. For embedded applications, we need fast and real-time inference on an embedded platform, for instance, mounted on a micro aerial vehicle or robotic system. The encoder is used for depth estimation; is commonly a network designed for image classification. Popular choices include VGG-16 [14] and ResNet-50 [15] because of their solid expressive power and high accuracy [2]. Such networks also suffer from increased complexity and latency, which lead to preventing being employed for applications running in real-time on embedded systems [2]. Fast Depth uses lightweight encoder-decoder network architecture, and the network can be further pruned to be even more efficient following latency constraints in embedded systems [2].

The Encoder of the network is MobileNet and uses depthwise decomposition (depth-wise separable convolution). This structure is more efficient than networks with standard convolution like ResNet and VGG, leading to reduced latency [2]. The decoder network (NNConv5) is formed using five cascading upsample layers and a single pointwise layer at the last layer [2]. Each upsample layer functions 5×5 convolution and decreases the number of output channels by 1/2, comparable to the number of input channels. Convolution is pursued by nearest neighbor interpolation that doubles the spatial resolution of intermediate feature maps. The decoder utilizes depthwise decomposition (depth-wise separable convolution) to reduce the complexity of all convolutional layers any further and to result in a slim and fast decoder [2]. The Encoder reduces the spatial resolution step by step and extracts higher-level features from the input. The output of the Encoder that feeds to the decoder evolves a group of low-resolution features in a way that multiple image details can be lost, making it more difficult for the decoder to recover pixel-wise (dense) data. Skip connections let image details from high-resolution feature maps in the Encoder be integrated into features within the decoder. This allows the decoding layers to reconstruct a more complex and more detailed dense output. Skip connections have previously been employed in networks for image segmentation, such as U-Net and Deeper Lab, showing that they can be advantageous in networks producing dense outputs. Skip connections are from the MobileNet Encoder to the outputs of the middle three layers in the decoder. Feature maps are fused via addition instead of concatenation to decrease the counts of feature map channels input to the decoding layers [2].

5 Analyses of the Predicted Mask

As mentioned, the output mask produced by the mask network (G) can be as sparse as possible. These set of sparse pixels are the most critical pixels for the depth estimator network to perceive depth. As a result, the mask can be further analyzed in different scenes to give us more insight into the inference of the CNN (in this case, a lightweight, energy-efficient network architecture).

Figure 5 shows RGB images and corresponding masks for different values of sparseness parameters. The following results can be concluded:

- **Vanishing points:** CNN uses vanishing points of the scene to estimate depth. Pixels near vanishing points are highlighted significantly in the produced masks. As it is clear from the images, the points near the horizon get more robust weights in CNN, and they are displayed brighter. The pixels of far objects are more highlighted and appear as Filled objects. CNN emphasizes more on the distant object to infer depth. Although this happens for different lambda values, for smaller values are more significant.

- **Edges and borders of objects:** Although the predicted masks differ from the edges, the most highlighted pixels are around the objects' edges.
- **Filled regions of objects:** Although the edges are important, they are not the only important points for the CNN to infer depth. The predicted masks tend to consist of filled regions. For example, in images 2 and 12, houses and, traffic signs appeared as filled objects. Filled regions are emerged more frequently for far objects.
- **Depth cues used by humans:** Some depth cues are used by the human visual system as monocular depth cues. The predicted masks for different scenes show the perception of some of these cues (Fig. 5). For example, we perceive linear perspective depth cues as parallel sides of the road meeting on the horizon. In many photos showing a long road, CNN gives us the perception of this cue by highlighting the edges of the roadsides (for example, in images 5, 6, and 7). As another example, lights and shades give us the perception of depth. This is also highlighted in resulted masks. For example, at different heights of the pedestrian path and road (images 9, 10, 11), it appeared as a strong highlighted line in the mask. The arrangements of pixels also give us some other monocular depth cues, such as aerial perspective, relative size, interposition, and texture gradient.

For the indoor dataset, edges and filled regions are the main features to infer depth (Fig. 4). Especially, edges and borders of the objects are the most critical features for CNN to infer depth.

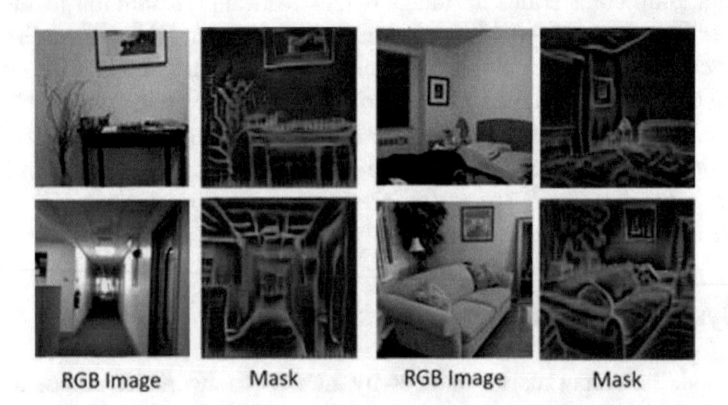

Fig. 4. RGB images and corresponding mask for indoor scenes

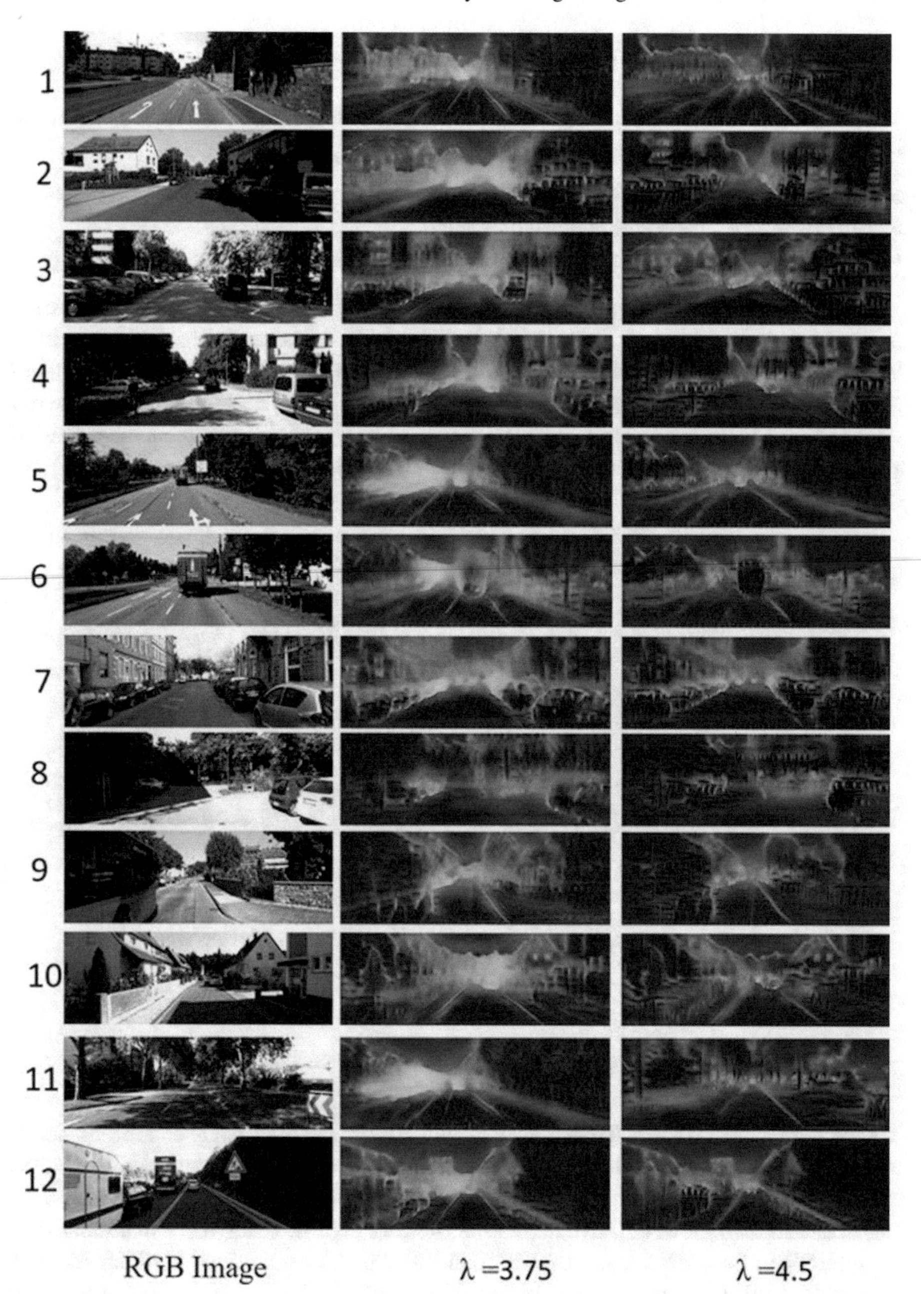

Fig. 5. RGB images and corresponding masks for different values (3.75, 4.5) of sparseness parameters

6 Other Network Architecture

The results of the inference analysis for different models (Encoder-Decoder based on ResNet 50 and SeNet 154) are pretty similar using the proposed method:

- Models highlighted more distant points in the mask.
- Models tend to use edges and filled objects.
- Models use some depth cues used by humans' visual system.

However, the sparseness behavior differs for larger models with more extensive parameters. A lighter model (fast-depth) tends to produce sparser masks. In comparison, larger models with more parameters remain denser even with more extensive sparseness parameters. Finally, the results have shown that the proposed method is not dependent on the selected models. It can be used as a general method for the inference analysis of any CNN model for monocular depth estimation task (Fig. 6).

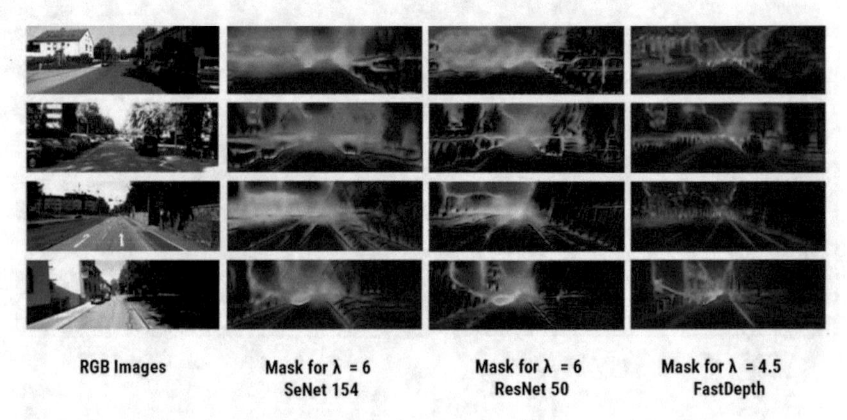

Fig. 6. The visualisation of the inferences for different model architectures

7 Summary and Conclusion

This Visualization of the inference of the CNN has proved that: CNN can infer a depth map accurately from a small number of image pixels in the monocular depth estimation task. CNNs trained on outdoor datasets use vanishing points (points near to horizon), edges, and filled regions as the main features to infer depths. Using these main features, the inference of a CNN for the task of monocular depth estimation is similar to what we see as depth to a large extent. In fact, this method shows it can produce features comparable to humans' visual depth cues. Human-like cues such as aerial perspective, linear perspective, the relative size of objects, interposition (overlap), texture gradients, light, and shades all have emerged in the mask representing the inference of the CNN.

The performance of CNN facing new scenes also can be evaluated by analyzing the produced mask. In fact, by using the produced mask, it is possible to examine the effect of different objects and features on the scene. In other words, when a feature has appeared in the mask, it shows that this feature is critical for the CNN to predict depth. For instance, some features, such as crosswalks, emerged in the mask for the test data, indicating that CNN effectively uses them to improve the accuracy of the predictions. On the contrary, when a feature is not included in the mask, it shows the CNN has no clue to use them, or they have no effect on the task of monocular depth estimation.

Finally, this method can be utilized as a complementary phase for final analyse of any trained depth estimator network to further evaluate the decision made by the network in different scenes.

References

1. Hu, J., Zhang, Y., Okatani, T.: Visualization of convolutional neural networks for monocular depth estimation. In: Proceedings of the IEEE/CVF International Conference on Computer Vision, pp. 3869–3878 (2019)
2. Wofk, D., Ma, F., Yang, T.J., Karaman, S., Sze, V.: FastDepth: fast monocular depth estimation on embedded systems. In: 2019 International Conference on Robotics and Automation (ICRA), pp. 6101–6108. IEEE (2019)
3. Vyas, P., Saxena, C., Badapanda, A., Goswami, A.: Outdoor monocular depth estimation: a research review. arXiv preprint arXiv:2205.01399 (2022)
4. Shadi, S., Hadi, S., Nazari, M.A., Hardt, W.: Outdoor navigation for visually impaired based on deep learning. In: Proc. CEUR Workshop Proc, vol. 2514, pp. 97– 406 (2019)
5. Saleh, S., Manoharan, S., Nine, J., Hardt, W.: Towards robust perception depth information for collision avoidance. In: 2020 IEEE Congreso Bienal de Argentina (ARGENCON), pp. 1–4. IEEE (2020)
6. de Queiroz Mendes, R., Ribeiro, E.G., dos Santos Rosa, N., Grassi, V., Jr.: On deep learning techniques to boost monocular depth estimation for autonomous navigation. Robot. Auton. Syst. **136**, 103701 (2021)
7. Saleh, S., Manoharan, S., Nine, J., Hardt, W.: Perception of 3D scene based on depth estimation and point-cloud generation. In: Czarnowski, I., Howlett, R.J., Jain, L.C. (eds.) Intelligent Decision Technologies. Smart Innovation, Systems and Technologies, vol. 309, pp. 495–508. Springer, Singapore (2022). https://doi.org/10.1007/978-981-19-3444-5_43
8. Ribeiro, M.T., Singh, S., Guestrin, C.: "Why should i trust you?" Explaining the predictions of any classifier. In: Proceedings of the 22nd ACM SIGKDD International Conference on Knowledge Discovery and Data Mining, pp. 1135–1144 (2016)
9. Zhou, B., Khosla, A., Lapedriza, A., Oliva, A., Torralba, A.: Learning deep features for discriminative localization. In: Proceedings of the IEEE Conference on Computer Vision and Pattern Recognition, pp. 2921–2929 (2016)
10. Smilkov, D., Thorat, N., Kim, B., Viégas, F., Wattenberg, M.: SmoothGrad: removing noise by adding noise. arXiv preprint arXiv:1706.03825 (2017)
11. Zintgraf, L.M., Cohen, T.S., Adel, T., Welling, M.: Visualizing deep neural network decisions: prediction difference analysis. arXiv preprint arXiv:1702.04595 (2017)
12. Hu, J., Ozay, M., Zhang, Y., Okatani, T.: Revisiting single image depth estimation: toward higher resolution maps with accurate object boundaries. In: 2019 IEEE Winter Conference on Applications of Computer Vision (WACV), pp. 1043–1051. IEEE (2019)
13. Laina, I., Rupprecht, C., Belagiannis, V., Tombari, F., Navab, N.: Deeper depth prediction with fully convolutional residual networks. In: 2016 Fourth International Conference on 3D Vision (3DV), pp. 239–248. IEEE (2016)

14. Simonyan, K., Zisserman, A.: Very deep convolutional networks for large-scale image recognition. arXiv preprint arXiv:1409.1556 (2014)
15. He, K., Zhang, X., Ren, S., Sun, J.: Deep residual learning for image recognition. In: Proceedings of the IEEE Conference on Computer Vision and Pattern Recognition, pp. 770–778 (2016)

Arabic Text-to-Speech Service with Syrian Dialect

Hadi Saleh[1], Ali Mohammad[1(✉)], Kamel Jafar[2,3], Monaf Solieman[4,5], Bashar Ahmad[6], and Samer Hasan[6]

[1] HSE University, Moscow, Russia
amokhammad_1@edu.hse.ru
[2] Russian Technological University - MIREA, Moscow, Russia
[3] Syrian Virtual University - SVU, Damascus, Syria
[4] Tartous University, Tartous, Syria
[5] Al-Andalus University for Medical Sciences, Tartous, Syria
[6] Infostrategic, Dubai, UAE

Abstract. This research aims to develop an Arabic text-to-speech (TTS) service with Syrian dialect, which is a variety of Arabic spoken in Syria and some neighboring countries, with easy access to it for people with disabilities or difficulty reading Arabic, such as people with visual impairments or learning disabilities. To achieve this goal, we employ two state-of-the-art Machine Learning (ML) approaches: Tactron2 and Transformers, which have achieved impressive results in various natural language processing tasks, including TTS. We compared the two approaches and evaluated the resulting TTS service using subjective measures. Our results show that both approaches can produce high-quality speech in the Syrian dialect, but transformers have the advantage of being more efficient and more flexible in handling different languages and accents.

Keywords: Arabic Text-To-Speech · Syrian Dialect · Tacotron2 · Transformers · TTS

1 Introduction

Natural Language, by definition, is a "System of Symbolic Communication (spoken, signed or written) without intentional human planning and design" [1]. Natural language processing (NLP) is a subfield of Artificial Intelligence (AI) that focus on modeling Natural Languages used in many applications such as Machine Translation, Optical Character Recognition (OCR), Speech Synthesis (Text-To-Speech), Speech Recognition (Speech-To-Text), Sentiment Analysis (SA), Question Answering, etc. NLP quality increased incredibly during the last decade (2010–2020) along with a rise in public and general usage. Nowadays there are many NLP models and tools that can be applied independently of the language using the same basic and general mechanisms. Unfortunately, it does not imply that all languages are getting the same attention. The Arabic Language is the

I. Czarnowski et al. (Eds.): KESIDT 2023, SIST 352, pp. 109–127, 2023.
https://doi.org/10.1007/978-981-99-2969-6_10

main language of the Arab World and the religious language of non-Arab Muslims. Standard Arabic Language is used for papers, articles, lectures, and formal letters exchanged among governmental agencies but is not the common dialect used among Arab People. In fact, there are five main Arabic dialects, from which several sub-dialects branch out, spread around Arab World countries. The Syrian Dialect is considered the simplest dialect and the closest to the Standard Arabic Language. It's widespread due to the cultural and human spread around the world. Above all, it was chosen by the Netflix platform to be the official Arabic dialect for dubbing their movies and series. Although Arabic NLP has many challenges, it has evolved during the last few years due to the research studies done in the Arab Gulf Countries (mainly in The United Arab Emirates, Qatar, and Saudi Arabia) which achieved great progress in this field, as many tools, specialized in Arabic language processing, have been proposed with the attempts to overcome the challenges facing researchers in the field of natural language, such as Farasa and Camel Tool. Text-To-Speech (TTS) allows greater population access to content, such as those with learning disabilities or reduced vision. It also opens doors to anyone else looking for easier ways to access digital content. TTS could be an advantageous way to teach elementary school pupils their native language or even help people learn a foreign language in a self-study way. TTS is an essential component in Digital Personal Assistance and it's important to generate a more human-like voice to make the users feel relief, for example listening to a GPS guide during a Car drive or getting answers to questions asked by chatbots. TTS researchers are now attempting to make machines speak with characteristics that are nearly indistinguishable from humans' characteristics, especially the expression of tones and emotions in synthesized speech. Proposing Machine Learning approaches to the already existing concatenative and parametric speech synthesis methods has demonstrated that higher-quality audio can be produced. In the long run, researchers aspire to develop TTS systems based on Machine Learning (ML) models that we can listen to without realizing the difference between their output speech and human speech. Even after years of effective experimentation and research, it still requires a lot of skills and resources for the regular user to be able to use some of the most evolved TTS models and technologies available. Here comes the importance of this research, to provide an analysis of TTS ML models, and based on that, build a TTS web service that is easy to access and use anytime via an API.

2 Problem Statement

The past years showed us that the availability of voices to develop our customized services or use cases relies directly on the products of big companies. Although they have advanced TTS models that support the Arabic language, these models (the free ones at least) are still facing a problem of concatenating (gluing) the words together to generate more natural Arabic speech due to the lack of Arabic Tajweed rules (pronunciation cases) knowledge. To overcome these problems, these big companies offered the option of building custom TTS models for clients with some cost impact. Amazon launched Brand Voice [3], which allows clients to collaborate with the Amazon Polly team of AI researchers and linguists to create exclusive and high-quality custom Neural TTS models. However, it is very hard for the average developer or user to accomplish this achievement with Amazon without incurring large costs. Google released the Custom Voice (Beta) [4], with similar functionality to Amazon Brand Voice, however, it is more limited and involves a lot of boilerplate processes. We propose in this research to generate an Arabic TTS model with Syrian Dialect based on available ML techniques to achieve efficient human-like synthesized speech and integrate it within an easy-to-implement web service that is accessed using an API.

2.1 Phonemes and Mel-spectrogram

A Phoneme, in phonology and linguistics, is a unit of sound that can differ one word from another in a particular language. The International Phonetic Association (IPA) provided these phonemes as a standardized representation of speech sounds in written form. Audio and speech signals have frequency content that varies over time, and are known as non periodic signals. To present the spectrum of a signal of such kind, several spectrums are computed by performing Fast Fourier Transform on several windowed segments of this signal. This transformation is called Short-Time Fourier Transform, and it's applied on overlapping windowed segments of the signal, to get the so-called spectrogram. Mel-spectrograms are spectrograms that visualize audio and speech on the Mel scale (a logarithmic transformation of a signal's frequency) as opposed to the frequency domain. The rest of this research is organized as the following: Sect. 3 gives a background of this research related to the related work in this field; A comparative analysis of two neural architectures, Tacotron2 and Transformers is demonstrated in Sect. 4; while Sect. 5 describes the main architecture of a neural TTS model based on Transformers called FastSpeech; Sect. 6 is related to the methodology used to develop and implement the corresponding Arabic TTS service; And Sect. 7 is dedicated to discuss the results of this research. Finally Sect. 8 provides conclusion and future work to improve this research.

3 Background

End-to-end generative TTS models, such as Tacotron [2] and Tacotron2 [16], have recently been proposed to simplify conventional speech synthesis pipelines by replacing the production of dialectal and audio features with a single neural network, thanks to the rapid development of neural networks. Tacotron and Tacotron2 generate mel-spectrograms directly from texts, then use a vocoder like the Griffin Lim method [5] or WaveNet [6] to synthesize the output speech. On some datasets, the quality of synthesized speech is substantially improved and even similar to human speech. These end-to-end models normally contain two components, namely Encoder and Decoder. The Encoder translates the input sequence of phonemes to semantic space in order to generate a sequence of encoder hidden states, and the Decoder also constructs a sequence of decoder hidden states then outputs the Mel-spectrogram frames. Both of them are built using Recurrent Neural Networks (RNNs), like Long Short-Term Memory (LSTM) [7] or Gated Recurrent Units GRU [8]. However, RNNs can only take input and generate output sequentially, because the current hidden state can't be built without using both of the previous hidden state and the current input, which limits the capability of parallelization in both training and inference process. The same reason also leads to, for a certain frame, biased information from many steps after multiple recurrent processing.

To tackle these two problems, Li et al. [9] proposed combining the advantages of Tacotron2 and Transformer, by replacing RNNs in Neural Machine Translation (NMT) models, and train an novel end-to-end TTS model, in which the multi-head attention mechanism is introduced to replace the RNN structures in the encoder and decoder, as well as the vanilla attention network.

Zerrouki et al. [10] have explored the possibility to adapt the existing TTS converters into Arabic language in eSpeak. They developed an Arabic TTS support to be integrated in eSpeak system, with structured new voices using existing phonemes as well as building TTS rule conversations concerning the main issue of discretization in Arabic, using available features like rules stemming and extra word list pronunciation and dictionary, to make a light morphology analyzer and vocalizer. They started by using a third party diacritizer to restore diacritics of texts before the process of TTS, and then to implement text to phonemes conversion rules, then a light morphological analyser was structured based on eSpeak rules, after that they started developing new voices for eSpeak through the process of adapting the existing phonemes or creating the missing ones, afterward, new rules were established to convert texts into phonemes.

Zine and Meziane [11] proposed an approach based on sub-segments for concatenation, and conducted a Diagnostic Rhyme Test (DRT) to measure the synthesized speech intelligibility. They found that the result of the DRT is approximately 95% for the word-level test, and 80% for the sentence-level test, which can be considered a satisfactory percentage. Then, in a later work, Zine et al. [12,14] developed a web application that implemented a TTS API, which was used to learn Arabic language and investigated the enhancement of Arabic TTS using the mentioned approach by adopting a lemma-based approach for concatenative Arabic TTS. They have presented a unit selection approach that combines long units as lemmas with sub-segments, and described the materials and methods used for generating the lemma frequency list, as well as the process of construction of the lemma database, to cope with the large number of Arabic lemmas. The results have shown significant improvement in both intelligibility and naturalness of the produced speech.

Abdelali et al. [13] presented automatic diacritization of Arabic Maghrebi dialects, used in Northern Africa, based on a character-level deep neural network architecture, in order to enable NLP to model conversational Arabic in dialog systems. They used a character-level DNN architecture that stacks two Bidirectional LSTM (biLSTM) layers over a Conditional Random Fields (CRF) output layer. Mono-dialectal training achieved Word Error Rate (WER) less than 3.6%. Though sub-dialects are phonetically divergent, our joint training model implicitly identifies sub-dialects, leading to small increases in WER.

Fahmy et al. [15] proposed an Arabic speech synthesizer using an Tacotron2, followed by Wave-Glow to synthesize Arabic speech. They described using Tacotron2 architecture to generate intermediate feature representation from Arabic diacritic text using a pre-trained English model and a total of 2.41 h of recorded speech, followed by WaveGlow as a vocoder to synthesize high-quality Arabic speech, along with applying transfer learning from English TTS to Arabic TTS successfully in spite of the fact that the two languages are quite different in terms of character level embedding and language phonemes.

Based on the previous researches, we are going to use the same approach, and depend on training a custom model, by investigating the architecture used in [15] and compare it with the Transformer architecture mentioned in [9] and [17] to choose the best option that serves our goal.

4 Comparative Analysis

Currently, there exist many Neural TTS models that achieve great results compared to traditional approaches. In this section, we compare the between two of the most used techniques, Tacotron2 and Transform. And according to this comparison we explain the reason for the chosen model.

4.1 Tacotron2

As it is stated in [16], "Tacotron2 is a fully neural TTS architecture that uses a sequence-to-sequence recurrent network (RNN) with attention to predict mel-spectrograms with a modified WaveNet vocoder". Tacotron2 [9,16,17] consists of two components (Fig. 1):

1. Feature prediction recurrent sequence-to-sequence network with attention to generate sequences of predicted mel-spectrogram frames from a character sequence input.
2. Modified version of WaveNet that generates time-domain waveform samples based on the predicted mel-spectrogram frames.

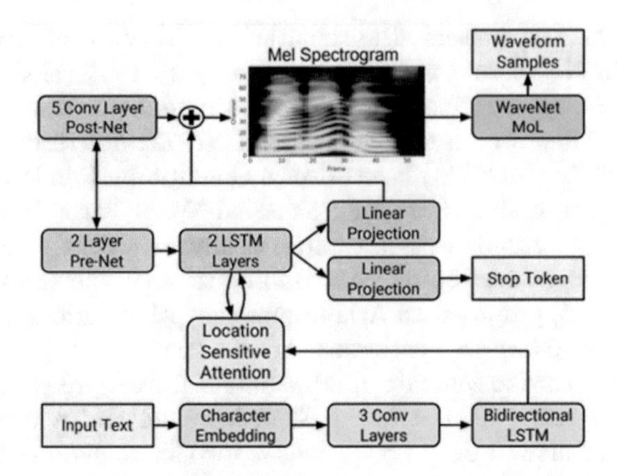

Fig. 1. Tacotron2 Architecture. [16]

First, character sequence input is processed with a 3-layer CNN to extract a longer-term context, then fed into a bidirectional LSTM encoder. While previous mel-spectrogram is processed first using decoder pre-net, a 2-layer fully connected network where output is concatenated with previous context vector, followed by a 2-layer LSTM. New context vector at this time step is calculated using the output, which is concatenated with the output of the 2-layer LSTM to predict the mel-spectrogram and stop token with two different linear projections respectively. Finally, to refine the predicted mel-spectrogram, it is fed into a 5-layer CNN with residual connections.

4.2 Transformers

It is stated in [9] "Transformer (Fig. 2) is a sequence to sequence network, based solely on attention mechanisms and dispensing with recurrences and convolutions entirely". Transform [9,17] demonstrated great results in the field of Neural Machine Translation compared to RNN-based models. Transform also consists of two components: an encoder and a decoder, those are built using layers of several identity blocks. Each encoder block contains two sub-networks: a multi-head attention and a feed forward network, while each decoder block contains an extra masked multi-head attention comparing to the encoder block. Both encoder and decoder blocks have residual connections and layer normalization.

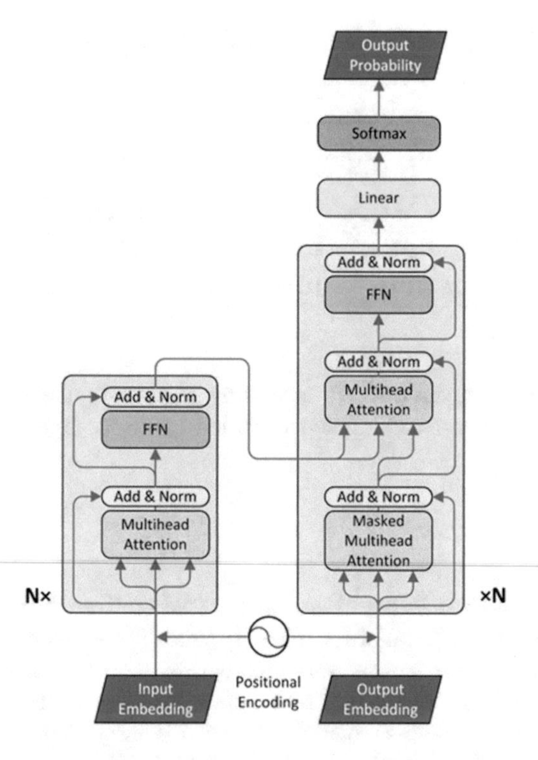

Fig. 2. System architecture of Transformer. [17]

Compared to Tactron2, which is an RNN-based model, using Transformer has two advantages in neural TTS:

- Parallel Training: eliminating recurrent connections provides in parallel the frames of an input sequence for the decoder;
- Self-attention: building long-range dependencies directly, by injecting the global context of the whole sequence into each input frame.

Transformer reduced the number of paths, where forward and backward signals travel between input and output sequence, down to 1. This reduction helps the prosody of synthesized waves, which requires several words in the neighborhood, and sentence-level semantics. The criteria for the comparison are the following:

- Mean Opinion Score (MOS): is a subjective measure of the perceived quality of a digital audio or video signal, typically given on a scale of 1 to 5, used to evaluate and diagnose problems that may be affecting the signal's quality;
- Accuracy: a percentage value represents the ratio of correctly synthesized words and sentences;
- Real Time Factor (RTF): is a metric value to measure the total time required to synthesize a speech after processing a text (Table 1).

Table 1. Comparison between Tacotron2 and Transformer. [9,17]

	Tacotron2	Transformer
MOS	3.86 ± 0.09	3.88 ± 0.09
Accuracy (1 GPU)	90.51%	93.01%
Accuracy (4 GPU)	89.57%	94.83%
RTF on CPU	0.216 ± 0.016	0.851 ± 0.076
RTF on GPU	0.094 ± 0.009	0.634 ± 0.025

Based on this compassion, We are going to investigate an available neural architecture that depend on Transform, called FastSpeech, to train the required TTS model.

5 FastSpeech

FastSpeech [18] is a Feed-Forward Network based on Transformer aims to generate mel-spectrogram in parallel for TTS. It takes a phoneme sequence as input and generates mel-spectrograms non-autoregressive. Its basic principle depends on a feed-forward network based on the self-attention in Transformer and 1D convolution. Since a mel-spectrogram sequence is larger than its corresponding phoneme sequence, FastSpeech adopts a Length Regulator that extracts attention alignments, to predict phonemes duration, by expanding the source phoneme sequence, to match the length of the target mel-spectrogram sequence for parallel mel-spectrogram generation. The Regulator is built on a phoneme duration predictor, which predicts the duration of each phoneme. FastSpeech speeds up mel-spectrogram generation by 270x and the end-to-end speech synthesis by 38x, using the following architecture:

5.1 Feed-Forward Transformer

Main component of FastSpeech is Feed-Forward Transformer (FFT) (Fig. 3a) [18], that stacks multiple FFT blocks for phoneme to mel-spectrogram transformation, with N blocks on both phoneme and mel-spectrogram sides, with a Length Regulator in the middle to reduce the length gap between the phoneme and mel-spectrogram sequences. A single FFT block (Fig. 3b) is composed of self-attention and 1D convolutional networks, both are followed by residual connections, layer normalization, and dropout. Cross-position information are extracted by the multi-head attention in self-attention networks. In contrast to Transformer, a 2-layer 1D convolutional network with ReLU activation instead of dense network because in speech tasks the adjacent hidden states are connected tightly to phoneme and mel-spectrogram sequences.

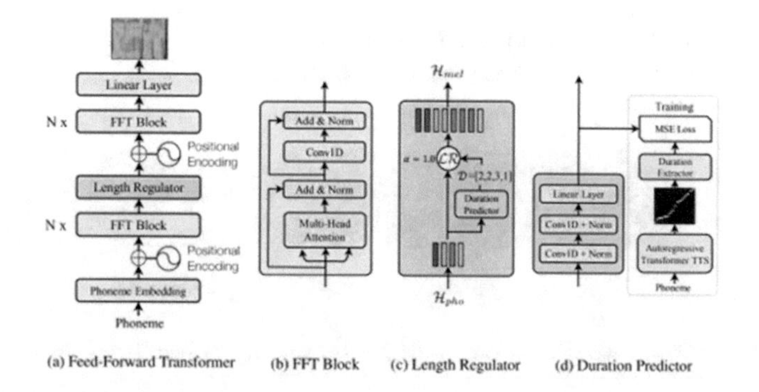

(a) Feed-Forward Transformer (b) FFT Block (c) Length Regulator (d) Duration Predictor

Fig. 3. System architecture for FastSpeech. (a). The feed-forward Transformer. (b). The feed-forward Transformer block. (c). The length regulator. (d). The duration predictor. [18].

5.2 Length Regulator

Length Regulator (Fig. 3c) [18] solves the problem of length mismatch between the phoneme and spectrogram sequence in FFT and controls speech speed and part of prosody. A single phoneme corresponds to several mel-spectrograms, referred to as phoneme duration, which mean that a mel-spectrogram sequence is larger than its phoneme sequence. The Length Regulator expands the hidden states of the phoneme sequence to match the length of the mel-spectrograms.

$$H_{mel} = LR(H_{pho}, D, \alpha) \tag{1}$$

H_{mel} is expanded sequence, LR is Length Regulator, $H_{pho} = [h_1, h_2, ...h_n]$ is phoneme sequence and n is its length, $D = [d_1, d_2, ...d_n]$ is phoneme duration sequence, where $d_1 + d_2 + ... + d_n = m$ is the length of mel-spectrogram sequence and α is a hyperparameter that controls voice speed by determining the length of the expanded sequence H_{mel}. An example of how LR works (Fig. 4), given $H_{pho} = [h_1, h_2, ...h_3]$ and D = [3 , 2 , 3]:

- Normal Speed: $\alpha = 1, H_{mel} = [h_1, h_1, h_1, h_2, h_2, h_3, h_3, h_3]$
- Slow Speed: $\alpha = 1.5, D = [5, 3, 5]$,
  $H_{mel} = [h_1, h_1, h_1, h_1, h_1, h_2, h_2, h_2, h_3, h_3, h_3, h_3, h_3]$
- Fast Speed: $\alpha = 0.5, D = [2, 1, 2]$,
  $H_{mel} = [h_1, h_1, h_2, h_3, h_3]$

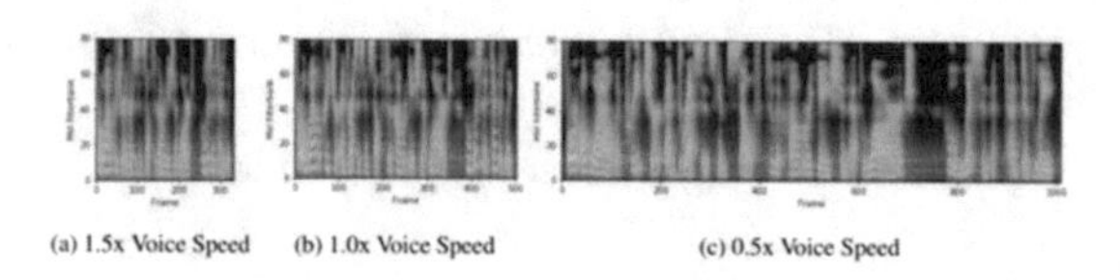

(a) 1.5x Voice Speed (b) 1.0x Voice Speed (c) 0.5x Voice Speed

Fig. 4. The mel-spectrograms of the voice with 1.5x, 1.0x and 0.5x speed respectively. [18]

The same approach is used to control the break between words (Fig. 5), where the duration of the space characters is accordingly adjusted.

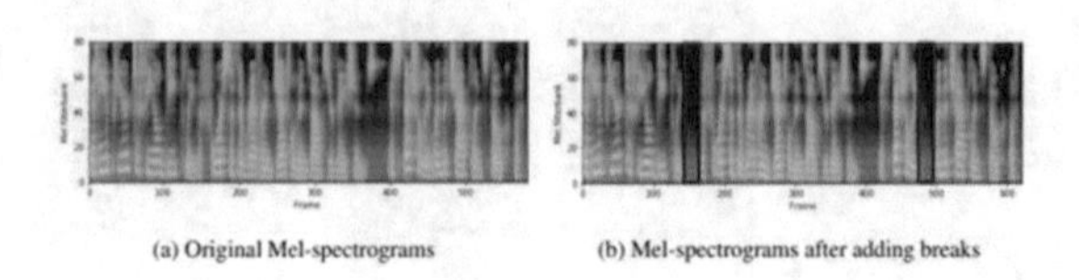

Fig. 5. The mel-spectrograms before and after adding breaks between words. [18].

5.3 Duration Predictor

Duration Predictor (Fig. 3d) [18] composed of a 2-layer 1D convolutional network with ReLU activation, each layer is followed by normalization and dropout layers, after that an extra linear layer is added to output a scalar that represent exactly the predicted phoneme duration. This module is trained simultaneously with the FastSpeech model and uses Mean Square Error (MSE) loss when it predicts mel-spectrograms length for each phoneme. The Duration Predictor training process starts with extracting the ground-truth phoneme duration using the auto-regressive teacher TTS model. This encoder-attention-decoder model is trained according to [17], where decoder-to-encoder attention alignments, for every training sequence pair, is extracted from the teacher model. Then predicted phonemes' duration is extracted using the duration extractor.

6 Service Development and Implementation

6.1 Architecture

From an architecture viewpoint, this research depend on a simple system that perform two main tasks: Recording dataset and Testing the trained model, and it is composed of the following components (Fig. 6):

1. User Interface: the system is a web-based interface where users can record their own dataset or test the available model.
2. RESTful API Server: responsible for handling the requests between the user interface, trained model, and internal business logic.
3. Database: simple database using MySQL to save dataset records and generate the main dataset file, metadata.csv, later.
4. Recording Proprieties Modules: to calculate the speech rate and volume of the speech recordings.
5. TTS Module: to handle testing available TTS model.

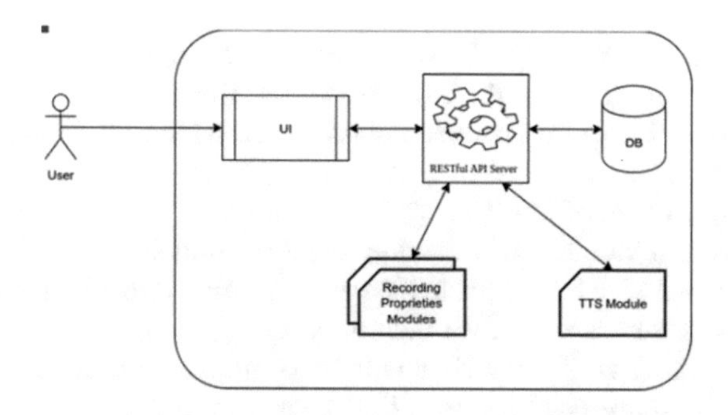

Fig. 6. Architecture of the proposed System

6.2 Criteria

There are rules (which will be mentioned later) that guarantee to get a more natural speech if they are followed and to make a good judgment on the results we take into consideration a group of criteria that characterize the synthesized speech. These criteria are as follows:

- Naturalness: Is the voice natural or artificial?
- Intelligibility (proper pronunciation): even if the words are logically erroneous, they are comprehensible;
- Comprehensibility: receiving and understanding the audio;
- Robotism: It is not acceptable to be a robot;
- Overall impression: What is the evaluators' overall impression of the speech?
- Acceptance: Will the audios be accepted and usable for the TTS project?

6.3 Limitations

The limitations of this work are as follows:

- The steps of downloading, reviewing, and feed-backing of the recordings were done manually;
- The input text when training and testing must be diacritized;
- The recording process requires a high-quality Microphone;
- The training process requires a minimum of 5 h of clean recordings;
- Overall impression: What is the evaluators' overall impression of the speech?
- The model needs resources to work effectively.

6.4 Technologies and Tools

This section includes a description of the tools and techniques used in the system's development. We describe in this section the techniques and tools which were used during the development process of the system.

- Python: an interpreted high-level programming language, which has multiple advantages in the fields of AI/ML and Computer Science.
- TensorFlow-GPU: an open-source platform provided by Google. It is widely used for the development of ML/DL applications.
- TensorBoard: a tool of the TensorFlow framework used normally to visualize the training and validation procedures of ML models.
- Phonemizer: an IPA-based library to convert text into phonemes.
- Scikit-learn library is used for machine learning in python.
- React.JS: JavaScript library that is used to implement the UI.
- MySQL: a database used to store the recordings data.

6.5 Model Training

After selecting the appropriate neural architecture, we start the procedure of training the model (Fig. 7). This procedure starts with building, recording, and reviewing the dataset, to generate the training metadata that consists of text/audio pairs, then training the Aligner model that is used to build the dataset that includes data about the duration and pitch of the recorded text data, which will be used also to train the TTS model.

Dataset. We build a text dataset that covers all the phonemes of the Arabic language using sentences from different Arabic books. This dataset was reviewed by a Linguistics Specialist to guarantee the correct spelling and diacritization of the words. Next, a native speaker, who has an excellent and appropriate Arabic pronunciation, recorded this dataset using a simple recording tool based on an open-source application called "The Mimic Recording Studio", which reduces the complexity of collecting the training data from users. This application is modified to reflect more accurate recording proprieties for users (Fig. 8):

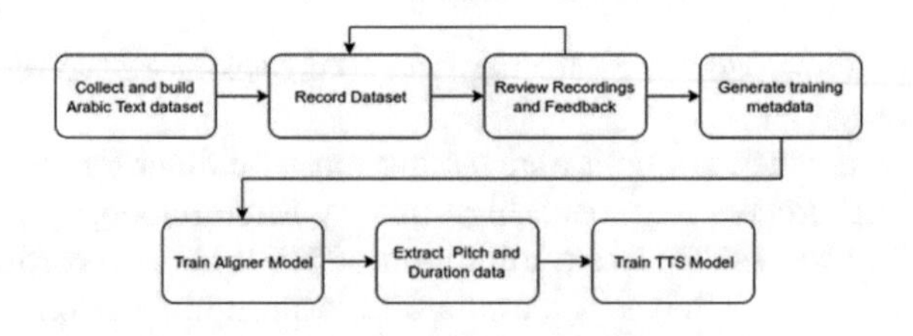

Fig. 7. Model Training Procedure

Fig. 8. Recording User Interface

- Speech rate [phoneme/sec]: measures the speech speed of the whole recording to help the users keep a steady speech speed for all the recordings.
- Speech rate (cut) [phoneme/sec]: measures the speech speed of the recording after removing the silent periods of it.
- Speech RMS [dB]: measures the volume of the speech to help the users keep a single speech loudness level for suitable and understandable speech.

The recording process was done under the following rules:

- Quiet recording environment (minimum noise), mostly at night.
- Good microphone and with fixed and suitable space between the mouth and the microphone.
- Sentences reading in a natural and neutral way and with correct formation.
- Recordings were reviewed when they are finished and, if needed, re-recorded multiple times.
- Only 150–200 sentences are recorded each day to maintain the speaker's vocals and guarantee speech quality.
- Recordings are saved in a WAV mono format with 22050 Hz sample rate.

Each day, a group of recordings is reviewed by the linguistics specialist to give feedback and improve this dataset. After selecting the best recordings, we build the training dataset based on these files structures:

- "wavs" is a folder that contains all the selected recording files.
- "metadata.csv" (Fig. 9) is the main dataset file, where each row represents a <wav—text> pair.

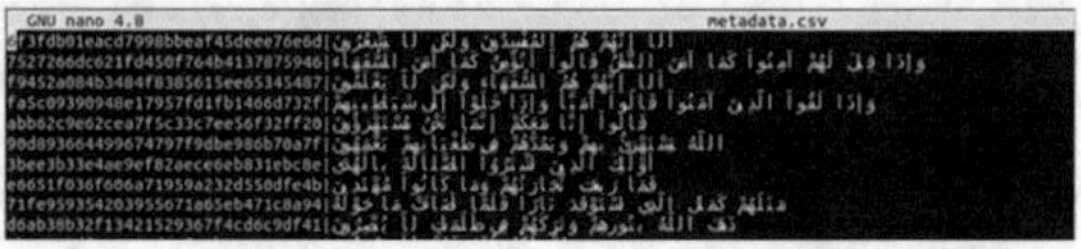

Fig. 9. Contents of metadata.csv

The final step before training starts is to pre-process the dataset to convert the text into phonemes (Fig. 10), and wav files into Mel-spectrograms (Fig. 11) with the same name but NumPy formatting, and divide the dataset into training and validating data, 80/20 respectively.

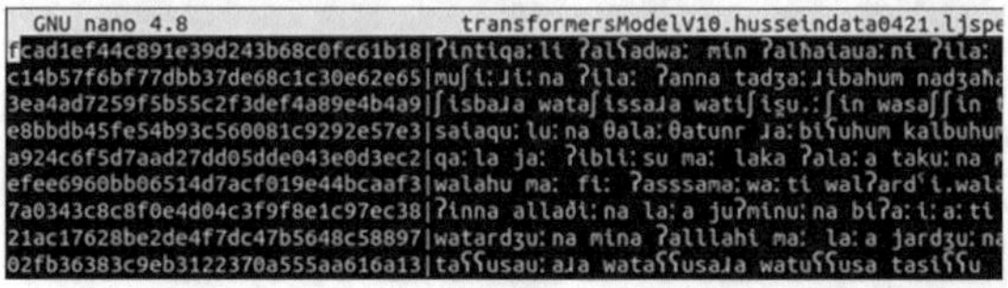

Fig. 10. Contents of metadata.csv with phonemes

Fig. 11. Generated Mel-spectrograms of corresponding wav files.

Models Training. After preparing the dataset, we start the training process. First, we train the Aligner model to produce the position of each phoneme as accurate as possible in order to align it with its corresponding mel-spectrogram frames, then we extract the predicted durations and pitch of the sentences in the dataset to be used as input when training the final TTS model. Training process has following configuration (Table 2):

Table 2. Configurations of training procedure

No. of recordings	Total duration	Aligner Training max steps	TTS training max steps
1200 wav file	117 min	260,000 steps	120,000 steps

The training procedure took place on a super server, and it has the following resources (Table 3):

Table 3. Training server information

Server Name	$LinuxFCS - GPU5.13.0 - 48 - generic54$ 20.04.1 $-$ $UbuntuSMPx86_64GNU/Linux$
Architecture	$x86_64$
CPU op-mode(s)	32-bit, 64-bit
Address sizes	46 bits physical, 48 bits virtual
CPU(s)	112
Model name	Intel(R) Xeon(R) Gold 6238R CPU @ 2.20 GHz
CPU(s)	4
GPU Model	NVIDIA® Tesla T4 GPU - 16 GB

First, we train the Aligner model (Fig. 12) to produce the position of each phoneme as accurately as possible in order to align it with its corresponding Mel-spectrogram frames, then we extract the predicted duration and pitch of the sentences in the dataset to be used as input when training the final TTS model (Fig. 13).

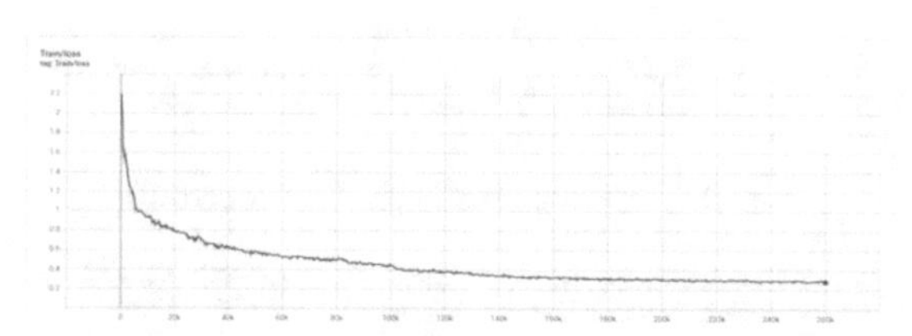

Fig. 12. Aligner Model Training Curve

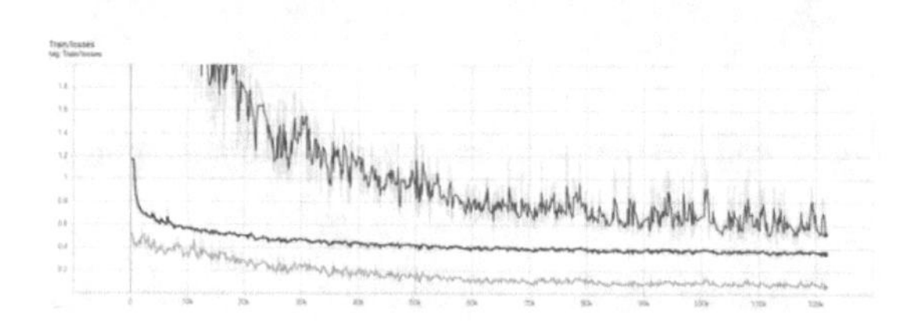

Fig. 13. TTS Model Training Curve

6.6 Service API

Simple service API is implemented to test the trained models (Fig. 14). The users have to specify only the diacritized text they want to synthesize and wait for the model to give them the resultant speech.

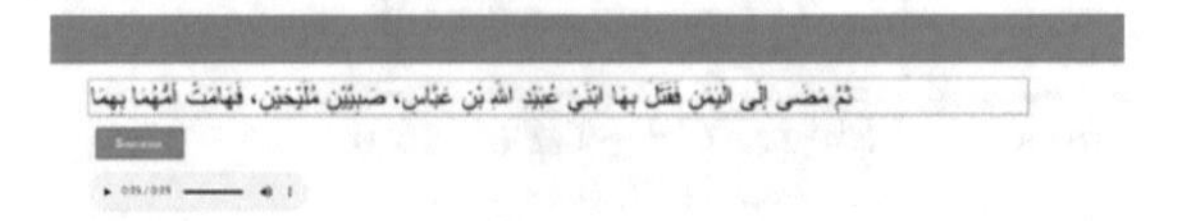

Fig. 14. User Interface to test the trained model

The procedure of speech synthesizing (Fig. 15) starts by taking the input text and converting it into the corresponding phonemes, based on IPA standards, and using it as input for the trained model that predicts the Mel-spectrograms frames corresponding to the phonemes sequence, then we convert these Mel-spectrograms into a waveform that represents the speech.

7 Evaluation

During the training procedure, we tracked the training process for the Mel-spectrogram (Fig. 16), the pitch of the phonemes (Fig. 17), and the duration (Fig. 18) of the phonemes. The predicted values are close to the trained target values.

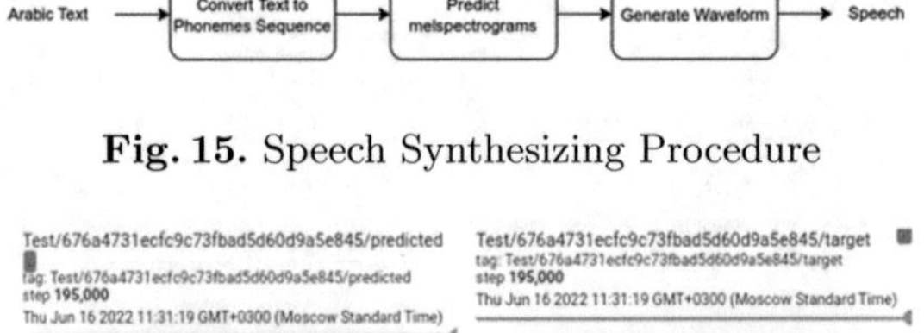

Fig. 15. Speech Synthesizing Procedure

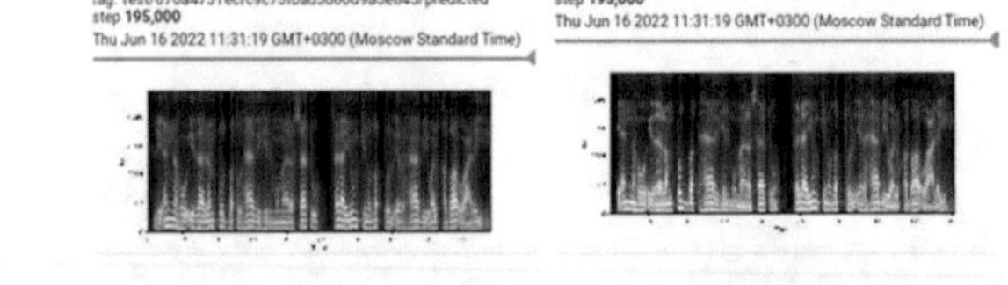

Fig. 16. Target and Predicted values for the mel-spectrogram training

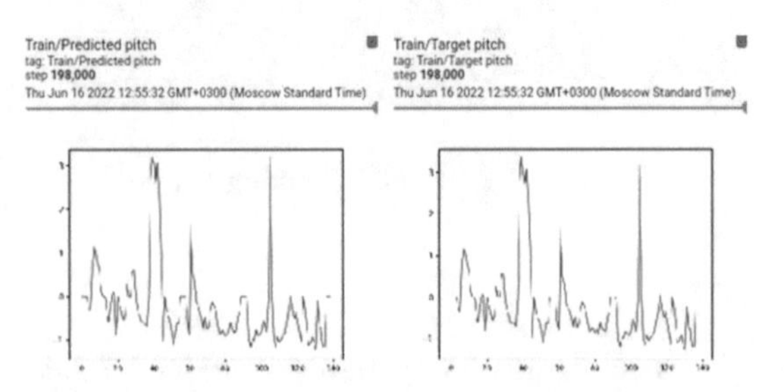

Fig. 17. Target and Predicted values for the pitch training

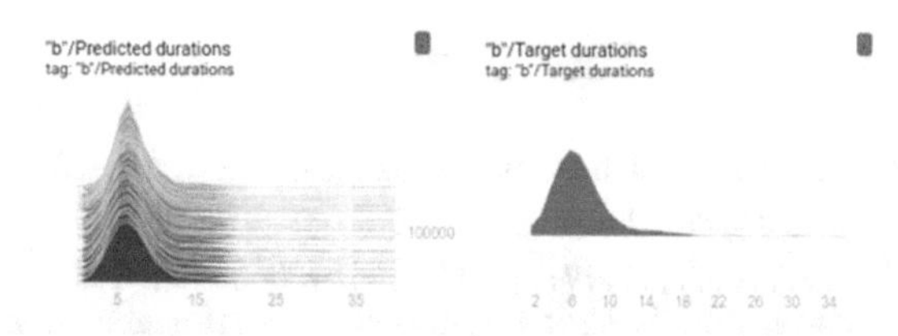

Fig. 18. Target and Predicted values for the duration of phoneme "b"

To evaluate the resultant TTS model we followed the subjective evaluation using MOS metric. We asked a group of 10 Syrian people from different cities to listen to a group of 20 audio files generated by our TTS model and evaluate them based on the previously mentioned criteria. The results of this evaluation are shown below (Table 4) according to each criterion:

For each criterion, we have 200 samples, 20 audio files x 10 evaluators, except for Overall Impression and Acceptance they have only 10 samples because these two questions were asked at the end of the survey after listening to audio files., the numbers of audio files. The MOS score for each criterion except Acceptance was calculated using the following formula:

Table 4. Training server information

Score	5	4	3	2	1	MOS
Naturalness (200)	93	94	13	0	0	4.4
Intelligibility (200)	136	64	0	0	0	4.68
Comprehensibility (200)	106	92	2	0	0	4.52
Robotism (200)	91	101	8	0	0	4.415
Overall Impression (10)	5	5	0	0	0	4.5
	Yes			No		
Acceptance (10)	8			2		4.0
Total						4.42

$$M_{OScr} = \frac{(i=1)^5 ixM_i}{N_{cr}} \tag{2}$$

where i is the scoring range (1–5), M_i is the number of samples corresponding to the score i and N_cr is the total number of samples for each criterion. for Acceptance is 80% and the corresponding value equals $(8/10) \times 5 = 4$. We considered that all the criteria have the same weight, so the overall evaluation value was calculated by taking the mean of all the values and we got a MOS equals 4.42, a very convenient value and consistent with the results of the chosen Architecture.

8 Conclusion and Future Prospects

In this research, we proposed an implementation of an Arabic-Syrian Dialect Text-To-Speech Service based on Machine Learning techniques. We reviewed and compared two different neural architectures, Tactron2 and Transformer, and investigated some models that depend on Transformer such as FastSpeech and FastPitch. We used an open-source application to record the dataset and trained the required TTS model and evaluated the results of the trained model in the area of naturalness. Although the model is quite good to be used in other applications, the text should be correctly discretized for proper pronunciation, otherwise, it would affect the accuracy of the model. Based on the Result we received during this research, The following aspects could be taken into consideration in the future:

- Improve Recording Admin Panel with better functionalities and better user experience.
- Enable reviewing and checking datasets within the Admin Panel.
- Train TTS Models online.
- Add more features to TTS models to handle speech speed and/or tone.

References

1. Darwish, K., et al.: A panoramic survey of natural language processing in the Arab world. Commun. ACM **64**(4), 72–81 (2021)
2. Wang, Y., et al.: Tacotron: towards end-to-end speech synthesis. arXiv preprint arXiv:1703.10135 (2017)
3. Amazon. Build a unique Brand Voice with Amazon Polly (2021). https://aws.amazon.com/blogs/machine-learning/build-a-unique-brand-voice-with-amazon-polly. Accessed 23 Sept 2021
4. Google. Custom Voice (Beta) Overview (2021). https://cloud.google.com/text-to-speech/custom-voice/docs. Accessed 23 Sept 2021
5. Griffin, D., Lim, J.: Signal estimation from modified short-time fourier transform. IEEE Trans. Acoust. Speech Signal Process. **32**(2), 236–243 (1984)
6. van den Oord, A., et al.: Wavenet: a generative model for raw audio. arXiv preprint arXiv:1609.03499 (2016)
7. Hochreiter, S., Schmidhuber, J.: Long short-term memory. Neural Comput. **9**(8), 1735–1780 (1997)
8. Cho, K., et al.: Learning phrase representations using RNN encoder-decoder for statistical machine translation. arXiv preprint arXiv:1406.1078 (2014)
9. Li, N., Liu, S., Liu, Y., Zhao, S., Liu, M.: Neural speech synthesis with transformer network. In: Proceedings of the AAAI Conference on Artificial Intelligence, vol. 33, no. 01, pp. 6706–6713 (2019)
10. Zerrouki, T., Abu Shquier, M.M., Balla, A., Bousbia, N., Sakraoui, I., Boudardara, F.: Adapting espeak to Arabic language: converting Arabic text to speech language using espeak. Int. J. Reason.-Based Intell. Syst. **11**(1), 76–89 (2019)
11. Zine, O., Meziane, A.: Novel approach for quality enhancement of Arabic text to speech synthesis. In: International Conference on Advanced Technologies for Signal and Image Processing (ATSIP) 2017, pp. 1–6 (2017)

12. Zine, O., Meziane, A., Boudchiche, M.: Towards a high-quality lemma-based text to speech system for the Arabic language. In: Lachkar, A., Bouzoubaa, K., Mazroui, A., Hamdani, A., Lekhouaja, A. (eds.) ICALP 2017. CCIS, vol. 782, pp. 53–66. Springer, Cham (2018). https://doi.org/10.1007/978-3-319-73500-9_4
13. Abdelali, A., Attia, M., Samih, Y., Darwish, K., Mubarak, H.: Diacritization of maghrebi Arabic sub-dialects, arXiv preprint arXiv:1810.06619 (2018)
14. Zine, O., Meziane, A., et al.: Text-to-speech technology for Arabic language learners. In: 2018 IEEE 5th International Congress on Information Science and Technology (CiSt), pp. 432–436 (2018)
15. Fahmy, F.K., Khalil, M.I., Abbas, H.M.: A transfer learning end-to-end Arabic text-to-speech (TTS) deep architecture. In: IAPR Workshop on Artificial Neural Networks in Pattern Recognition, pp. 266–277 (2020)
16. Shen, J., et al.: Natural TTS synthesis by conditioning wavenet on MEL spectrogram predictions. In: 2018 IEEE International Conference on Acoustics, Speech and Signal Processing (ICASSP), pp. 4779–4783 (2018)
17. Karita, S., et al.: A comparative study on transformer vs RNN in speech applications. In: 2019 IEEE Automatic Speech Recognition and Understanding Workshop (ASRU), pp. 449–456 (2019)
18. Ren, Y., et al.: Fastspeech: fast, robust and controllable text to speech. In: Advances in Neural Information Processing Systems, vol. 32 (2019)

A Systematic Review of Sentiment Analysis in Arabizi

Sana Gayed[1,2(✉)], Souheyl Mallat[2], and Mounir Zrigui[2]

[1] Higher Institute of Computer Science and Communication Techniques ISITCom, University of Sousse, 4011 Hammam Sousse, Tunisia
sana.gaied@gmail.com

[2] Research Laboratory in Algebra, Numbers Theory and Intelligent Systems RLANTIS, University of Monastir, 5019 Monastir, Tunisia
mounir.zrigui@fsm.rnu.tn

Abstract. Sentiment analysis (SA), also called opinion mining, is concerned with the automatic extraction of opinions conveyed in a certain text. Many studies have been conducted in the area of SA especially on English texts, while other languages such as Arabic received less attention. Recently, Arabic Sentiment Analysis (ASA) has received a great deal of interest in the research community. Several studies have been conducted on Arabic and especially in arabizi. This survey presents a systematic review of Arabic sentiment analysis research related to arabizi.

Keywords: Sentiment Analysis · Arabic sentiment analysis · Arabizi · social media · review

1 Introduction

Sentiment Analysis (SA) is concerned with the extraction of the sentiments conveyed in a piece of text based on its content. Reading this description, one would assume that SA is a sub-field of Natural Language Processing (NLP) and, due to the long and rich history of NLP, there must be several old papers addressing it. However, SA did not come to existence and flourish till the start of the new millennium. The reason for this is simply because that time witnessed the rise of the Web 2.0, which is to turn more towards Internet users, by making them move from the status of spectator of different web pages to that of actor of these. This means that users can post whatever content they want including their views and opinions. Thus, SA can be used to analyze this massive raw textual data in order to be able to summarize what the general public think of a certain issue or a certain product [1].

The natural languages that are more ubiquitous online (such as English and Chinese) are the best target for applying SA. This is verified by the large number of SA papers and tools for these languages. On the other hand, other languages lack a level of interest proportional to their "online ubiquity". One example is the Arabic language. Despite being one of the fastest growing languages in terms of online users, the field of Arabic NLP, and especially the part related the Arabic online content, is still not as mature as its

I. Czarnowski et al. (Eds.): KESIDT 2023, SIST 352, pp. 128–133, 2023.
https://doi.org/10.1007/978-981-99-2969-6_11

English counterpart. However, the past few years have witnessed a surge of interested in ASA. The purpose of this paper is to shed light onto the existing works on ASA. It also helps in directing new research efforts in order to bridge the gap between Arabic and other languages, which would allow for successful SA applications [2].

The literature is indicative of a few survey papers, which summarize the latest research in Arabic SA. The surveys of comparatively few references are offered by the authors of [3–5], who, between the years 2012 and 2016, explained methods to establish SA systems and Arabic subjectivity. These methods were defined, when there was limited research on Arabic SA. Moreover, a survey of Arabic SA tweets citing only nine journals and conferences was presented by the authors of [6]. This paper aims to commence a systematic review of sentiment analysis from Arabic text dialect. The most recent studies undertaken and published after 2012 will be highlighted in this paper.

This paper has the following composition. Section 2 contains the details of the informal Arabic text and sentiments analysis. Section 3 covers the methodology used. The survey results are presented in Sect. 4. The last section shows the conclusion.

2 The Informal Arabic Text and Sentiment Analysis

The size of Arabic users on the internet is over 226 million, and the figure on Facebook is above 141 million. Furthermore, Arabic is considered the 4th most language used in the web [7]. Literature consists of 22 Arabic Dialects (AD) illustrating the official Arabic speaking countries [3]. However, the formal language to convey the formal communication, for instance, the news, magazines, books, and official statements, is referred to as the Modern Standard Arabic (MSA). On the other hand, the Holy Quran language is known as Classical Arabic [4]. Either Arabic letters or Romanian ones (Arabizi) can be used to write the AD and MSA. The Arabizi explains the Arabic words through Latin characters, signs, and numbers, and it is because of the intricacy of the Arabic language [5]. There are six classifications of MSA and AD: Khaliji (used in the Arabic Gulf area), Egyptian (used in Egypt), Maghrebi (northern of Africa), Shami (Levantine) is spoken in (Jordan, Lebanon, Palestine, and Syria), Iraqi and Sudanese Arabic [7]. There are a number of sub-Arabic vernaculars found in each of these groups [5]. Arabic is a fertile language in vocabulary and morphological. It can be revealed in many forms.

After the rise of social networks and the growing users leading to a substantial unstructured format of the text published on the internet, the formal-informal dialects MSA and AD handling turn out to be more critical for communication. MSA is far different from the informal AD, and the former is a kind of Arabic language, which is not native to any country. Moreover, identifying the description of the text is becoming difficult because of AD and MSA complexity. In contrast, certain shortcomings are associated with Arabic sentiment analysis. The possible causes could be the restricted number of NLP tools and the diverse types of contents on the web, as a dialectal scheme is used to write such content(s) [8].

Techniques employed by computer machines to analyze the unprocessed natural spoken language among people is referred to as the NLP and it is carried out by extracting meaning and grammar from the input [9].

In recent years, the volume of data related to the daily activities of Internet users has increased significantly. Different information such as user reviews make this data. As a result, a new area of research known as opinion mining (OM) or sentiment analysis (SA) is emerging from such a large volume of data. As for SA, this domain mines texts using advanced tools. Internet users generate large volumes of unstructured content that is processed at using NLP, information retrieval and machine learning techniques. Sentiment analysis explores the text carrying the opinion according to its semantic orientation. It is possible to have a natural, negative or positive semantic orientation. Text can be treated differently by SA: several opinions are present in a document; or the whole document has only one opinion.

There are different variations of the Arabic language, including Classical Arabic (CA), Modern Standard Arabic (MSA), and Dialectal Arabic (DA) [10]. MSA and DA could be written either in Arabic or in Roman script (Arabizi), which corresponds to Arabic written with Latin letters, numerals and punctuation. Arabic is considered a more challenging language to handle by NLP systems than other languages, such as English, for various reasons. Among these reasons are the complexity of its morphology, the inconsistent and ambiguous orthography [11–15]. Once social networks emerged and large volumes of unstructured text were published to the growing number of users, communication began to become more reliant on DA and MSA [14–16].

Arabizi is a form of writing Arabic text which relies on using Latin letters rather than Arabic letters. This form of writing is common with the Arab youth. Tunisian Arabizi represents the Tunisian arabic text written using Latin characters and numerals rather than Arabic letters.

In [17], a lexicon-based sentiment analysis system was used to classify the sentiment of Tunisian tweets. A Tunisian morphological analyzer developed to produce a linguistic features. The author used 800 Arabic script tweets (TAC dataset) and achieved an accuracy of 72.1%.

A supervised sentiment analysis system was presented in [18] for Tunisian Arabic script tweets. The support vector machine achieved the best results for binary classification with an accuracy of 71.09% and an F-measure of 63%. Different bag-of-word schemes used as features, binary and multiclass classifications were conducted on a Tunisian Election dataset (TEC) of 3,043 positive/negative tweets combining MSA and Tunisian dialect.

In [19], a study is conducted on the impact on the Tunisian sentiment classification performance when it is combined with other Arabic based preprocessing tasks (Named Entity Tagging, stopwords removal…). A lexicon-based approach and the support vector machine model were used to evaluate the performances on two datasets; TEC (Tunisian Election dataset) and TSAC (Tunisian Sentiment Analysis Corpus).

In [19], authors evaluate, three deep learning methods (convolution neural networks (CNN), long short-term memory (LSTM), and bidirectional long-short-term-memory (BiLSTM)) on a corpus containing comments posted on the official Facebook pages of Tunisian supermarkets to conduct to an automatic sentiment analysis. In their evaluation, authors wanted to show that the gathered features could lead to very encouraging performances through the use of CNN and BiLSTM neural networks.

A robustly optimized BERT approach was used to establish sentiment classification from the Tunisian corpus in [20] due to the urgent need to sentiment analysis by marketing and business firms, government organizations, and society as a whole. Authors proposed a Tunisian Robustly optimized BERT approach model called TunRoBERTa which outperformed Multilingual-BERT, CNN, CNN combined with LSTM and RoBERTa. Their proposed model was pretrained on seven unlabelled Tunisian datasets publicly available.

In [21], to produce document embeddings of Tunisian Arabic and Tunisian Romanized alphabet comments, authors used the doc2vec algorithm. The generated embeddings were fed to train a Multi-Layer Perceptron (MLP) classifier where both the achieved accuracy and F-measure values were 78% on the TSAC (Tunisian Sentiment Analysis Corpus) dataset. This last dataset combines 7,366 positive/negative Tunisian Arabic and Tunisian Romanized alphabet Facebook comments.

In [22], syntax-ignorant n-gram embeddings representation composed and learned using an unordered composition function and a shallow neural model was proposed, it helps to relieve hard work due to the hand-crafted features. The proposed model, called Tw-StAR, was evaluated to predict the sentiment on five Arabic dialect datasets including the TSAC dataset.

doc.

3 Methodology Used

To be more concrete, the focus of this article is to increase the criteria such as inclusion, and this has been listed below.

- Papers published from the year 2012 till 2022,
- Papers using the following keywords: Arabic, dialects, opinion analysis, Sentiment Analysis on arabizi.

The following electronic databases were searched: IEEE, ACM, Scopus, Web of Science, Science Direct, Google Scholar, etc. To achieve better and more qualitative results, several keywords presented in Table 1 are used.

Table 1. The search process to collect papers.

Database	Number of Papers	Keywords
IEEE & IEEE Xplore	43	Sentiment analysis, opinion mining, Naïve Bayes, Support Vector Machine, NLP, arabizi,tunizi, etc.
ACM	30	
Scopus	28	
Web of Science	32	
Springer	49	
Google Scholar	70	
Others	18	
Total	270	

Natural Language Processing is an implementation and needs experiment to apply it [23–26]. However, any experiment also needs a prior literature review or survey to have a breadth knowledge about the previous researches. Also, LR needed to deepen the knowledge, to find out the gap, and to specify the specific area that needs enhancement. Moreover, a literature review/survey needed to determine and examine the variables needed for the experiment and to know the algorithms used before and the performance for each one.

4 Results

We can observe that the predominant techniques in arabizi studies are: Naive Bayes, Long Short-Term Memory (LSTM), Convolutional Neural Network (CNN), Support Vector Machine (SVM) and; K-Nearest Neighbor (KNN).

5 Conclusion

The objective of this review to analyze Arabic sentiment analysis studies that involved arabizi script. We found 270 research articles varied between 2012 and 2022. The main research findings indicate that ML algorithms NB and SVM are used on sentiment analysis of Arabizi. Deep learning approaches, like LSTM and CNN have satisfactory results. The most studies in arabizi are Egyptian and Jordanian, Algerian, Saudi and recently Tunisian, while other dialects like Mauritanian, Yemeni were neglected.

References

1. Liu, B.: Sentiment analysis and opinion mining. Synth. Lect. Hum. Lang. Technol. **5**(1), 1–167 (2012)
2. Abdulla, N.A., Al-Ayyoub, M., Al-Kabi, M.N.: An extended analytical study of Arabic sentiments. Int. J. Big Data Intell. **1**(1–2), 103–113 (2014)
3. Mustafa, H.H., Mohamed, A., Elzanfaly, D.S.: An enhanced approach for arabic sentiment analysis. Int. J. Artif. Intell. Appl. **8**(5), 1–14 (2017)
4. Assiri, A., Emam, A., Al-Dossari, H.: Saudi twitter corpus for sentiment analysis. World Acad. Sci. Eng. Technol. Int. J. Comput. Electr. Autom. Control Inf. Eng. **10**(2), 272–275 (2016)
5. Baly, R., Khaddaj, A., Hajj, H., El-Hajj, W., Shaban, K.B.: ArSentD-LEV: a multi-topic corpus for target-based sentiment analysis in Arabic levantine tweets. arXiv Prepr, arXiv1906.01830 (2019)
6. Alawya, A.: Aspect terms extraction of Arabic dialects for opinion mining using conditional random fields. In: Gelbukh, A. (ed.) CICLing 2016. LNCS, vol. 9624, pp. 211–220. Springer, Cham (2018). https://doi.org/10.1007/978-3-319-75487-1_16
7. Oussous, A., Benjelloun, F.-Z., Lahcen, A.A., Belfkih, S.: ASA: a framework for Arabic sentiment analysis. J. Inf. Sci. **46**, 544–559 (2019)
8. Darwish, K., Magdy, W.: Arabic information retrieval. Foundations and Trends® in Information Retrieval, **7**(4), 239–342 (2014)
9. Alotaiby, T.N., Alshebeili, S.A., Alshawi, T., Ahmad, I., Abd El-Samie, F.E.: EEG seizure detection and prediction algorithms: a survey. EURASIP J. Adv. Sign. Proces. **2014**(1), 1–21 (2014). https://doi.org/10.1186/1687-6180-2014-183

10. Alhumoud, S.O., Altuwaijri, M.I., Albuhairi, T.M., Alohaideb, W.M.: Survey on Arabic sentiment analysis in twitter. Int. Sci. Index **9**(1), 364–368 (2015)
11. Rozovskaya, A., Sproat, R., Benmamoun, E.: Challenges in processing colloquial Arabic. In Proceedings of the International Conference on the Challenge of Arabic for NLP/MT (pp. 4–14) (2006)
12. Habash, N., Soudi, A., Buckwalter, T.: On Arabic Transliteration. Arabic Computational Morphology: Knowledge-Based and Empirical Methods, pp. 15–22 (2007)
13. Abdellaoui, H., Zrigui, M.: Using Tweets and emojis to build TEAD: an Arabic dataset for sentiment analysis. Computación y Sistemas **22**(3), 777–786 (2018)
14. Sghaier, M.A., Zrigui, M.: Tunisian dialect-modern standard Arabic bilingual lexicon. In: 14th IEEE/ACS International Conference on Computer Systems and Applications, AICCSA, pp. 973–979. IEEE, Hammamet, Tunisia (2017)
15. Gayed, S., Mallat, S., Zrigui, M.: Exploring word embedding for arabic sentiment analysis. In: Szczerbicki, E., Wojtkiewicz, K., Nguyen, S.V., Pietranik, M., Krótkiewicz, M. (eds.) Recent Challenges in Intelligent Information and Database Systems. ACIIDS 2022. Communications in Computer and Information Science, vol. 1716, pp. 92–101. Springer, Singapore (2022). https://doi.org/10.1007/978-981-19-8234-7_8
16. Sghaier, M.A., Zrigui, M. : Rule-based machine translation from Tunisian dialect to modern standard Arabic. In: 24th International Conference on Knowledge-Based and Intelligent Information & Engineering Systems, pp. 310–319. Elsevier, a virtual conference (2020)
17. Sayadi, K., Liwicki, M., Ingold, R., Bui, M.: Tunisian dialect and modern standard Arabic dataset for sentiment analysis: Tunisian election context. In: 2nd International Conference on Arabic Computational Linguistics. Turkey (2016)
18. Mulki, H., Haddad, H., Bechikh Ali, C., Babaoglu, I.: Tunisian dialect sentiment analysis: a natural language processing-based approach. Computaci´on y Sistemas **22**(4), 1223– 1232 (2018)
19. Medhaffar, S. Bougares, F., Estève, Y. Hadrich-Belguith, L.: Sentiment analysis of Tunisian dialects: linguistic resources and experiments. In: 3rd Arabic Natural Language Processing Workshop, pp. 55–61. Association for Computational Linguistics, Valencia (2017)
20. Antit, C., Mechti, S., Faiz, R.: TunRoBERTa: a Tunisian robustly optimized BERT approach model for sentiment analysis. In: 2nd International Conference on Industry 4.0 and Artificial Intelligence (ICIAI) (2021)
21. Mulki, H., Haddad, H., Gridach, M., Babaoglu, I.: Syntax-Ignorant N-gram embeddings for sentiment analysis of Arabic dialects. In: 4th Arabic Natural Language Processing Workshop, pp. 30–39. Association for Computational Linguistics, Florence (2019)
22. Zahran, M.A., Magooda, A., Mahgoub, A.Y., Raafat, H.M., Rashwan, M., Atyia, A.: Word representations in vector space and their applications for arabic. CICLing **1**, 430–443 (2015)
23. Mahmoud, A., Zrigui, M.: Semantic similarity analysis for corpus development and paraphrase detection in Arabic. Int. Arab J. Inf. Technol. **18**(1), 1–7 (2021)
24. Haffar, N., Hkiri, E., Zrigui, M.: Using bidirectional LSTM and shortest dependency path for classifying Arabic temporal relations. In: 24th International Conference Knowledge-Based and Intelligent Information & Engineering Systems (KES), pp. 370–379. Elsevier, a virtual conference (2020)
25. Mahmoud, A., Zrigui, M.: Arabic semantic textual similarity identification based on convolutional gated recurrent units. In : International Conference on INnovations in Intelligent SysTems and Applications (INISTA), pp. 1–7. IEEE, Kocaeli (2021)
26. Haffar, N., Ayadi, R., Hkiri, E., Zrigui, M.: Temporal ordering of events via deep neural networks. In: 16th International Conference on Document Analysis and Recognition (ICDAR), pp. 762–777. Lausanne (2021)

Reasoning-Based Intelligent Applied Systems

Optimize a Contingency Testing Using Paraconsistent Logic

Liliam Sayuri Sakamoto(✉), Jair Minoro Abe, Aparecido Carlos Duarte, and José Rodrigo Cabral

Graduate Program in Production Engineering, Paulista University, UNIP, Rua Dr. Bacelar, 1212-Vila Clementino, São Paulo 04026002, Brazil
liliam.sakamoto@aluno.unip.br

Abstract. In this paper, a risk analysis was approached in Contingency Testing with the purpose of serving as a support for the prevention of the disaster situation in an IT environment by a Data Protection Officer (DPO) supported using the Logic Paraconsistent Annotated Evidential Eτ. When observing the aspect of risk, that is, of non-compliance or adherence to ISO 22.301/2013 - Security and resilience—Business continuity management systems—Requirements and Brazilian General Data Protection Law (LGPD), there is a focus on global verification, and it is not always respected in its details interns. In this proposal, simple decision-making is contrasted with those based on Paraconsistent Annotated Logic Eτ. Only a few risk factors were selected for this study concerning contingency tests. Some indications have been analyzed in this context and may present a higher or lower degree of the positive sign, however, situations that lead to uncertainty percentages may occur. It can be concluded that a lot of information has ambiguous and incomplete tendencies. However, this information, instead of being wholly discarded because it is considered totally inconclusive, can be analyzed and optimized decisions can be made.

Keywords: Paraconsistent Logic · Contingency Testing · Data Protection Officer (DPO) · Brazilian General Data Protection Law (LGPD)

1 Introduction

The importance of information systems for companies is growing, as each day they depend on technology to manage their decision-making. This dependency also causes concern about the availability of this resource, which can be supported by its own IT structure or by the provision of services by a specialized company.

Although there are several formats for structuring this IT support structure, companies must increasingly be concerned with ensuring this availability, which brings about the need for a contingency plan, which can be combined with a Crisis Management Plan and the Disaster Recovery Plan [1]. The use of Logic Paraconsistent Annotated Evidential Eτ [7] with the practical application of the Para-Analyzer algorithm for decision-making is one of the important points for the innovation of the work of prevention and detection

I. Czarnowski et al. (Eds.): KESIDT 2023, SIST 352, pp. 137–146, 2023.
https://doi.org/10.1007/978-981-99-2969-6_12

of situations or scenarios of probable risks of events that lead to a possible supportive or monitoring tools generally take into account only the classic approach, that is, they are based on the duality of whether it is a risky situation or not, however, no study involves the contradictions and the doubts that most of the time they are despised.

By using the Logic Paraconsistent Annotated Evidential Eτ [7], it is possible to:

- arrive at more filtered conclusions, where countless contexts at internal levels are analyzed, even those considered contradictory.
- another possibility is centered on the joint analysis of other contexts previously neglected and that becomes part of the active analysis, constantly updated.

2 Backgrounds

2.1 Compliance and Risk Considerations

According to the Committee of Sponsoring Organizations of the Treadway Commission - COSO [2], risk management can be applied in all types of activities, in any company, public or private, or, in any form of company, regardless of its size [3]. One of the areas covered by COSO is compliance which is so important for risk assessment: Compliance - (compliance - if it complies with an obligation or standard) with laws and regulations to which the entity is subject [2].

Other frameworks also prove that there is an alignment of Governance with the need for risk analysis or called risk management, COBIT, a framework for internal IT controls [4]. According to Kant [5]: "... Everything that happens has a cause", in this case, he addresses the materialization of risk in which an agent is needed to provoke a certain situation, that is, this makes the risk happen, which results in the non-realization of a given objective, The relationship between the cause and effect that is the consequence of the risk, because a certain risk can happen because of a given situation that will have an alleged consequence.

In the aspect of cause and effect, when investors buy shares, surgeons perform operations, engineers design bridges, entrepreneurs open their businesses and politicians run for elected office, the risk is an inevitable partner [6]. Risk management is understood as a general procedure for risk reduction, that is, when it is applied in any instance, the possible consequences are all acceptable, and there may be coexistence with the worst expected result [6]. Risk is presented in some way and to some degree in most human activities and is characterized by being partially known, changing over time, and being manageable in the sense that human action can be applied to change its shape and the degree of its It is made.

The risk management process starts with uncertainties, concerns, doubts, and unknowns that turn into acceptable risks. But each company needs to define its risk appetite. Although the legislation tells municipalities that they must have a disaster recovery plan, they do not present evidence of tests, they only despair when the worst happens. According to Bernstein [5]: "In the 20th century, risk management was disseminated, studied, and used mainly in the areas of health, finance, life insurance, and so on. All projects in these areas deal with risks, as profits depend on attractive opportunities, balanced by well-calculated risks".

Risk is defined: Firstly, risk affects future events. Present and past are not a concern, because what we harvest today has already been sown by our previous actions [14]. Risk involves change, such as a change in thinking, opinion, actions, or places…, and thirdly, risk involves choice and uncertainty that the choice itself involves, so, paradoxically, risk, like death and taxes, is one of the few certainties in life. The GDPR - General Protection Data Regulation in the European Union presents specific documentation for personal data protection privacy [12], while the LGPD [11] is the Brazilian legislation about this issue.

2.2 ISO 22.301/2013 - Business Continuity Management System

This standard specifies requirements for establishing and managing an effective Business Continuity Management System [1].The organization must carry out exercises and tests that:

a) Are consistent with the scope and objectives of BCMS - Business Continuity Management System [1].
b) are based on appropriate scenarios that are well planned with clearly defined objectives and objectives,
c) taken over time to validate all their business continuity agreements, involving interested parties,
d) minimize the risk of interruption of operations,
e) produce formal post-exercise reports containing results, recommendations, and actions to implement improvements,
f) are reviewed in the context of promoting continuous improvement and
g) are conducted at planned intervals and when there are significant changes within the organization or the environment in which it operates.

2.3 Logic Paraconsistent Annotated Evidential Eτ

Abe et al. [7] mention that Logic Paraconsistent Annotated Evidential Eτ is a family of non-classical logic that appeared in the late 90s of the last century in logical programming. Annotated logic constitutes a class of paraconsistent logic. Such logic is related to certain complete lattices, which play an important role. A knowledge specialist on the subject to be addressed issues a quantitative opinion ranging from 0.0 to 1.0. These values are respectively the favorable evidence that is expressed by the symbol μ and the contrary evidence λ.

Abe et al. [7]: programs can now be built using paraconsistent logic, making it possible to treat inconsistencies in a direct and elegant way. With this feature, it can be applied in specialist systems, object-oriented databases, representation of contradictory knowledge, etc. with all the implications in artificial intelligence. In Abe et al. [7]: "Logic Paraconsistent Annotated Evidential Eτ has an Eτ language, and the atomic propositions are of the type p ((μ, λ) where p is a proposition and d $\mu, \lambda \in [0, 1]$ (closed unit real interval). Intuitively, μ indicates the degree of unfavorable evidence1 of p and λ the degree of contrary evidence of p. The reading of the values μ, λ depends on the applications considered and may change in fact, μ it may be the degree of belief favorable and λ it may be the degree of belief contrary to proposition p; also, μ can indicate the probability expressed by p occurring and λ the improbability expressed by p occurring. The atomic propositions p (μ, λ) of logic Eτ can intuitively be read as: I believe in p with the degree of favorable belief μ and the degree of contrary belief λ, or the degree of favorable evidence of p is μ and the degree of evidence to the contrary of p is λ".

Paraconsistent logics are logics that can serve as the underlying logic of theories in which A and ¬A (the negation of A) are both true without being trivial [6]. There are many types of paraconsistent systems. In this article, we consider the Paraconsistent Annotated Evidential Logic Eτ. The formulation in Logic Eτ is of the type p (μ, λ), in which p is a proposition and e (μ, λ) ϵ [0, 1] is the real unitary closed interval.

A proposition p (μ, λ) can be read as: "The favorable evidence of p is μ and the unfavorable evidence is λ" [7]. For instance, p (1.0, 0.0) can be read as a true proposition, p (0.0,1.0) as false, p (1.0, 1.0) as inconsistent, p (0.0, 0.0) as paracomplete, and p (0.5, 0.5) as an indefinite proposition [8]. Also, we introduce the following concepts: Uncertainty degree: Gun $(\mu, \lambda) = \mu + \lambda - 1$ $(0 \leq \mu, \lambda \leq 1)$ and Certainty degree: Gce $(\mu, \lambda) = \mu - \lambda$ $(0 \leq \mu, \lambda \leq 1)$ [9].

An order relation is defined on [0, 1]: $(\mu 1, \lambda 1) \leq (\mu 2, \lambda 2) \leftrightarrow \mu 1 \leq \mu 2$ and $\lambda 2 \leq \lambda 1$, constituting a lattice that will be symbolized by τ.

With the uncertainty and certainty degrees, we can get the following 12 output states (Table 2): extreme states, and non-extreme states. It is worth observing that this division can be modified according to each application [10].

2.4 Para-Analyser Algorithm

In this proposed algorithm, there is a set of information obtained, which can sometimes seem contradictory, making it difficult to analyze the scenario for risk analysis. Generally, in such situations, this information is discarded or ignored, that is, they are considered "dirty" of the system, however at best they may even receive different treatment. Silva Filho, Abe, and Torres [10] quote: "However, the contradiction most of the time contains decisive information, as it is like the encounter of two strands of opposing truth values. Therefore, to neglect it is to proceed in an anachronistic way, and that is why we must look for languages that can live with the contradiction without disturbing the other information. As for the concept of uncertainty, we must think of a language that can capture the 'maximum' of 'information' of the concept". In this line of reasoning for the analysis based on Paraconsistent Logic, situations of Inconsistency and Paracompleteness will be considered together with the True and False, represented according to Table 1:

The set of these states or objects ($\tau = \{F, V, T, \perp\}$) can also be called annotation constants and can be represented using the Hasse diagram as shown in Fig. 1:

The operator about τ é: $\sim: |\tau| \rightarrow |\tau|$ that will operate, intuitively, like this:

~ T = T (the 'negation' of an inconsistent proposition is inconsistent)
~ V = F (the 'negation' of a true proposition is false)
~ F = V (the 'negation' of a false proposition is true)
~ ⊥ = ⊥ (the 'negation' of a paracomplete proposition is paracomplete)

Annotated Paraconsistent Logic will be used, this type must be composed of 1, 2, or "n" values. With the calculations of the values of the axes that make up the representative

Table 1. Extreme States Fonte: Abe et al. [7].

Extreme States	Symbol
True	V
False	F
Inconsistent	T
Paracomplete	⊥

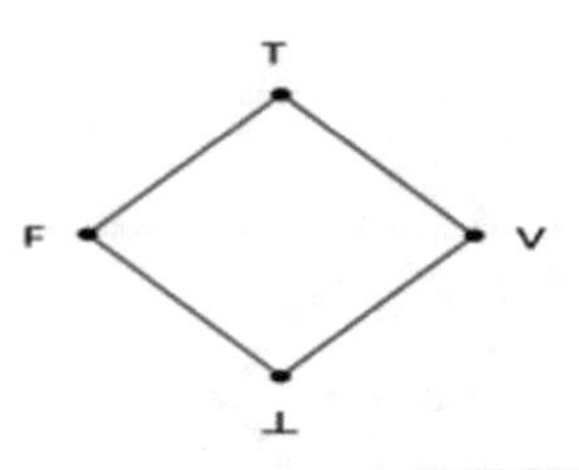

Fig. 1. [10] Hasse diagram.

figure of the lattice, it can be divided or internally delimited into several regions of different sizes and formats, thus obtaining a discretization of the same. From the bounded regions of the lattice, it is possible to relate the resulting logical states, which, in turn, will be obtained by interpolating the Degrees of Certainty Gc and Contradiction Gct. Thus, for each point of interpolation between the Degrees of Certainty and Contradiction, there will be a single delimited region that the lattice will be divided depending on the accuracy desired in the analysis [10].

The representation of Fig. 2 shows a representation of the lattice constructed with values of Degrees of Certainty and Contradiction and sectioned into 12 regions. Thus, at the end of the analysis, one of the 12 possible resulting logical states will be obtained as an answer for decision-making.

Table 2. Non-extreme States

Non-extreme States	Symbol
Quasi-true tending to Inconsistent	QV → T
Quasi-true tending to Paracomplete	QV → ⊥
Quasi-false tending to Inconsistent	QF → T
Quasi-false tending to Paracomplete	QF → ⊥
Quasi-inconsistent tending to True	QT → V
Quasi-inconsistent tending to false	QT → F
Quasi-paracomplete tending to True	Q⊥ → V
Quasi-paracomplete tending to false	Q⊥ → F

Some additional control values are:

- Vscct = maximum value of uncertainty control = Ftun
- Vscc = maximum value of certainty control = Ftce
- Vicct = minimum value of uncertainty control = -Ftun
- Vicc = minimum value of certainty control = -Ftce

All states are represented in the next Fig. 2.

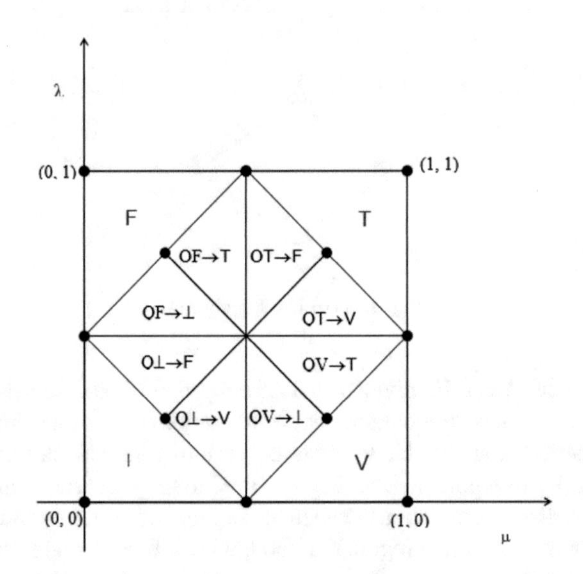

Fig. 2. Extreme and Non-extreme states that represents Table 2

3 Methodology

3.1 The Proposal of Data Analysis for Decision-Making

For decision-making regarding contingency tests, the analysis is based on the following steps: setting control values, choosing influencing factors, establishing sections for each factor, building the database, field research, calculating the constants of resulting annotation, determination of the barycenter, and decision-making.

3.1.1 Setting Control Values

Knowledge Engineers (KE) were chosen, that is, specialists in the field with the capacity to adequately discern the Contingency Test in IT. Therefore, the setting of control values must be high, this being a matter of relevance.

Each KE determines the level of requirement of the decision, consequently, it is fixing the decision regions, as well as the decision rule and the para-analyzer algorithm.

3.1.2 Choice of Influencing Factors and Sections

The premise is that the KE must know in detail the analyzed environment and list the influencing factors that detect signs of Risk of disaster in IT or not. This KE must be based on ISO 22.301/2013 [1]. Each factor being listed must be well delimited to create favorable conditions for detection and must be tested to enable analysis for decision-making.

Another important point is the allocation of weights for the different factors listed. These weights will compensate for the degree of inference of each factor in this decision. In the first phase: only factors based on contingency tests and exercises will be analyzed (item 8.5 of ISO 22.301 / 2013 [1].

The analyzed factor can have "n" sections that will be called "s", and it is established that the "s" sections Sj (1 < j < s), that is, they can translate conditions of the factors, which depend on the refinement of the analyze.

In this study, weights will be balanced, and the sections are defined as follows (Table 4):

Table 3. Weights, Factors, and sections.

Weights	Factors	Sections
1	F1 – The organization carries out exercises and tests consistent with the scope and objectives of the Contingency Plan	S1. Periodic exercises
2	F2 - The exercises and tests are based on appropriate scenarios that are well planned with clearly defined objectives	S1. Data Center scenario
3	F3 - The exercises and tests minimize the risk of interruptions in IT operations	S1. Interruption of telecom link

Table 4. Authors: Averege over μ and λ factors and sections

Result	Factor	Section	Conclusion		Definition
			μ	λ	
A	F1	S1	0,2	0,8	False
B	F2	S1	0,7	0,5	Inconsist
C	F3	S1	1	0,2	True

4 Discussions and Results

The analysis of the values is by the Annotated Evidenced Paraconsistent Logic Eτ, which would be presented in Table 3 results in the para-analyzer algorithm graphic in (Fig. 3).

Figure 3A (F1-S1): A consensus is reached on this section within the scope factor, where there is no conclusive situation regarding the evidence of a possible IT disaster. The average of favorable evidence is 0.2 and favorable evidence 0.8, in this case, the result is false tending to be inconsistent, a situation that indicates that monitoring is necessary. Law 12.340/2010 establishes one year for the Municipalities included in the registry to prepare the Civil Protection and Defense Contingency Plan because the Brazilian Atlas of Natural Disasters points out that the 31,909 disasters recorded in the period considered in the survey affected 96,494,755 people and caused 3,404 deaths between 1991 and 2010 [13]. In comparing "A" to Law 12.340 there are no corrective actions to do about the situation.

Figure 3B (F2-S1): represents the analysis carried out by the experts regarding the appropriate scenario factor, in which case the chosen one was the Data Center. For in this section, where there is no way to prove when and where this may occur. The average of favorable evidence is 0.7 and favorable evidence is 0.5, in this case, the result is inconsistent since the context should have been more detailed. In Federal Emergency Management Agency (FEMA), required every Florida county to identify potential locations of disaster recovery centers (DRCs), and there is no definition of a Data Center [14]. But in this type of situation, there are no exact actions to do.

Figure 3C (F3-S1): represents the analysis carried out by the specialists regarding the risk minimization factor in the telecom link test, there is evidence of the possible risk of IT disaster, as an indicator in the contingency test.

This works represents the analysis performed by the para-analyzer algorithm by 09 different specialists.

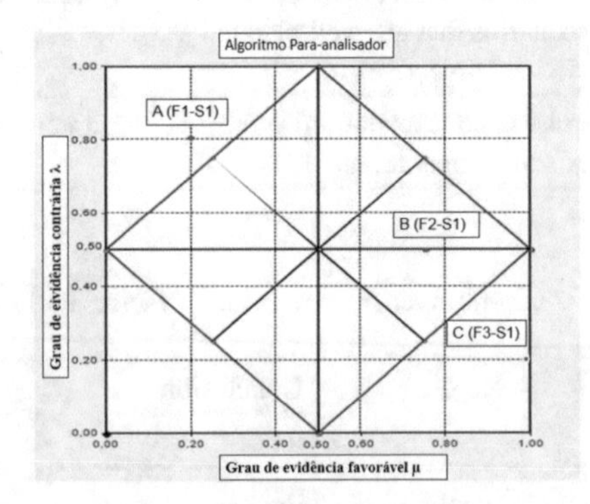

Fig. 3. Authors: Grafical representation of Table 3

Law 9.472, of July 16, 1997, which provides for the organization of telecommunications services, determines that the National Telecommunications Agency establishes a free emergency service and maintains a plan with the attribution, distribution, and allocation of radio frequencies and the provision of radio frequency bands for emergency and public security services [13]. In comparison with the "C," we could verify that is the more relevant disaster.

Decision-making that involves a larger number of people tends to more qualified results, increasing the knowledge of the decision situation, and mitigating, through the aggregation of information and knowledge, the distortions of the individualistic view [14]. Decision-making in a world that changes dynamically and quickly is a challenge, and decision-aid techniques help decision-makers to face this challenge [14].

5 Conclusion

These situations lead companies to make decisions that are nothing more than, a complex and very comprehensive process, where we must analyze several factors and make the combination of the most diverse and varied possibilities. As several applications are developed to improve this risk analysis and often do not show consistency, verification was made possible here through non-classical logic such as paraconsistent. The heterodox Logics entered the scene with great impetus: no one could predict where the multipurpose, relevant, and paraconsistent Logics will take us. Perhaps, in the coming years, we will have a new change in the idea of logic, impossible to imagine now. The analysis of the identified risks is the activity that allows the characterization of the most important aspects of each risk, aiming to explore the best mitigation strategies. Usually, risks are classified, according to some established criterion, to make management focus on the risks considered to be priorities. What can be evidenced is that the first scenario does not indicate certainty of a disaster risk situation, the second scenario presents a doubtful risk, and the third scenario already induces a possible situation of greater risk.

There may be several ways of making decisions about risky situations, however, through the Logic Paraconsistent Annotated Evidential Eτ shows a differentiated form of more correct decision because it involves the knowledge of several specialists. This article presented three analyses with conclusions that converge to the simplified trends in three scenarios addressed. Other studies on contingency management have been verified, such as in power distribution operators [15], and in Northern Ireland with governance scenarios very close to the concepts of ISO 22.301/2013 [16], which will be focuses of future studies. For, the challenge of this study is to deepen the maturity of the changes of the corporate world and study contingency management scenario.

Acknowledgements. We thank the research group Paraconsistent logic and artificial intelligence maintained by the Paulista University and conducted by researcher Dr. Abe.

References

1. ISO 22.301/2013 - Security and resilience—Business continuity management systems—Requirements. https://www.iso.org/standard/75106.html. Accessed 23 Apr 2021
2. Cocurullo, A.: Gestão de riscos corporativos: riscos alinhados com algumas ferramentas de gestão: um estudo de caso no setor de celulose e papel. São Paulo. 2°. Ed. ABDR (2004)
3. Fernandes, F.C., et al.: Mecanismos de Gerenciamento de Riscos na Cadeia de Suprimentos e Logística de Micro e Pequenas Empresas do Médio Vale do Itajaí: Discussão Exploratória sobre oportunidades de pesquisa. VIII Encontro de Estudos em Empreendedorismo e Gestão de Pequenas Empresas (EGEPE). Goiânia (2014)
4. Giampaoli, R.Z., et al.: Contribuições do modelo COBIT para a Governança Corporativa e de Tecnologia da Informação: desafios, problemas e benefícios na percepção de especialistas e CIOs. Análise a Revista Acadêmica da FACE **22**(2), 120–133 (2011)

5. Kant, I.: Crítica de la razón pura. editorial Lozada, 7a. edn, p. 156 (1977)
6. Bernstein, P.L.: Desafio aos Deuses: A fascinante história do risco. 3ª. Edição. São Paulo. Campus, p. Vii (1996)
7. Abe, J.M., et al.: Lógica Paraconsistente – Anotada Evidencial Eτ – Santos: Editora Comunnicar (2011)
8. Carvalho, F.R., Abe, J.M.: Tomadas de decisão com ferramentas da lógica paraconsistente anotada – Método Paraconsistente de Decisão – MPD – São Paulo: Blucher (2011)
9. Da Costa, N.C.A.: Ensaios sobre os fundamentos da lógica. Hucitec, São Paulo (1980)
10. Silva Filho, J.I., Abe, J.M., Torres, G.L.: Inteligência Artificial com as redes de análises paraconsistentes. LTC, Rio de Janeiro (2008)
11. Brazil. Lei Geral de Proteção de Dados Pessoais (LGPD). Lei nº 13.709, de 14 de Agosto de 2018. http://www.planalto.gov.br/ccivil_03/_ato2015-2018/2018/lei/l13709.htm. Accessed 21 Dec 2022
12. European Commission GDPR - General Data Protection Regulation 2016. https://gdpr-info.eu/. Accessed 21 Apr 2022
13. Ganem, R.S.: Gestão de desastres no Brasil. Estudo. Consultoria Legislativa. Câmara dos (2012)
14. Wold, G.H.: Disaster recovery planning process. Disaster Recovery J. **5**(1), 10–15 (2006)
15. Fattaheian-Dehkordi, S., Tavakkoli, M., Abbaspour, A., Fotuhi-Firuzabad, M., Lehtonen, M.: Optimal energy management of distribution networks in post-contingency conditions. Int. J. Electr. Power Energy Syst. **141**, 108022 (2022). https://doi.org/10.1016/j.ijepes.2022.108022. ISSN 0142-0615
16. Schimmel, F.: of Thesis: an expanded contingency framework for mediated. J. Conflict Resolut. **25**(1), 157–180 (2023)

Age-Group Estimation of Facial Images Using Multi-task Ranking CNN

Margarita N. Favorskaya(✉) and Andrey I. Pakhirka

Reshetnev Siberian State University of Science and Technology, 31 Krasnoyarsky Rabochy Ave., Krasnoyarsk 660037, Russian Federation
{favorskaya,pahirka}@sibsau.ru

Abstract. Age as one of the facial attributes plays a significant role in surveillance, web content filtering in electronic customer relationship management, face recognition, among others. The process of ageing changes the colour and texture of skin, as well as the facial skeleton lines with other additional attributes. This process is burdened by such features as ethnicity, gender, emotion, illumination, pose, makeup, and other artifacts that make the task of age estimation non-trivial. One of the promising approaches is to consider the problem as multi-task in order to improve accuracy. We propose a new multi-task CNN for identity verification in real-time access systems. It includes a ranking sub-CNN that groups facial images into predefined age ranges and two sub-CNNs that binary gender classification and multi-class facial expression classification. Even a small number of additional attributes shows more accurate estimates of the age grouping. We have tested our multi-task CNN using five public datasets with promising results.

Keywords: Age-group · Facial image · Multi-task ranking · Deep learning

1 Introduction

Face-based age estimation is one aspect of the face recognition problem and can find various applications such as driving license renewal system, passport renewal system, finding missing children, finding criminals, web content filtering, and so on. Even the datasets of employers in organizations need age-invariant face information in order to keep the datasets from being updated. Ageing affects people differently due to ethnicity, biological factors (gender, genetics, and diseases), social lifestyle factors (smoking, alcohol, emotional stress, and changes in weight), and environmental factors (extreme climate). On the one hand, all these factors make the task of face-based age estimation non-trivial. On the other hand, age-invariant face recognition systems can accurately identify a person and allow to avoid updating large facial databases.

Age estimation is usually referred to as either a classification problem according to discrete values [1], or as a regression problem estimating the age as a scalar [2]. Classification-based approaches utilize analysis of facial features or hyperplanes as classifiers under the assumption that class labels are basically uncorrelated and inherently unrelated to each other. The rapid development of CNN models makes this approach a

I. Czarnowski et al. (Eds.): KESIDT 2023, SIST 352, pp. 147–156, 2023.
https://doi.org/10.1007/978-981-99-2969-6_13

promising solution for age estimation, including age-group classification. However, the age labels themselves are ordinal by nature. At the same time, regression approaches consider the labels as numerical values using the ordering information for age estimation. For the age estimation problem, nonlinear regression approaches such as quadratic regression, Gaussian process and support vector regression (SVR) are recommended. Age regression schemes often use CNN in combination with a regression model using SVR based on radial basis functions. However, human ageing patterns are non-stationary in the feature space (e.g., bone growth in childhood and skin wrinkles in adulthood). Thus, stationary kernel functions are not directly suitable for measuring pair-wise similarities between ages.

Recently, the concept of ordinal ranking for age estimation has become attractive, since the effects of human aging associated, for example, with the shape of the face in childhood or the texture of the skin in adulthood, are different in different age ranges. Hence, the decision based on the "larger than" relationship provides a more reliable property for age estimation than differences between the labels.

Our contribution is following. We propose a method for estimating age-group for large facial datasets to enable fast search and identification using the predicted age value, gender and expression as the additional attributes. It will be useful for identity verification in real-time access systems. Unlike other ranking models, our age ranking model for age-group estimation is non-uniformly distributed over the intervals according to natural changes in facial biological features and social lifestyles. Proposed model was evaluated on the five publicly available datasets.

The rest of the paper is organized as follows. Section 2 provides a brief overview of related work. Section 3 presents a proposed method for age-group classification using capsule network. Section 4 details the experimental results, and Sect. 5 concludes the paper.

2 Related Work

The facial age grouping can be considered as a separate problem and as a problem related to age prediction. Many authors note that the effectiveness of age estimation needs to be studied depending on ethnicity and gender. Traditional approaches first utilized ageing patterns extraction and then machine learning methods for classification or regression. Shape-based ageing features and texture features of the face are two main sources of age information. However, shape-based ageing patterns, such as landmark extraction or triangulation methods, are sensitive to pose variations and insufficient for age estimation. Many global and local features such as local binary patterns or biologically inspired features have been proposed to describe skin texture. Techniques such as ordinal classification for age estimation, ranking SVM scheme, SVM and SVR classifiers can be used as age estimators. Since the 2010 s, a typical way has been a hierarchical approach for automatic age estimation based on hand-crafted global and local facial features.

Recently, deep learning models have become a very popular for implementing face recognition as well as related applications. There are two types of deep learning methods depending on the input fed to the CNNs. Some CNN models process the full image, while other models divide the face image into many local patches, each of which is fed into independent convolutional sub-networks. The responses from sub-networks are then combined at a fully connected layer. Architectures with sub-networks tend to be less deep.

Depending on the applications, different strategies can be used, such as coarse classification (age-group), per-year classification, regression, label distribution learning (as an intermediate approach between the discrete classification and continuous regression), or ranking. The training pipelines of the deep learning models can also have the sequential or parallel architectures. Thus, in [3] image pre-processing, face classification network, age regression network, metric learning, and SVR are the main chain in training. On the one hand, most work supposes a separate deep model for each face attribute (e.g., age) [4, 5]. On the other hand, deep multi-task learning takes into account multiple CNN models predicting one category of homogeneous attributes. Thus, deep multi-task learning network proposed in [6] included a shared feature extraction stage followed by a multi-task estimation stage with category-specific feature learning.

A wide variety of facial attributes can be both correlated (either positive or negative) and heterogeneous. Some local gender-related attributes are correlated, such as beard, mustache, and makeup. At the same time, individual attributes (age, gender, race, unibrow, big lips, etc.) are heterogeneous in terms of data type, scale and semantic meaning and depend on each other. Many proposed deep architectures differ in these factors. Antipov et al. [7] analyzed four factors of the CNN training: the target age encoding and loss function, the CNN depth, the need for pre-training on public datasets, and the mono-task (separate gender recognition and age estimation) or multi-task (simultaneous gender recognition and age estimation) training strategy.

Human age and facial expression such as anger, annoyed, disgusted, grumpy, happy, neutral, sad, and surprised are also closely related features that require additional attention, especially for face age recognition in the wild. The fact is that facial expressions can change the perception of the face, like the human ageing. Thus, smiling or crying of a young person lead to skin folds from both sides of the noise to the corners of the mouth. The algorithm can detect these folds as nasolabial folds commonly seen in older people. Such facial features that resemble wrinkles are associated with many types of expression. For example, Yang et al. [8] combined the scattering and convolutional neural networks for joint automatic age estimation and face expression recognition from a single face image.

Ranking CNN for age estimation demonstrate smaller estimation errors compared to multi-class classification approaches because multiple basic CNNs are trained with ordinal age labels. DeepRank+, which was proposed by Yang et al. [9], included a 3-layer ScatNet, a dimensionality reduction component by principal component analysis and a 3-layer fully-connected network. Chen et al. [10] suggested a ranking-CNN model that contained a series of basic CNNs. Then, their binary outputs were aggregated to make the final age prediction. The main advantage was that each basic CNN was trained using all the labeled data, which led to better performance of feature learning and preventing overfitting. Cao et al. [11] have studied the consistent rank logits framework with strong theoretical guarantees for rank-monotonicity and consistent confidence scores applied to the age estimation problem.

3 The Proposed Method

Different people have different rates of the ageing process depending on the people's gene, health condition, living style, working environment, and sociality. In addition, ageing shows different features at different periods of life. Men and women usually have different patterns of facial aging due to varying degrees of makeup and anti-aging treatments, and female face images can potentially show younger appearance.

We are looking for a trade-off between single-task and full multi-task learning in order to get the best age grouping results on large facial datasets. We limited our study to two main additional attributes, such as gender and face expression. However, our approach can be extended by other heterogeneous attributes (for example, race or individual features). We also assume a limited number of facial expressions, consisting of neutral, smiling and sad expressions. The main idea is to simultaneously train a CNN model that includes rank, binary and multi-class classifications, instead of independently training sub-networks responsible for different estimates. Moreover, we propose a non-uniform ranking due to the change in face with age, when the skin becomes thicker, its color and texture change, the composition of the tissues becomes more subcutaneous, and wrinkles and lines of the facial skeleton appear.

Figure 1 illustrates the overview of the proposed CNN for age group estimation, where mnemonics denote sub-CNNs. Face and landmarks detection is a well-elaborated process that is applied at the first step of face analysis. Also, face normalization and cropping are necessary steps. We solve the age-group estimation problem as a multi-task problem, developing a system with interconnections between subnets. In this case, we have the dual benefit of increased accuracy in age group estimation taking into account gender and expression attributes and the resulting multi-attributes showing predicted gender, age group, and expression. Gender prediction is a binary classification problem, for which we use a CNN with one neuron at the output layer. Gender sub-network was trained using cross-entropy function, which is mathematically equivalent to logistic regression loss function for binary classification.

Age prediction is usually modelled as a multi-class classification problem with age labels in the interval [1…100]. Since this task remains difficult, we suggest splitting the ages into groups as a preliminary step. Setting age estimation as a ranking problem provides a better solution. Ordinal Hyperplanes Ranker (OHRank) proposed by Chang et al. [12], which used a divide-and-conquer strategy for ordinal regression, remains one of the promising solutions applicable for age-group estimation. Our approach inspired by OHRank differs in that the label encoding scheme is tuned to $1 \ldots K$ subproblems with K output nodes of K age-group sub-networks in such manner the $1 \ldots K$ intervals have different values. The non-uniform distribution of age grouping helps to carefully consider the degree of natural changes in the face. At that time, each output of each age-group sub-network is a binary classifier, simplifying age-group prediction.

Recognition of facial expressions as a separate problem includes six basic states, such as anger, disgust, fear, happiness, sadness, and surprise. For simplicity, we have limited this set to only smile, neutral, and sad states. Thus, the expression sub-network has three outputs, acting as three binary classifies. The expression sub-network consists of CNN and ScatNet inspired by Yang et al. [8]. It should be noted that the ScatNet outputs are

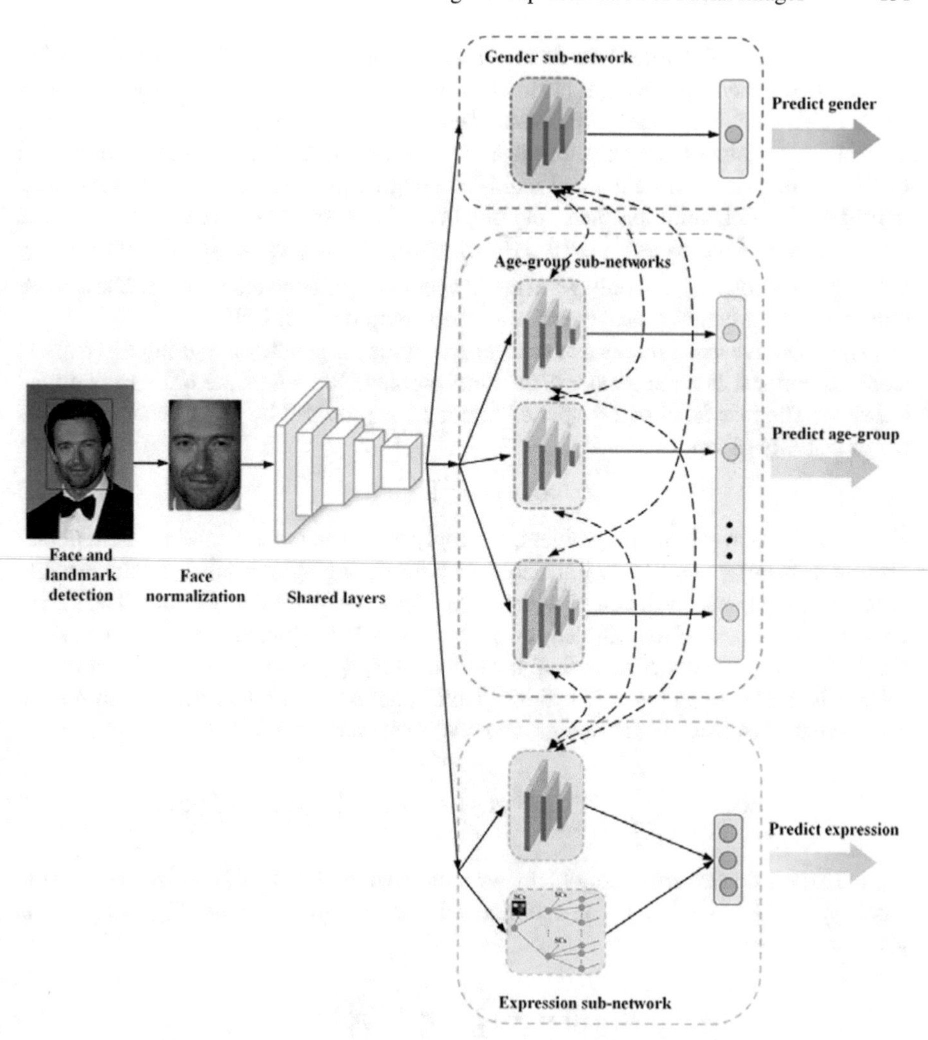

Fig. 1. The proposed CNN overview for age group estimation.

reduced using PCA algorithm. All sub-networks are trained simultaneously that allows to take into account the mutual influence of gender, age and expression attributes.

Our multi-task CNN consists of three branches, and in addition, the age-group branch includes several sub-networks according to the selected number of intervals. As a result, the *D*-dimensional output feature vector $\mathbf{X} = \{\mathbf{x}_i, y_i\},\ i = 1, \ldots, N$, where $\mathbf{x}_i \in \mathrm{R}^D$ is the feature vector, *N* is the training samples, is fed to the inputs of the three branches after the shared layers of the initial sub-CNN. The output y_j is a composition of the gender label, the age-group labels and the expression labels, denoted by $(y_j{}^G, y_j{}^A, y_j{}^E)$ with predicted gender label $(y_j{}^G = [1])$, age-group labels from the set $(y_j{}^A = [2 \ldots k_A])$ and expression labels from the set $(y_j{}^E = [k_A + 1 \ldots k_E])$, where k^A and k^E are the numbers of age-group labels and expression labels, respectively. Such *D*-dimensional output feature vector is available if all three branches are tuned to solve only classification or regression problems. Inspired by OHRank [12], we converted the age-group regression problem into a multi-label classification problem. Thus, gender, age and expression predictions are handled via binary classifications in a unified CNN or, as is sometimes mentioned in the literature, joint multi-class and multi-level-regression coding. Due to the accepted multi-label classification problem, the outputs of three branches are encoded as three binary vectors, denoted as $\mathbf{c}_j^G \in \{0, 1\}^1$, $\mathbf{c}_j^A \in \{0, 1\}^{kA}$ and $\mathbf{c}_j^E \in \{0, 1\}^{kE}$, where only $y_j{}^A$th element equals 1 and only $y_j{}^E$th element equals 1 and the others are 0. These three binary vectors are then concatenated into a total output of our CNN.

The objective loss functions have different forms for gender, age-group and expression estimations. For the gender estimation problem, we adopt the ordinary softmax classifier. The loss function L^G applied to node 1 is provided by Eq. 1, where $p^G(\mathbf{x}_i)$ is the probability of input $\mathbf{x}_i$.

$$L^G(\mathbf{x}_i) = -c^G \log p^G(\mathbf{x}_i) \tag{1}$$

For the age-group estimation problem, we apply the setting of age labels, where binary classifiers from $y_j{}^A = [2 \ldots k_A]$ are used to infer the age-group label. Each classifier focuses on the binary decision of whether the input is smaller than j or not. The binary cross-entropy loss is often adopted for the multi-label classification problem. Consider the jth $(2 \le j \le k_A)$ node in the final layer and let $F_j{}^A(\mathbf{x}_i) = \sigma(z_j)$ be the output of this node, where $\sigma(z) = 1/(1 + \mathrm{e}^{-z})$ is the sigmoid function and z_j the activation of node j. The summarized loss of each age-group node is calculated by follows:

$$L^A(\mathbf{x}_i) = -\sum_{j=2}^{k_A} \left(c_j^i \log\left(F_j^A(\mathbf{x}_i)\right) + \left(1 - c_j^i\right) \log\left(1 - F_j^A(\mathbf{x}_i)\right)\right). \tag{2}$$

For the expression estimation problem, we solve a single-label problem with expression labels $y_j{}^E = [k_A + 1 \ldots k_E]$. The summarized loss function for nodes $[k_A + 1 \ldots k_E]$ is given as

$$L^E(\mathbf{x}_i) = -\sum_{j=k_A+1}^{k_E} c_j^E \log p_j^E(\mathbf{x}_i), \tag{3}$$

where $p^E(\mathbf{x}_i)$ is the probability of input $\mathbf{x}_i$ belonging to class j, which is obtained by the softmax function with the activation z_j of node j:

$$p_j^E(\mathbf{x}_i) = e^{z_j} \Big/ \sum\nolimits_m e^{z_m}.$$

Taking into account Eqs. 1–3, the overall objective function L_{loss} for joint multi-level ranking regression and classification is determined by Eq. 4, where $\mathbf{W}$ denotes CNN parameters including the weights of interconnections, α, β and γ are positive constants that control the relative contribution between three tasks, λ is a L2-norm regularization.

$$L_{loss} = \underset{\mathbf{W}}{\arg\min} \sum_{\mathbf{x}_i \in \mathbf{X}} \left(\alpha L^G(\mathbf{x}_i) + \beta L^A(\mathbf{x}_i) + \gamma \sum_{j=k_A+1}^{k_E} L_j^E(\mathbf{x}_i) + \lambda \sum_q \mathbf{W}_q^2 \right) \tag{4}$$

Adam optimization was used to minimize Eq. 4.

4 Experimental Results

In experiments, we have used five imbalanced facial datasets with age, gender and ethnicity metadata:

- IMDB dataset and WIKI dataset are often combined as the IMDB-WIKI dataset [13]. IMDB-WIKI dataset is a large-scale facial dataset with age and gender metadata. It contains 523,051 face images: 460,723 face images from 20,284 celebrities from IMDb and 62,328 from Wikipedia. The age labels are distributed from 0 to 100 years.
- UTKFace dataset [14] is a facial dataset with age, gender and ethnicity labels, which ranges from 0 to 116 years. It contains over 20,000 facial images with large variation in facial expression, illumination, occlusion, etc.
- FG-NET dataset [15] contains 1,002 facial images of 82 people with age range from 0 to 69 and age gap up to 45 years.
- MORPH II dataset [16] is a large-scale longitudinal face database, collected in real-world conditions with variations in age, pose, expression and lighting conditions. It contains 55,134 facial images of 13,617 subjects ranging from 16 to 77 years old.

For our study, we used two different CNN, widely adopted in several face analysis tasks: SENet [17] and DenseNet [18]. We applied SENet and DenseNet with the pre-trained weights while our own CNN was trained from scratch. Training was done on the machine with a GTX 2060 s graphics card, AMD Ryzen 7 1700X CPU, and 32 GB RAM. For each model, the input image size was 224 × 224 pixels. Each dataset was divided into the training, validation and test sets as 70%, 15% and 15% of images, respectively. All models were trained for 80 epochs. The loss function graphs for training and validation processes of the three networks (DenseNet, SENet and our CNN) on the WIKI, IMDB, UTKFace, FG-NET, and MORPH II datasets are shown in Fig. 2.

The comparative loss values and mean accuracy values for age-group prediction on the five datasets are given in Table 1 and Table 2. Three types of network models were used. The proposed multi-task CNN achieved the mean accuracy value 96.8% for multi-task age-group prediction with ranges [18–25], [26–40], [41–65] (Table 2).

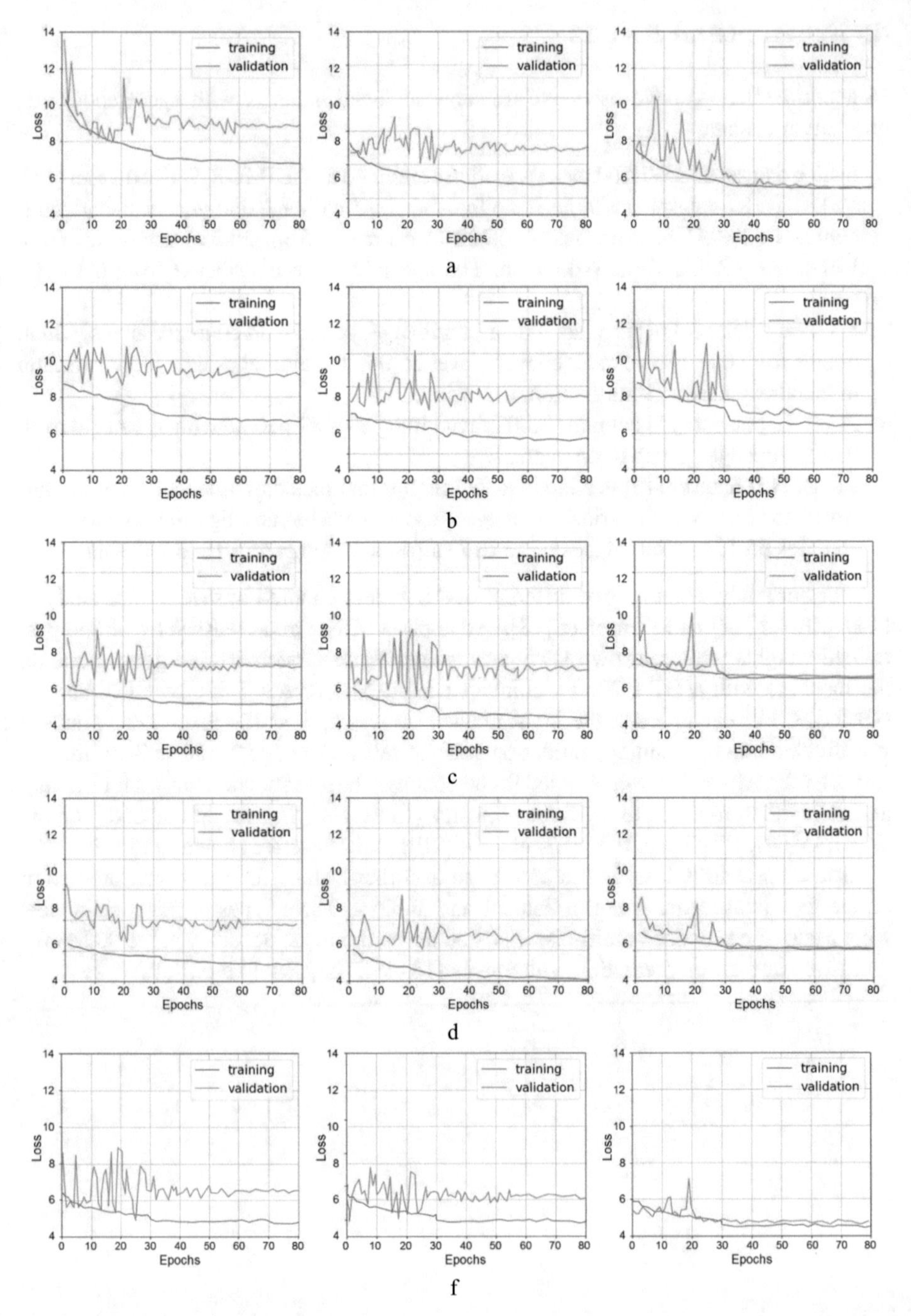

Fig. 2. The loss function graphs for training and test processes of the DenseNet, SENet and our proposed CNN (from left to right) on the datasets: **a** WIKI, **b** IMDB, **c** UTKFace, **d** FG-NET, **f** MORPH II.

Table 1. The comparative loss values for age-group prediction.

Model	WIKI	IMDB	UTKFace	FG-NET	MORPH II
DenseNet	8.87	9.32	6.78	7.68	6.29
SENet	7.54	7.34	6.60	7.05	6.14
Our CNN	5.66	6.95	6.20	6.02	5.04

Table 2. The comparative mean accuracy values for age-group prediction.

Model	WIKI	IMDB	UTKFace	FG-NET	MORPH II
DenseNet	91.5	90.4	95.5	94.3	96.2
SENet	96.1	94.9	97.0	95.4	96.4
Our CNN	96.7	95.2	98.7	95.2	98.5

5 Conclusions

In this study, we proposed a novel multi-tasking CNN deep model for age-group estimation under consideration gender and expression estimates using facial images. Our deep model includes three branches presented by CNNs with the relatively simple architectures after the initial feature shearing. We consider the age-group estimation problem as a ranking problem, not a regression problem. The branches of our multi-tasking CNN have interconnections, which allow to obtain better age-group classification. In experiments, we have used five imbalanced facial datasets with age, gender and ethnicity metadata such as the IMDB dataset, the WIKI dataset, the UTKFace dataset, the FG-NET dataset, and the MORPH II dataset. We compared the proposed multi-task CNN architecture with the DenseNet and SENet models and achieved the mean accuracy value 96.8% for multi-task age-group prediction with ranges [18–25], [26–40], [41–65].

References

1. Eidinger, E., Enbar, R., Hassner, T.: Age and gender estimation of unfiltered faces. IEEE Trans. Inf. Forensics Secur. **9**(12), 2170–2179 (2014)
2. Chao, W.-L., Liu, J.-Z., Ding, J.-J.: Facial age estimation based on label-sensitive learning and age-oriented regression. Pattern Recognit. **46**(3), 628–641 (2013)
3. Sendik, O., Keller, Y.: DeepAge: deep learning of face-based age estimation. Signal Process. Image Commun. **78**, 368–375 (2019)
4. Iqbal, M.T.B., Shoyaib, M., Ryu, B., Abdullah-Al-Wadud, M., Chae, O.: Directional age-primitive pattern (DAPP) for human age group recognition and age estimation. IEEE Trans. Inf. Forensics Secur. **12**(11), 2505–2517 (2017)
5. Duan, M., Li, K., Ouyang, A., Win, K.N., Li, K., Tian, Q.: EGroupNet: a feature-enhanced network for age estimation with novel age group schemes. ACM Trans. Multimedia Comput. Commun. Appl. **16**(2), 1–23 (2020)

6. Han, H., Jain, A.K., Wang, F., Shan, S., Chen, X.: Heterogeneous face attribute estimation: a deep multi-task learning approach. IEEE Trans. Pattern Anal. Mach. Intell. **40**(11), 2597–2609 (2018)
7. Antipov, G., Baccouche, M., Berrani, S.-A., Dugelay, J.-L.: Effective training of convolutional neural networks for face-based gender and age prediction. Pattern Recognit. **72**, 15–26 (2017)
8. Yang, H.-F., Lin, B.-Y., Chang, K.-Y., Chen, C.-S.: Joint estimation of age and expression by combining scattering and convolutional networks. ACM Trans. Multimedia Comput. Commun. Appl. **14**(1), 1–18 (2017)
9. Yang, H.-F., Lin, B.-Y., Chang, K.-Y., Chen, C.-S.: Automatic age estimation from face images via deep ranking. In: British Machine Vision Conference (BMVC), pp. 55.1–55.11. BMVA Press, Swansea, UK (2015)
10. Chen, S., Zhang, C., Dong, M., Le, J., Rao, M.: Using ranking-CNN for age estimation. In: 2017 IEEE Conference on Computer Vision and Pattern Recognition (CVPR), pp. 5183–5192. IEEE, Honolulu, HI, USA (2017)
11. Cao, W., Mirjalili, V., Raschka, S.: Rank consistent ordinal regression for neural networks with application to age estimation. Pattern Recognit. Lett. **1402**, 325–331 (2020)
12. Chang, K.Y., Chen, C.-S., Hung, Y.-P.: Ordinal hyperplanes ranker with cost sensitivities for age estimation. In: 2011 IEEE Conference on Computer Vision and Pattern Recognition (CVPR), pp. 585–592. IEEE, Colorado Springs, CO, USA (2011)
13. Rothe, R., Timofte, R., Van Gool, L.: Deep expectation of real and apparent age from a single image without facial landmarks. Int. J. Comput. Vis. **126**(2–4), 144–157 (2016)
14. Zhang, Z., Song, Y., Qi, H.: Age progression/regression by conditional adversarial autoencoder. In: 2017 IEEE Conference on Computer Vision and Pattern Recognition (CVPR), pp. 5810–5818. IEEE, Honolulu, HI, USA (2017)
15. Fu, Y., Guo, G., Huang, T.S.: Age synthesis and estimation via faces: a survey. IEEE Trans. Pattern Anal. Mach. Intell. **32**(11), 1955–1976 (2010)
16. Ricanek, K., Tesafaye, T.: Morph: a longitudinal image database of normal adult age-progression. In: 7th International Conference on Automatic Face and Gesture Recognition (FGR06), pp. 341–345 (2006)
17. Hu, J., Shen, L., Sun, G.: Squeeze-and-excitation networks. In: 2018 IEEE/CVF Conference on Computer Vision and Pattern Recognition, pp. 7132–7141. IEEE, Salt Lake City, UT, USA (2018)
18. Huang, G., Liu, Z., Van Der Maaten, L., Weinberger, K.Q.: Densely connected convolutional networks. In: 2017 IEEE Conference on Computer Vision and Pattern Recognition (CVPR), pp. 4700–4708. IEEE, Honolulu, HI, USA (2017)

Evaluation Instrument for Pre-implementation of Lean Manufacturing in SMEs Using the Paraconsistent Annotated Evidential Logic Eτ Evaluation Method

Nilton Cesar França Teles(✉), Jair Minoro Abe, Samira Sestari do Nascimento, and Cristina Corrêa de Oliveira

Paulista University, São Paulo, Brazil
{nilton.teles,samira.nascimento3}@aluno.unip.br, jair.abe@docente.unip.br, crisolive@ifsp.edu.br

Abstract. The Lean Manufacturing methodology provides competitive advantages, greater productivity, better product quality, customer and worker satisfaction. This research aims to develop an evaluation instrument to be used by consultants in the evaluation of Lean Manufacturing concepts by employees of a company. This instrument uses indicators defined in the literature, in order to improve the implementation of Lean Manufacturing in small and medium-sized companies, which do not benefit from this implementation due to the lack of methods that assess how much they are prepared for the process. This article presents the results of an evaluation simulation of assimilation of these indicators using the Paraconsistent Annotated Evidential Logic Eτ. The results showed that measuring the participants' degree of assimilation of Lean Manufacturing concepts and tools can improve the success of Lean Manufacturing implementation in small and medium-sized companies.

Keywords: Lean Manufacturing · Paraconsistent Logic · Implementation · Non-Classical Logic · Framework

1 Introduction

While many benefits have been documented by companies using the Lean Manufacturing (LM) methodology, only 30% of large US companies have achieved the potential benefits. [3]. In the UK only 25% of large companies were successful with LM [4]. However, the number of small and medium-sized enterprises (SMEs) represent 90% of all companies [5] that have not benefited from the LM [1]. The failure rate in the adoption of LM occurs due to the lack of appropriate structures [2] as an example, maturity assessment models at the beginning of the implementation.

Therefore, there is a need for a structure based on indicators and training for the implementation of LM in SMEs. However, just developing this framework will not be

I. Czarnowski et al. (Eds.): KESIDT 2023, SIST 352, pp. 157–170, 2023.
https://doi.org/10.1007/978-981-99-2969-6_14

useful until it is verified how well the implementation project participants have assimilated the LM concepts and tools. In the first stage of this research, a literature review was carried out to identify the indicators that help the adoption of LM in SMEs.

In the second stage, an evaluation instrument was defined, using the Paraconsistent Annotated Evidential Logic Eτ, and an interview script for the Lean specialists to apply to the professionals selected for the implementation project. Finally, a simulation of the participants' assimilation of the concepts of the LM [6] was carried out.

2 Theoretical Reference

2.1 Lean Manufacturing

Competition is making customer acquisition increasingly fierce, and to stay ahead, companies must find effective methods, technologies, and performance tools to deliver the value customers need. Among the methods and tools that help companies to remain in the market, LM stands out, which is a continuous improvement program that aims to reduce production costs, manufacturing time and increase productivity. It can be said that as a standardized production system aimed at eliminating waste, it has been practiced in the industrial and economic sectors [7].

2.2 Paraconsistent Annotated Evidential Logic Eτ

Paraconsistent logics are a class of non-classical logic built to find ways of giving non-trivial treatment to contradictory situations, demonstrating that paraconsistent logics are more propitious in framing problems caused by situations on contradictions, which appear when we deal with descriptions from the real world [8]. For [6], consistencies appear when two or more specialists give opinions on the same subject in the real world. Intuitively, in Paraconsistent Annotated Evidential Logic Eτ what we do is assign an annotation (μ; λ), with μ and λ belonging to the closed interval [0;1], to each elementary (atomic) proposition p in such a way that μ translates the degree of belief (or favorable evidence) one has in p and λ, the degree of disbelief (or contrary evidence). In [6] the authors elaborate a representation where the Cartesian Unitary Square Plane (CUSP) presents X and Y values varying in a real closed interval [0, 1], so these values represent the degrees of belief, μ, and disbelief, λ, respectively. The CUSP is divided into twelve regions, according to Fig. 1.

2.3 Degrees of Contradiction and Certainty

Given the definitions in Fig. 1, it is understood, for example, that if a point "falls" in the region of the CPQ lines, the result is true as shown in Table 1. In this context, the CUSP can be divided into 12 regions.

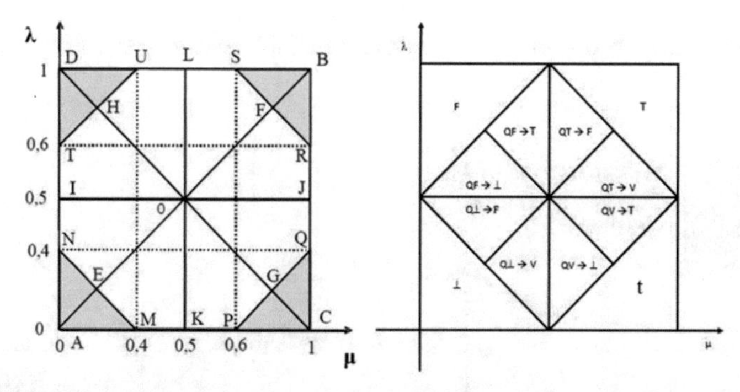

Fig. 1. Division of the CUSP into twelve regions, adopting as limits |G*contr*| = 0.60 and |H*cert*| = 0.60, source: [6].

Table 1. Values for μ and λ, G*contr* and H*cert*, from each CUSP region, source: [6].

Region	μ	λ	G_{contr}	H_{cert}	Description	Symbol
AMN	[0;0,4]	[0;0,4]	[-1;-0,6]	[-0,4;0,4]	Paracomplete	⊥
BRS	[0,6;1]	[0,6;1]	[0,6;1]	[-0,4;0,4]	Inconsistent	T
CPQ	[0,6;1]	[0;0,4]	[-0,4;0,4]	[0,6;1]	True	t
DTU	[0;0,4]	[0,6;1]	[-0,4;0,4]	[-1;-0,6]	False	F
OFSL	[0,5;0,8[	[0,5;1]	[0;0,6[	[-0,5;0[	Quasi-inconsistent tending to false	QT → F
OHUL	]0,2;0,5[	[0,5;1]	[0;0,5[	]-0,6;0[	Quasi-false tending to inconsistent	QF → T
OHTI	[0;0,5[	[0,5;0,8[	[-0,5;0[	]-0,6;0[	Quasi-false tending to paracomplete	QF → ⊥
OENI	[0;0,5[	]0,2;0,5[	]-0,6;0[	]-0,5;0[	Quasi-paracomplete tending to false	Q⊥ → F
OEMK	]0,2;0,5[	[0;0,5[	]-0,6;0[	[0;0,5[	Quasi-paracomplete tending to true	Q⊥ → t
OGPK	[0,5;0,8[	[0;0,5[	[-0,5;0[	[0;0,6[	Quasi-true tending to paracomplete	Qt → ⊥
OGQJ	[0,5;1]	]0,2;0,5[	[0;0,5[	[0;0,6]	Quasi-true tending to inconsistent	Qt → T
OFRJ	[0,5;1]	[0,5;0,8[	[0;0,6[	[0;0,5]	Quasi-inconsistent tending to true	QT → t

2.4 ProKnow-C Method

In this work, ProKnow-C was used as an intervention tool [9]. The ProKnow-C process consists of four steps: i) selecting a portfolio of articles on the research topic; ii) bibliometric analysis of the portfolio; iii) systemic analysis and iv) definition of the research question and research objective. For bibliographical research, the first two steps are illustrated in Fig. 2.

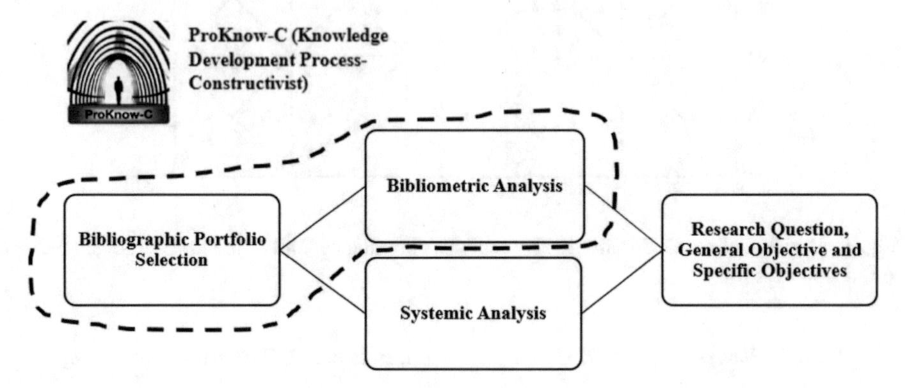

Fig. 2. Systemic review process steps (ProKnow-C), source: [10].

3 Methodology

In this work, three axes were used for the research in English, justified by the fact of exploring articles in international databases: (i) Lean, (ii) SAE J4000 and (iii) SMEs; from the axes, the keywords were defined. Keys relevant to the topic of interest as well as the Boolean operator, as described below: "Lean Manufacturing" OR "Toyota Production System" AND "Best Practice" OR "Implementation" OR "Framework" OR "Implantation" OR "SAE J4000" AND "Small and Medium Enterprises" OR "SMEs" OR "Small Enterprises" OR "Medium Enterprises". For the development of this article, the literature search procedure was used to identify gaps and potential topics of interest related to the evaluation of the implementation of the LM in SMEs, in order to justify the research, identify existing models and evaluation methods, and propose an evaluation model based on dimensions and indicators of maturity evaluation for the LM pre-implementation.

3.1 Raw Article Bank Selection

The first phase of the selection focused on forming a bank of raw articles. First, it was necessary to determine the international research axes. The axes were defined as: (i) Lean, (ii) SAE J4000 and (iii) SMEs as follows in Fig. 3.

In this research, we chose the Web of Science and Scopus databases to search the bibliographic portfolio. When performing the searches, titles, keywords, abstracts, and recent articles with a maximum of ten years between 2013 to 2022 and limited to open-access articles were selected. The articles from the database of raw articles were filtered in different ways in order to originate the final bibliographic portfolio: (i) filtering for redundancy: duplicate articles were excluded, resulting in the elimination of 32 articles;

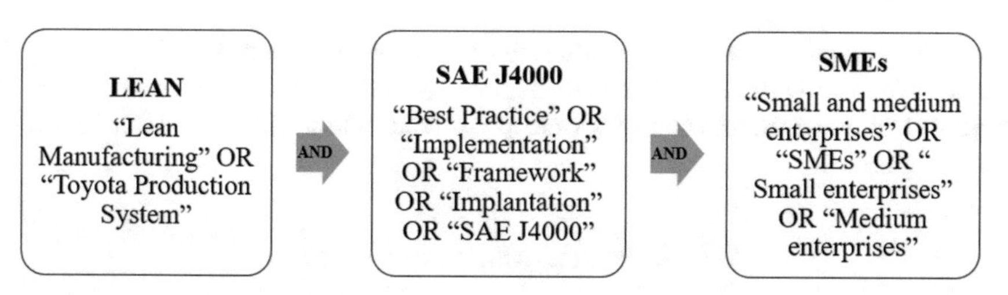

Fig. 3. Research axes and keywords, source: Authors.

(ii) filtering regarding the alignment of the titles: the titles related to the research topic were checked, which led to the elimination of 6 articles and the permanence of 58 articles; (iii) filtering regarding the alignment of abstracts: then, the abstracts of the articles were read and resulted in the elimination of 40 articles, leaving 18 articles in the database of raw articles; (iv) filtering for scientific recognition: the scientific representativeness of the articles was verified by inspecting the number of citations they have in Google Scholar. The article with the highest scientific recognition had 165 citations, as shown in Fig. 4, while some works had no citations. Thus, a cut-off point was determined according to the pertinence of scientific recognition of the articles. A new category was constituted, excluding articles that had less than three citations, which were deposited in "Repository B", and analysed later. Articles with more than three citations constituted "Repository A", being candidates for direct inclusion in the final portfolio of articles. The details of the quotes are illustrated in Fig. 4.

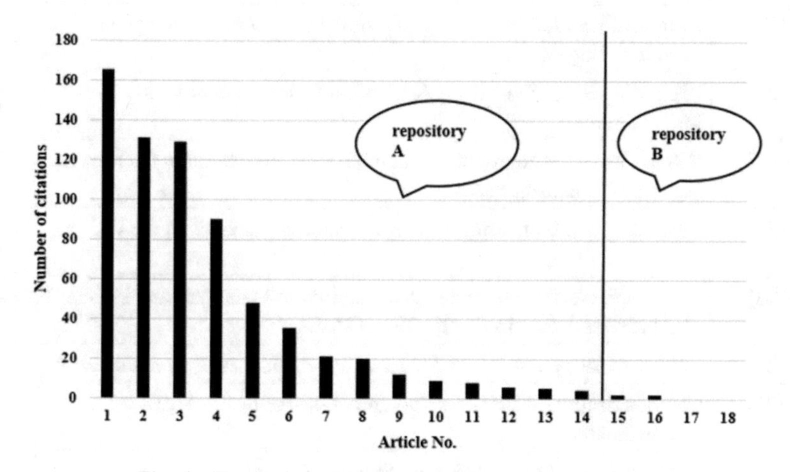

Fig. 4. Number of citations of articles, source: Authors.

After deepening the number of citations, 14 articles remained in "Repository A", which were read in full. In order to verify the pertinence of articles published less than two years ago or with the same author as the selected articles, "Repository B" was analysed. After analysing "Repository B", the four articles published less than two years ago were selected because they were aligned with the research theme but with little scientific recognition, which is justified by the recent publication. After reading the articles in total, 18 articles remained in the final portfolio, as shown in Table 2, on which bibliometric and systemic analyses were performed, whose titles are described in the following topics.

Table 2. Portfolio Articles, source: Authors.

No.	Author	Title	Year
1	[11]	Lean implementation frameworks: the challenges for SMEs	2017
2	[5]	Development of lean manufacturing systems to improve acceptance of manufacturing in developing countries	2020
3	[12]	Lean in small companies: Literature review	2019
4	[13]	Strengths and weaknesses of companies that implement lean manufacturing	2016
5	[14]	Key Success Factor for Lean Adoption in the Automotive Industry	2014
6	[15]	Critical factors for the implementation of Lean in small companies	2019
7	[16]	Lean excellence business management for manufacturing companies with a focus on KRI	2020
8	[17]	Similarities with lean manufacturing success in SMEs	2017
9	[18]	An approach to analyze barriers in the implementation of Lean	2021
10	[19]	Barriers to Lean Adoption in Indian SMEs	2014
11	[20]	Lean readiness factors and organizational readiness for change in manufacturing companies	2020
12	[21]	Examining the barriers to successful implementation of Lean Manufacturing	2020
13	[12]	A framework for improving Lean implementation through the leveraged aspects of organizational culture	2019
14	[22]	Evaluating and comparing lean manufacturing methods in the UK and France	2018
15	[23]	Lean implementation enables sustainability in manufacturing systems for small and medium-sized enterprises (SMEs)	2022
16	[24]	Lean transformation and implementation challenges in manufacturing	2022
17	[25]	Characterisation of the production processes of textile SMEs in Cundinamarca	2019
18	[26]	Lean Tools Selector: A Decision Support System	2021

3.2 Criteria for Composing the Indicators

The studies in Table 2 show LM implementation components at a tactical level. It comprises customers, suppliers, and investments. Furthermore, finally, at the operational level, the criteria that allow the composition and measurement of indicators stand out. This composition has over 50 criteria, which are divided according to each area of interest. It was found that the criteria assigned by the author [5] in Table 3 flow harmoniously in the composition of a conceptual model that aims, through performance indicators, to cover the main evaluation metrics of the LM maturity model.

4 Model Building

To develop the proposed model, a systematic analysis of international articles related to the topic of interest was carried out to develop a questionnaire with dimensions and performance indicators according to the LM methods used in previous studies. The construction of the model follows the steps shown in Fig. 5.

4.1 Definition of Lean Maturity Categories and Indicators

The categories can be considered axes that guide the decision-maker evaluation, allowing comparisons between alternatives. Regarding the adopted criteria, it can be stated that they refer to the first items defined in the composition of the questionnaire, for which six qualitative criteria were established, which are shown in Table 3 and are described as: (i) shop floor management; (ii) manufacturing strategy; (iii) quality management; (iv) manufacturing process; (v) supplier and customer management and (vi) workforce management.

4.2 Definition of the Questionnaires Made up of the Indicators

Next, the questionnaires for the experts were developed, which are basically the metrics to examine the performance of each indicator. The questions were defined qualitatively. Concerning this study, the composition of the indicators was established based on the adaptation of the indicators used by the study by [5], which are relevant to the objective of this work. Two dimensions were established, with the first dimension (μ) addressing issues of favorable evidence and the second dimension referring to unfavorable evidence (λ). In Table 4, as an example, one can observe the composition of all indicators where: EF(μ), favourable evidence questionnaire (μ), "how much do you believe the training was assimilated?"; UE (λ), unfavourable evidence questionnaire (λ), "how much do you believe that the training was not assimilated?". Experts should ask respondents for a brief description of their understanding of the indicators described to assess how much they assimilated knowledge from pre-implementation training for the LM.

Table 3. LM indicators applicable to SMEs, source: [5].

Nº	LM categories	Nº ID	Global Weight	LM Indicators
1	Factory floor Management (SFM)	*SFM1*	0.0436	Production schedule
		SFM2	0.0118	Effective use of resources
		SFM3	0.0824	Effective inventory management
		SFM4	0.0147	Product flow control
		SFM5	0.0709	Cycle time reduction
		SFM6	0.0471	Security improvement initiatives
		SFM7	0.0263	Reduced setup time
2	Manufacturing Strategy (MS)	*MS1*	0.0630	Quality product design
		MS2	0.0805	Network-aligned distribution management
		MS3	0.0283	Effective strategies for marketing management
		MS4	0.0380	Standardise product development approaches
3	Quality Management (QM)	*QM1*	0.0612	5S
		QM2	0.0219	Value stream mapping
		QM3	0.0784	Residue analysis
		QM4	0.0589	Total quality management
		QM5	0.0205	Total productive maintenance
4	Manufacturing Process (MP)	*MP1*	0.0413	Cell manufacturing
		MP2	0.0291	Technology management
		MP3	0.0335	Standardisation of work
		MP4	0.0167	Process focus
		MP5	0.0081	Continuous improvement approach
		MP6	0.0057	Visual management
5	Supplier and Customer Management (SCM)	*SCM1*	0.0129	Appropriate supplier evaluation strategies

(*continued*)

Table 3. (*continued*)

Nº	LM categories	Nº ID	Global Weight	LM Indicators
		SCM2	0.0046	Mapping delivery performance
		SCM3	0.0159	Supplier development
		SCM4	0.0089	Identifying customer requirements
		SCM5	0.0055	Follow-up on customer feedback
6	Workforce Management (WM)	*WM1*	0.0202	Appropriate workforce assessment strategies
		WM2	0.0296	Workforce training and education system
		WM3	0.0081	Employee proficiency
		WM4	0.0125	Workforce training and engagement

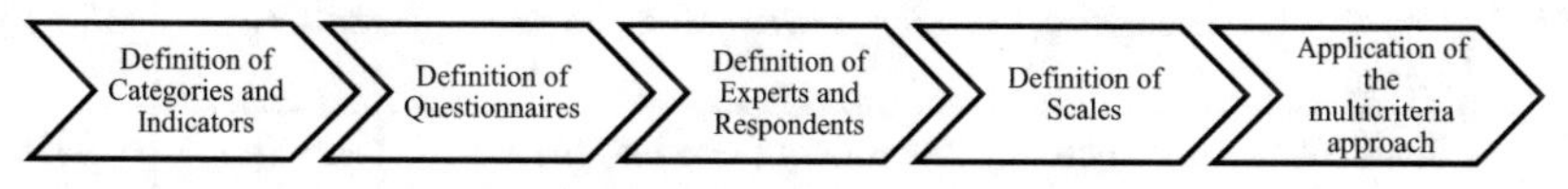

Fig. 5. Model building, source: Authors.

4.3 Definition of Experts and Interviewees

Four experts and eight interviewees proposed to participate in the simulation. In an instrument evaluation, it was necessary to present the conceptual basis of the construct and the steps for the evaluation process [27]. Table 5 describes the roles of the participants.

4.4 Definition of Measurement Scales for Assessing Lean Maturity

It was based on the Likert scale to define qualitative values, which were normalised to quantitative values. The scalar distribution method was used, with values ranging from zero to one: No Assimilation (0.00); Low Assimilation (0.25); Partially Assimilated (0.50); Assimilated (0.75); Fully Assimilated (1.00).

4.5 Expert Evaluation Stage

The purpose of this research is to carry out two evaluation rounds, where each expert individually interviews the project professionals. In the first round, the "Favorable Evidence Questionnaire (μ)" is used and in the second round, the "Unfavorable Evidence Questionnaire (λ)" is used. Each question receives eight answers (eight respondents)

Table 4. Questionnaires made up of the indicators, source: Authors.

KPIs	Description of indicators	FE (μ)	UE (λ)
SFM1	Implementation of continuous flow processes	SFM1(μ)	SFM1(λ)
SFM2	Effective allocation and monitoring of resources across the system	SFM2(μ)	SFM2(λ)
SFM3	Use of Kanban and 'just-in-time' practices within the company	SFM3(μ)	SFM3(λ)
SFM4	Adoption of take time (average production time per unit) in production	SFM4(μ)	SFM4(λ)
SFM5	Adoption of Poka Yoke (failsafe systems)	SFM5(μ)	SFM5(λ)
SFM6	Use of safety equipment that increases employee protection	SFM6(μ)	SFM6(λ)
SFM7	Use of standardised templates and fixtures to guide quick setups	SFM7(μ)	SFM7(λ)
MS1	The goal of zero defects, detect and provide solutions by eliminating the root cause	MS1(μ)	MS1(λ)
MS2	Shared distribution adoption, optimisation for transportation systems	MS2(μ)	MS2(λ)
MS3	Use of sales forecasting techniques to improve the pull system	MS3(μ)	MS3(λ)
MS4	Standardisation of parts by integrating them into the production system	MS4(μ)	MS4(λ)
QM1	Adoption of the 5S program	QM1(μ)	QM1(λ)
QM2	Mapping of the current state of activities and suggestions for the future state	QM2(μ)	QM2(λ)
QM3	Identification of types of waste from non-value-added activities	QM3(μ)	QM3(λ)
QM4	Innovative quality practices and top management commitment	QM4(μ)	QM4(λ)
QM5	Maintenance practices that prevent production system failures	QM5(μ)	QM5(λ)
MP1	Mapping of similar activities, development of product families	MP1(μ)	MP1(λ)
MP2	Adoption of advanced technologies that can help reduce lead time	MP2(μ)	MP2(λ)
MP3	Alignment of process operations and adoption of standardised work	MP3(μ)	MP3(λ)
MP4	Observation of fluctuating demands and standardisation of processes	MP4(μ)	MP4(λ)

(*continued*)

Table 4. (*continued*)

KPIs	Description of indicators	FE (μ)	UE (λ)
MP5	Adoption of Kaizen, continuous diagnosis of activities that do not add value	MP5(μ)	MP5(λ)
MP6	Adoption of visual management for waste identification and information systems	MP6(μ)	MP6(λ)
SCM1	Mapping of suppliers' cost reduction and adoption of 'just-in-time.'	SCM1(μ)	SCM1(λ)
SCM2	Measures to supply the products within the delivery period desired by the customer	SCM2(μ)	SCM2(λ)
SCM3	Integrate suppliers with processes, building a long-term relationship	SCM3(μ)	SCM3(λ)
SCM4	Modify products considering customer requirements	SCM4(μ)	SCM4(λ)
SCM5	Effective customer feedback system, responding to customer complaints	SCM5(μ)	SCM5(λ)
WM1	Evaluation programs and performance incentives for employees	WM1(μ)	WM1(λ)
WM2	Effective training for understanding and implementing LM activities	WM2(μ)	WM2(λ)
WM3	Enhancing employee multi-functionality for new challenges	WM3(μ)	WM3(λ)
WM4	Encouraging employees to get involved in the implementation of the LM	WM4(μ)	WM(λ)

Table 5. Job description of participants, source: Authors.

ID No.	No. of Employees	Office
MP	1	Managing Partner
FM	1	Financial Manager
CM	1	Commercial Manager
PM	1	Production Manager
IL	4	Industry Leaders

ID No.	Office
S1	Lean Six Sigma Consultant
S2	Lean Consultant
S3	Lean Six Sigma Consultant
S4	Lean Consultant

and the arithmetic mean of the evaluation is the value used in the para-analyzer algorithm. Once the interview script has been established, a simulation of possible results demonstrates the application of the method, as can be seen in Table 6.

5 Results and Discussion

The answers of the interviewees and the evaluation of the specialists indicate the global result (average of the degrees resulting from the Para-Analyser Algorithm) received the coordinates (0.78; 0.23) and the result of the feasibility evaluation was (0.55; 0.02). Considering that a requirement level of (0.50) was defined for this research, the result

Table 6. Result of the evaluation of the factory floor management (SFM), source: Authors.

		S1		S2		S3		S4	
KPIs	**Weight**	μ	λ	μ	λ	μ	λ	μ	λ
SFM1	0.0436	1.00	0.10	0.90	0.20	1.00	0.10	0.90	0.10
SFM2	0.0118	0.60	0.30	0.70	0.30	0.80	0.20	0.90	0.20
SFM3	0.0824	0.70	0.40	0.80	0.30	0.90	0.20	0.90	0.30
SFM4	0.0147	0.20	0.80	0.20	0.80	0.20	0.80	0.20	0.80
SFM5	0.0709	1.00	0.20	0.90	0.20	0.90	0.20	0.80	0.30
SFM6	0.0471	1.00	0.20	0.90	0.20	0.90	0.20	0.80	0.30
SFM7	0.0263	1.00	0.10	0.90	0.20	1.00	0.10	0.90	0.10

is favorable regarding the preliminary training process for the implementation of the LM, since the candidates demonstrated that they consistently assimilated most of the concepts applied in the training. In Fig. 6, it is possible to visualise that the coordinates of the indicators and barycenter (a point highlighted in Red) are in the viability or true area (t) of the CUSP.

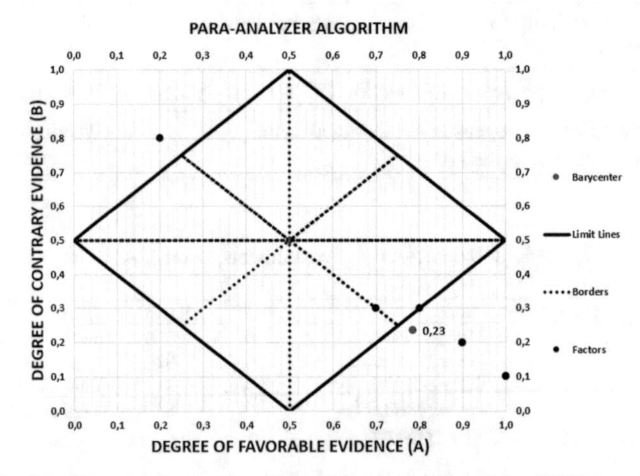

Fig. 6. Result of the Para-Analyzer Algorithm with the division of the CUSP into twelve regions, adapted from [6].

6 Conclusion

A Lean maturity assessment model was developed in its initial implementation using the Paraconsistent Logic method. In the first round of the interview, it was possible to tabulate the degree of favourable evidence evaluated by the specialists. In the second round, it was possible to tabulate the degree of unfavourable evidence. An essential step was aligning the logic concepts with the participants, promoting a theoretical basis for the application. The critical evaluation of the specialists improved the results, allowing the algorithm to a quantitative result based on qualitative evaluations. The percentage of agreement between the first round and the second round was more significant than 50%, meeting the pre-defined requirement level. It is expected that this evaluation method applied in the initial stage of LM implementation will help companies decide on more significant investments in training so that project participants are widely engaged with the success and expected results of LM.

Acknowledgement. This study was supported partially by the Coordenação de Aperfeiçoamento de Pessoal de Nível Superior - Brasil (CAPES) - Finance Code 001 Number Process 88887.663537/2022–00.

References

1. Henao, R., Sarache, W., Gómez, I.: Lean manufacturing and sustainable performance: trends and future challenges. J. Clean. Prod. **208**, 99–116 (2019)
2. Jadhav, J.R., Mantha, S.S., Rane, S.B.: Exploring barriers in lean implementation. Int. J. Lean Six Sigma **5**(2), 122–148 (2014). https://doi.org/10.1108/IJLSS-12-2012-0014
3. Gandhi, N.S., Thanki, S.J., Thakkar, J.J.: Ranking of drivers for integrated lean-green manufacturing for Indian manufacturing SMEs. J. Clean. Prod. **171**, 675–689 (2018). https://doi.org/10.1016/j.jclepro.2017.10.041
4. Hofer, C., Eroglu, C., Hofer, A.R.: The effect of lean production on financial performance: the mediating role of inventory leanness. Int. J. Prod. Econ. **138**(2), 242–253 (2012). https://doi.org/10.1016/j.ijpe.2012.03.025
5. Yadav, G., Luthra, S., Huisingh, D., Mangla, S.K., Narkhede, B.E., Liu, Y.: Development of a lean manufacturing framework to enhance its adoption within manufacturing companies in developing economies. J. Clean. Prod. **245**, 118726 (2020)
6. Abe, J.M. (ed.): Paraconsistent Intelligent-Based Systems. ISRL, vol. 94. Springer, Cham (2015). https://doi.org/10.1007/978-3-319-19722-7
7. Womack, J.P., Jones, D.T.: Beyond Toyota: How to root out waste and pursue perfection. Harvard Bus. Rev. **74**, 140–151 (1996)
8. Da Costa, N.C.A.: ea: Lógica Paraconsistente Aplicada (Applied paraconsistent logic). Atlas, Sao Paulo **7**, 25 (1999)
9. de Carvalho, G.D.G., et al.: Bibliometrics and systematic reviews: a comparison between the Proknow-C and the Methodi Ordinatio. J. Inf. **14**, 101043 (2020)
10. Ensslin, L., Ensslin, S.R., Lacerda, R.T.D.O., Tasca, J.E.: ProKnow-C, Knowledge Development Process–Constructivist: processo técnico com patente de registro pendente junto ao INPI. Brasil **21**, 54–66 (2010)
11. AlManei, M., Salonitis, K., Xu, Y.: Lean implementation frameworks: the challenges for SMEs. In: Manufacturing Systems 4.0, Sara Burgerhartstraat 25, Po Box 211, 1000 Ae Amsterdam, Netherlands (2017)
12. Alkhoraif, A.A., McLaughlin, P., Rashid, H.: A framework to improve lean implementation by review leveraging aspects of organisational culture: the case of Saudi Arabia. Int. J. Agile Syst. Manag. **12**, 124–179 (2019)
13. Moeuf, A., Tamayo, S., Lamouri, S., Pellerin, R., Lelievre, A.: Strengths and weaknesses of small and medium sized enterprises regarding the implementation of lean manufacturing. Ifac Papersonline **49**, 71–76 (2016)

14. Rose, A.N.M., Deros, B.M., Rahman, M.N.A.: Critical success factors for implementing lean manufacturing in Malaysian automotive industry. Res. J. Appl. Sci. Eng. Technol. **8**, 1191–1200 (2014)
15. Elkhairi, A., Fedouaki, F., El Alami, S.: Barriers and critical success factors for implementing lean manufacturing in SMEs. Ifac Papersonline **52**, 565–570 (2019)
16. Mohammad, I.S., Oduoza, C.F.: Lean-excellence business management for manufacturing SMEs focusing on KRI. Int. J. Prod. Perf. Manag. **69**, 519–539 (2020)
17. Rose, A.N.M., Ab Rashid, M.F.F., Nik Mohamed, N.M.Z., Ahmad, H.: Similarities of lean manufacturing approaches implementation in SMEs towards the success: case study in the automotive component industry. In: 9th International Unimas Stem Engineering Conference (Encon 2016) Innovative Solutions For Engineering And Technology Challenges, 17 Ave Du Hoggar Parc D Activites Coutaboeuf Bp 112, F-91944 Cedex A, France (2017)
18. Abu, F., et al.: An SEM approach for the barrier analysis in lean implementation in manufacturing industries. Sustainability **13**(4), 1978 (2021). https://doi.org/10.3390/su1304 1978
19. Ravikumar, M.M., Marimuthu, K., Parthiban, P., Abdul Zubar, H.: Critical issues of lean implementation in Indian micro, small and medium enterprises-an analysis. Res. J. Appl. Sci. Eng. Technol. **7**, 2680–2686 (2014)
20. Inuwa, M., Rahim, S.B.A.: Lean readiness factors and organizational readiness for change in manufacturing SMEs: the role of organizational culture. J. Crit. Rev. **7**, 56–67 (2020)
21. Yuik, C.J., Perumal, P.A., Feng, C.J.: Exploring critical success factors for the implementation of lean manufacturing in machinery and equipment SMEs. Eng. Manag. Prod. Serv. **12**, 77–91 (2020)
22. Demirbas, D., Holleville, L., Bennett, D.: Evaluation and comparison of lean manufacturing practices in britain and france: a case study of a printing solutions organisation. J. Econ. Cult. Soc. **57**, 93–150 (2018)
23. Qureshi, K.M., Bhavesh, G., Mewada, S.Y., Alghamdi, N.A., Mohamed, R.N., Qureshi, M.M.: Accomplishing sustainability in manufacturing system for small and medium-sized enterprises (smes) through lean implementation. Sustainability **14**(15), 9732 (2022). https://doi.org/10.3390/su14159732
24. Maware, C., Parsley, D.M.: The challenges of lean transformation and implementation in the manufacturing sector. Sustainability **14**(10), 6287 (2022). https://doi.org/10.3390/su1410 6287
25. Arteaga Sarmiento, W.J., Villamil Sandoval, D.C., Jesus Gonzalez, A.: Characterization of the production processes of textile SMEs in Cundinamarca. Logos Ciencia & Tecnologia 11, 60–77 (2019)
26. Mendes, A., Lima, T.M., Gaspar, P.D.: Lean tools selector - a decision support system. In: 2021 International Conference On Decision Aid Sciences And Application (Dasa), 345 E 47th St, New York, Ny 10017 USA (2021)
27. Grant, J.S., Davis, L.L.: Selection and use of content experts for instrument development. Res. Nurs. Health **20**, 269–274 (1997)

Recent Development of Multivariate Analysis and Model Selection

Modified C_p Criterion in Widely Applicable Models

Hirokazu Yanagihara[1(✉)], Isamu Nagai[2], Keisuke Fukui[1], and Yuta Hijikawa[3]

[1] Mathematics Program, Graduate School of Advanced Science and Engineering, Hiroshima University, Higashi-Hiroshima, Hiroshima 739-8526, Japan
yanagi-hiro@hiroshima-u.ac.jp

[2] Faculty of Liberal Arts and Science, Chukyo University, Nagoya, Aichi 466-8666, Japan

[3] Kaita Junior High School, Kaita-cho, Aki-gun, Hiroshima 736-0026, Japan

Abstract. A risk function based on the mean square error of prediction is a widely used measure of the goodness of a candidate model in model selection. A modified C_p criterion referred to as an MC_p criterion is an unbiased estimator of the risk function. The original MC_p criterion was proposed by Fujikoshi and Satoh (1997) in multivariate linear regression models. Thereafter, many authors have proposed MC_p criteria for various candidate models. A purpose of this paper is to propose an MC_p criterion for a wide class of candidate models, including results in previous studies.

Keywords: GMANOVA model · Mean square error of prediction · Model selection criterion · Unbiased estimator

1 Introduction

A risk function based on the mean square error of prediction is a widely used measure of the goodness of a candidate model in model selection. However, the risk function must be estimated because it contains several unknown parameters. The C_p criterion proposed by [3,4] is an asymptotic unbiased estimator of the risk function that can be used as a measure of the goodness of a candidate model instead of the unknown risk function in actual data analyses. The original C_p criterion is a model selection criterion proposed for variable selection in multiple regression, and its extension to multivariate regression models was proposed by [7].

Since the C_p criterion is an asymptotic unbiased estimator of the risk function, there may be a non-negligible bias in the C_p criterion against the risk function for sample sizes that are not sufficiently large. The larger the bias tends to be, the more remarkable it becomes as the number of dimensions of a vector of response variables increases in multivariate models. Therefore, it is important to completely remove the bias of the C_p criterion against the risk function.

I. Czarnowski et al. (Eds.): KESIDT 2023, SIST 352, pp. 173–182, 2023.
https://doi.org/10.1007/978-981-99-2969-6_15

The C_p criterion that has been modified to completely remove the bias against the risk function is called the modified C_p (denoted by MC_p) criterion. The original MC_p criterion was proposed by [1] for selecting variables in multivariate linear regression models. Thereafter, MC_p criteria have been proposed for selecting the degrees of polynomials in GMANOVA models (the GMANOVA model was proposed by [5]), as well as for selecting the ridge parameter in multivariate ridge regression and multiple ridge parameters in multivariate generalized ridge regression, by [6,8,9], respectively. The purpose of this paper is to propose an MC_p criterion for a wide class of candidate models, including results in previous studies. The derivation of our MC_p criterion is basically the same method as in [6]. However, their model was more limited than ours, so we could not directly use their lemma to our calculations. Hence, we have prepared new lemmas and used them to derive our MC_p criterion.

This paper is organized as follows. In Sect. 2, we give the formula of our new MC_p criterion. In Sect. 3, we show several examples of applying our MC_p criterion to a concrete model and perform a simple numerical study to verify whether this MC_p criterion completely removes the bias of the C_p criterion. Technical details are provided in the Appendix.

2 Main Result

Let $\boldsymbol{Y}$ be an $n \times p$ matrix of response variables, which is distributed according to the matrix normal distribution as

$$\boldsymbol{Y} \sim \mathcal{N}_{n \times p}(\boldsymbol{\Gamma}, \boldsymbol{\Sigma} \otimes \boldsymbol{I}_n), \tag{1}$$

where $\boldsymbol{\Gamma}$ is an $n \times p$ unknown location matrix, $\boldsymbol{\Sigma}$ is a $p \times p$ unknown covariance matrix, $\boldsymbol{I}_n$ is an $n \times n$ unit matrix, and $\otimes$ denotes the Kronecker product (see, e.g., chap. 16 in [2]). Furthermore, let $\boldsymbol{M}$ and $\boldsymbol{H}$ be $n \times n$ constant symmetric matrices satisfying

$$\boldsymbol{M}^2 = \boldsymbol{M},\ \mathrm{rank}(\boldsymbol{M}) = m < n,\ (\boldsymbol{I}_n - \boldsymbol{M})\boldsymbol{\Gamma} = \boldsymbol{O}_{n,n},\ (\boldsymbol{I}_n - \boldsymbol{M})\boldsymbol{H} = \boldsymbol{O}_{n,n}, \tag{2}$$

where $\boldsymbol{O}_{a,b}$ is an $a \times b$ matrix of zeros. Examples of $\boldsymbol{M}$ and $\boldsymbol{H}$ under specific models are given in Sect. 3. By using $\boldsymbol{M}$ and a $p \times q$ full column rank matrix $\boldsymbol{X}$ ($\mathrm{rank}(\boldsymbol{X}) = q \le p$), we define two $p \times p$ random matrices as

$$\boldsymbol{S} = \frac{1}{n-m}\boldsymbol{Y}'(\boldsymbol{I}_n - \boldsymbol{M})\boldsymbol{Y}, \quad \boldsymbol{G} = \boldsymbol{S}^{-1}\boldsymbol{X}\left(\boldsymbol{X}'\boldsymbol{S}^{-1}\boldsymbol{X}\right)^{-1}\boldsymbol{X}'. \tag{3}$$

To ensure the existence of $E[\boldsymbol{S}^{-1}]$, we assume that $n - m - p - 1 > 0$. It is clear that $\boldsymbol{S}$ is an unbiased estimator of $\boldsymbol{\Sigma}$ because of the assumption of $\boldsymbol{M}$. Let $\hat{\boldsymbol{Y}}$ be an $n \times p$ matrix of fitted values for $\boldsymbol{Y}$,

$$\hat{\boldsymbol{Y}} = \boldsymbol{H}\boldsymbol{Y}\boldsymbol{G}, \tag{4}$$

and let $d(\boldsymbol{A}, \boldsymbol{B})$ and $\hat{d}(\boldsymbol{A}, \boldsymbol{B})$ be the distance function and estimated distance function for matrices $\boldsymbol{A}$ and $\boldsymbol{B}$, which are defined as

$$\begin{aligned} d(\boldsymbol{A}, \boldsymbol{B}) &= \operatorname{tr}\left\{(\boldsymbol{A}-\boldsymbol{B})\boldsymbol{\Sigma}^{-1}(\boldsymbol{A}-\boldsymbol{B})'\right\}, \\ \hat{d}(\boldsymbol{A}, \boldsymbol{B}) &= \operatorname{tr}\left\{(\boldsymbol{A}-\boldsymbol{B})\boldsymbol{S}^{-1}(\boldsymbol{A}-\boldsymbol{B})'\right\}. \end{aligned} \tag{5}$$

The above distances are the sum of the squares of the Mahalanobis distances for the rows of $\boldsymbol{A}$ and $\boldsymbol{B}$. Then, we define the underlying risk function based on the multivariate MSE of prediction as proposed in [1]

$$R_p = E\left[d\left(\boldsymbol{Y}_{\mathrm{F}}, \hat{\boldsymbol{Y}}\right)\right], \tag{6}$$

where $\boldsymbol{Y}_{\mathrm{F}}$ is an $n \times p$ random matrix that is independent of $\boldsymbol{Y}$ and has the same distribution as $\boldsymbol{Y}$. The random matrix $\boldsymbol{Y}_{\mathrm{F}}$ can be regarded as a future observation or imaginary new observation. The following two lemmas for R_p are derived (the proofs are given in Appendices A.1 and A.2, respectively):

Lemma 1. *The risk function R_p in (6) can be computed as*

$$R_p = (n-m)p + 2q\operatorname{tr}(\boldsymbol{H}) + E\left[\hat{d}(\boldsymbol{M}\boldsymbol{Y}, \hat{\boldsymbol{Y}})\right]. \tag{7}$$

Lemma 2. *The bias of $d(\boldsymbol{Y}, \hat{\boldsymbol{Y}})$ against R_p is expressed as*

$$\begin{aligned} B_p &= R_p - E\left[\hat{d}(\boldsymbol{Y}, \hat{\boldsymbol{Y}})\right] \\ &= 2q\operatorname{tr}(\boldsymbol{H}) + \left\{E\left[d(\boldsymbol{M}\boldsymbol{Y}, \hat{\boldsymbol{Y}})\right] - E\left[\hat{d}(\boldsymbol{M}\boldsymbol{Y}, \hat{\boldsymbol{Y}})\right]\right\}. \end{aligned} \tag{8}$$

Taking $E[d(\boldsymbol{M}\boldsymbol{Y}, \hat{\boldsymbol{Y}})] \approx E[\hat{d}(\boldsymbol{M}\boldsymbol{Y}, \hat{\boldsymbol{Y}})]$ in Lemma 2 yields the asymptotic unbiased estimator of R_p, C_p, as

$$C_p = \hat{d}(\boldsymbol{Y}, \hat{\boldsymbol{Y}}) + 2q\operatorname{tr}(\boldsymbol{H}).$$

Next, we prepare formulas of expectations of matrices needed to evaluate the bias in (8) (the proof is given in Appendix A.3):

Lemma 3. *Let $\boldsymbol{C}_1$, $\boldsymbol{C}_2$, and $\boldsymbol{C}_3$ be $p \times p$ random matrices consisting of $\boldsymbol{S}$, defined as*

$$\begin{aligned} \boldsymbol{C}_1 &= \boldsymbol{\Sigma}^{-1/2}\boldsymbol{X}\left(\boldsymbol{X}'\boldsymbol{S}^{-1}\boldsymbol{X}\right)^{-1}\boldsymbol{X}'\boldsymbol{S}^{-1}\boldsymbol{\Sigma}^{1/2}, \\ \boldsymbol{C}_2 &= \boldsymbol{\Sigma}^{1/2}\boldsymbol{S}^{-1}\boldsymbol{X}\left(\boldsymbol{X}'\boldsymbol{S}^{-1}\boldsymbol{X}\right)^{-1}\boldsymbol{X}'\boldsymbol{S}^{-1}\boldsymbol{\Sigma}^{1/2}, \\ \boldsymbol{C}_3 &= \boldsymbol{\Sigma}^{1/2}\boldsymbol{S}^{-1}\boldsymbol{X}\left(\boldsymbol{X}'\boldsymbol{S}^{-1}\boldsymbol{X}\right)^{-1}\boldsymbol{X}'\boldsymbol{\Sigma}^{-1}\boldsymbol{X}\left(\boldsymbol{X}'\boldsymbol{S}^{-1}\boldsymbol{X}\right)^{-1}\boldsymbol{X}'\boldsymbol{S}^{-1}\boldsymbol{\Sigma}^{1/2}. \end{aligned} \tag{9}$$

The expectations of the above three matrices are given by

$$E\left[\boldsymbol{C}_1\right] = \boldsymbol{Q}_1\boldsymbol{Q}_1', \; E\left[\boldsymbol{C}_2\right] = c_2\boldsymbol{Q}_1\boldsymbol{Q}_1' + c_1c_3\boldsymbol{I}_p, \; E\left[\boldsymbol{C}_3\right] = \frac{1}{c_1}E\left[\boldsymbol{C}_2\right].$$

where $\boldsymbol{Q}_1$ *is a* $p \times q$ *matrix defined by*

$$\boldsymbol{Q}_1 = \boldsymbol{\Sigma}^{-1/2}\boldsymbol{X}\left(\boldsymbol{X}'\boldsymbol{\Sigma}^{-1}\boldsymbol{X}\right)^{-1/2}, \tag{10}$$

and c_1, c_2, *and* c_3 *are constants given by*

$$c_1 = \frac{n-m}{n-m-p-1}, \; c_2 = \frac{n-m}{n-m-p+q-1}, \; c_3 = \frac{q}{n-m-p+q-1}. \tag{11}$$

By using Lemma 3, an unbiased estimator of $E[d(\boldsymbol{MY}, \hat{\boldsymbol{Y}})]$ can be obtained as in the following lemma (the proof is given in Appendix A.4):

Lemma 4. *Let* $\boldsymbol{R}$ *be a* $p \times p$ *random matrix defined by*

$$\boldsymbol{R} = \boldsymbol{Y}'\boldsymbol{HY}\left(\boldsymbol{I}_p - \boldsymbol{G}\right)\boldsymbol{S}^{-1}. \tag{12}$$

An unbiased estimator of $E[d(\boldsymbol{MY}, \hat{\boldsymbol{Y}})]$ *can be given by*

$$\hat{D} = \left(1 - \frac{p+1}{n-m}\right)\hat{d}(\boldsymbol{MY}, \hat{\boldsymbol{Y}}) + \frac{2q}{n-m}\mathrm{tr}(\boldsymbol{R}). \tag{13}$$

It follows from simple matrix calculations that $\boldsymbol{S}^{1/2}(\boldsymbol{I}_p - \boldsymbol{G})\boldsymbol{S}^{-1/2}$ is a symmetric idempotent matrix. This implies that $\mathrm{tr}(\boldsymbol{R}) \geq 0$.

By using the unbiased estimator in Lemma 4, the new MC_p criterion can be defined as

$$MC_p = \left(1 - \frac{p+1}{n-m}\right)\hat{d}(\boldsymbol{Y}, \hat{\boldsymbol{Y}}) + p(p+1) + 2q\left\{\mathrm{tr}(\boldsymbol{H}) + \frac{\mathrm{tr}(\boldsymbol{R})}{n-m}\right\}. \tag{14}$$

The unbiasedness of the MC_p criterion is ensured by the following main theorem (the proof is given in Appendix A.5):

Theorem 1. *The* MC_p *defined in* (14) *is an unbiased estimator of the risk function* R_p *in* (6), *i.e.,* $E[MC_p] = R_p$.

3 Examples of MC_p

If we remark that $\boldsymbol{R} = \boldsymbol{O}_{p,p}$ holds in the case of multivariate linear regression, because of $q = p$, our MC_p includes the MC_p criteria of [1,8,9] as special cases. When $\boldsymbol{H} = \boldsymbol{M}$, the following result is derived (the proof is given in Appendix A.6):

$$\hat{d}(\boldsymbol{Y}, \hat{\boldsymbol{Y}}) = (n-m)p + \mathrm{tr}(\boldsymbol{R}). \tag{15}$$

This implies that

$$\left(1 - \frac{p+1}{n-m}\right)\hat{d}(\boldsymbol{Y}, \hat{\boldsymbol{Y}}) + \frac{2q\mathrm{tr}(\boldsymbol{R})}{n-m} = \left(1 - \frac{p-2q+1}{n-m}\right)\hat{d}(\boldsymbol{Y}, \hat{\boldsymbol{Y}}) - 2qp.$$

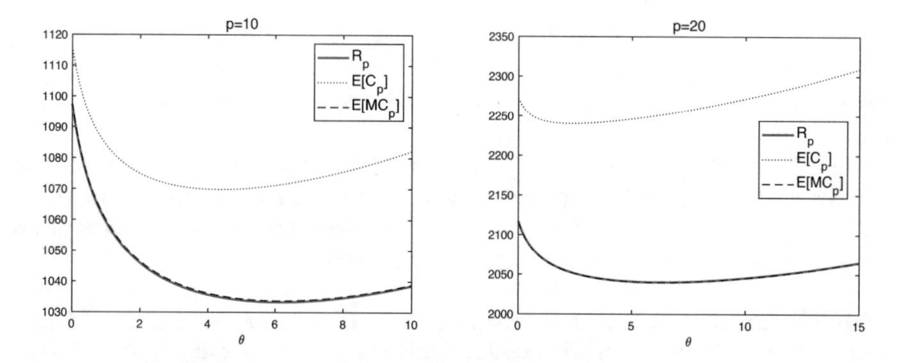

Fig. 1. R_p, $E[C_p]$, and $E[MC_p]$ in the cases $p = 10$ and 20

Notice that $\mathrm{tr}(\boldsymbol{H}) = m$ if $\boldsymbol{H} = \boldsymbol{M}$. Hence, MC_p with $\boldsymbol{H} = \boldsymbol{M}$ is rewritten as

$$MC_p = \left(1 - \frac{p - 2q + 1}{n - m}\right) \hat{d}(\boldsymbol{Y}, \hat{\boldsymbol{Y}}) + 2mq + p(p - 2q + 1).$$

This coincides with the MC_p of [6]. Moreover, when $\boldsymbol{H}$ is an idempotent matrix, i.e., satisfying $\boldsymbol{H}^2 = \boldsymbol{H}$, then MC_p coincides with one for selecting simultaneously within and between individual explanatory variables in the GMANOVA models. Next, we consider the following matrices as concrete $\boldsymbol{M}$ and $\boldsymbol{H}$:

$$\boldsymbol{M} = \boldsymbol{J}_n + \boldsymbol{A}(\boldsymbol{A}'\boldsymbol{A})^{-1}\boldsymbol{A}', \quad \boldsymbol{H} = \boldsymbol{J}_n + \boldsymbol{A}(\boldsymbol{A}'\boldsymbol{A} + \theta\boldsymbol{I}_k)^{-1}\boldsymbol{A}', \tag{16}$$

where θ is a non-negative ridge parameter, $\boldsymbol{A}$ is an $n \times k$ centralized matrix, i.e., $\boldsymbol{A}'\boldsymbol{1}_n = \boldsymbol{0}_k$, and $\boldsymbol{J}_n = \boldsymbol{1}_n\boldsymbol{1}_n'/n$. Here, $\boldsymbol{1}_n$ and $\boldsymbol{0}_n$ are n-dimensional vectors of ones and zeros, respectively. Then, the MC_p criterion with $\boldsymbol{M}$ and $\boldsymbol{H}$ in (16) coincides with one for selecting the ridge parameter in multivariate ridge regression for the GMANOVA model. Additionally, by changing $\boldsymbol{H}$ in the above MC_p criterion, we can also express one for selecting multiple ridge parameters in multivariate generalized ridge regression for the GMANOVA model.

Before concluding this section, a simple numerical study is conducted to verify how large the bias of the C_p criterion is against R_p and whether any bias against R_p remains in the MC_p criterion. In the numerical study, we deal with selection of the ridge parameter θ in multivariate ridge regression for the GMANOVA model, i.e., $\boldsymbol{M}$ and $\boldsymbol{H}$ given by (16) were used. The $n \times k$ matrix $\boldsymbol{A}$ was $\boldsymbol{A} = (\boldsymbol{I}_n - \boldsymbol{J}_n)\boldsymbol{A}_0\boldsymbol{\Phi}_k(0.9)^{1/2}$, where $\boldsymbol{\Phi}_k(\rho)$ is a $k \times k$ autoregressive correlation matrix in which the (a, b)th element is $\rho^{|a-b|}$, and $\boldsymbol{A}_0$ was an $n \times k$ matrix in which elements were generated independently from the uniform distribution $\mathcal{U}(-1, 1)$. Moreover, we set the (a, b)th element of $\boldsymbol{X}$ equal to $\{-1 + 2(a - 1)/(p - 1)\}^{b-1}$ $(a = 1, \ldots, p; b = 1, \ldots, q)$. Simulation data were generated from (1) with $\boldsymbol{\Gamma} = \boldsymbol{A}\boldsymbol{1}_k\boldsymbol{1}_k'\boldsymbol{X}'/2$, $\boldsymbol{\Sigma} = \boldsymbol{\Phi}_p(0.8)$, and $(n, k, q) = (100, 30, 3)$. Figure 1 shows R_p, $E[C_p]$, and $E[MC_p]$ evaluated from 1,000 repetitions. The left and right figures show the results when $p = 10$ and 20, respectively. In each figure, the horizontal axis represents the value of θ, and the solid red line, the dotted line, and the dashed line represent R_p, $E[C_p]$, and $E[MC_p]$, respectively. The expected values were computed as $\theta_j = 10(j-1)/99$ $(j = 1, \ldots, 100)$ when $p = 10$, and as $\theta_j = 15(j-1)/99$ $(j = 1, \ldots, 100)$ when $p = 20$. From the figure we can see that the bias of the C_p criterion against R_p was large even when n was moderate. In particular, the bias increased as p increased. It is also clear that the bias of the MC_p criterion against R_p was completely removed.

Acknowledgments. The authors wish to thank two reviewers for their helpful comments. This research was supported by JSPS Bilateral Program Grant Number JPJSBP 120219927 and JSPS KAKENHI Grant Number 20H04151.

Appendix: Mathematical Details

A.1 The Proof of Lemma 1

Let $\boldsymbol{\mathcal{E}}$ be an $n \times p$ random matrix defined by

$$\boldsymbol{\mathcal{E}} = (\boldsymbol{Y} - \boldsymbol{\Gamma})\boldsymbol{\Omega}, \quad \boldsymbol{\Omega} = \boldsymbol{\Sigma}^{-1/2}. \tag{A.1}$$

Notice that

$$d(\boldsymbol{Y}_{\mathrm{F}}, \hat{\boldsymbol{Y}}) = d(\boldsymbol{Y}_{\mathrm{F}}, \boldsymbol{M}\boldsymbol{Y}) + d(\boldsymbol{M}\boldsymbol{Y}, \hat{\boldsymbol{Y}}) + 2\mathcal{L}, \tag{A.2}$$

where $\hat{\boldsymbol{Y}}$ is given by (4), and $\mathcal{L} = \mathrm{tr}\{(\boldsymbol{Y}_{\mathrm{F}} - \boldsymbol{M}\boldsymbol{Y})\boldsymbol{\Omega}^2(\boldsymbol{M}\boldsymbol{Y} - \hat{\boldsymbol{Y}})'\}$. Since $\boldsymbol{Y}_{\mathrm{F}}$ is independent of $\boldsymbol{Y}$ and distributed according to the same distribution as $\boldsymbol{Y}$, and $\boldsymbol{M}$ satisfies (2), we have

$$E\left[d(\boldsymbol{Y}_{\mathrm{F}}, \boldsymbol{M}\boldsymbol{Y})\right] = E\left[d(\boldsymbol{Y}_{\mathrm{F}}, \boldsymbol{\Gamma})\right] + E\left[\mathrm{tr}(\boldsymbol{\mathcal{E}}'\boldsymbol{M}\boldsymbol{\mathcal{E}})\right] = (n+m)p. \tag{A.3}$$

Notice that $\boldsymbol{Y}_{\mathrm{F}}$, $\boldsymbol{M}\boldsymbol{\mathcal{E}}$, and $\boldsymbol{G}$ in (3) are mutually independent, and

$$\mathcal{L} = \mathrm{tr}\left\{\left((\boldsymbol{Y}_{\mathrm{F}} - \boldsymbol{\Gamma})\,\boldsymbol{\Omega} - \boldsymbol{M}\boldsymbol{\mathcal{E}}\right)\left(\boldsymbol{M}\boldsymbol{\mathcal{E}} - \boldsymbol{H}\boldsymbol{\mathcal{E}}\boldsymbol{\Omega}^{-1}\boldsymbol{G}\boldsymbol{\Omega} + (\boldsymbol{\Gamma} - \boldsymbol{H}\boldsymbol{\Gamma}\boldsymbol{G})\,\boldsymbol{\Omega}\right)'\right\}.$$

The above equations and (2) imply

$$\begin{aligned} E[\mathcal{L}] &= -E\left[\mathrm{tr}\left\{\boldsymbol{M}\boldsymbol{\mathcal{E}}\left(\boldsymbol{M}\boldsymbol{\mathcal{E}} - \boldsymbol{H}\boldsymbol{\mathcal{E}}\boldsymbol{\Omega}^{-1}\boldsymbol{G}\boldsymbol{\Omega}\right)'\right\}\right] \\ &= -E\left[\mathrm{tr}(\boldsymbol{\mathcal{E}}'\boldsymbol{M}\boldsymbol{\mathcal{E}})\right] + E\left[\mathcal{L}_1\right] = -mp + E\left[\mathcal{L}_1\right], \end{aligned} \tag{A.4}$$

where $\mathcal{L}_1 = \mathrm{tr}(\boldsymbol{\mathcal{E}}'\boldsymbol{H}\boldsymbol{\mathcal{E}}\boldsymbol{\Omega}\boldsymbol{G}'\boldsymbol{\Omega}^{-1})$. Since $\boldsymbol{\mathcal{E}}'\boldsymbol{H}\boldsymbol{\mathcal{E}}$ is independent of $\boldsymbol{G}$ and $\mathrm{tr}(\boldsymbol{G}) = q$, the following equation is obtained.

$$E\left[\mathcal{L}_1\right] = \mathrm{tr}(\boldsymbol{H})E\left[\mathrm{tr}\left(\boldsymbol{\Omega}\boldsymbol{G}'\boldsymbol{\Omega}^{-1}\right)\right] = \mathrm{tr}(\boldsymbol{H})E\left[\mathrm{tr}\left(\boldsymbol{G}'\right)\right] = q\mathrm{tr}(\boldsymbol{H}). \tag{A.5}$$

From (A.2), (A.3), (A.4), and (A.5), Lemma 1 is proved.

A.2 The Proof of Lemma 2

Notice that $\mathrm{tr}\{(\boldsymbol{Y} - \boldsymbol{MY})\boldsymbol{S}^{-1}(\boldsymbol{MY} - \hat{\boldsymbol{Y}})'\} = 0$ because of

$$\mathrm{tr}\{(\boldsymbol{Y} - \boldsymbol{MY})\boldsymbol{S}^{-1}(\boldsymbol{MY} - \hat{\boldsymbol{Y}})'\} = \mathrm{tr}\left\{(\boldsymbol{I}_n - \boldsymbol{M})\boldsymbol{YS}^{-1}(\boldsymbol{Y} - \boldsymbol{HYG})'\boldsymbol{M}\right\}.$$

It follows from this result and a calculation similar to the one used to find (A.2) that

$$\hat{d}(\boldsymbol{Y}, \hat{\boldsymbol{Y}}) = \hat{d}(\boldsymbol{Y}, \boldsymbol{MY}) + \hat{d}(\boldsymbol{MY}, \hat{\boldsymbol{Y}}). \tag{A.6}$$

It is easy to see from the definition of $\boldsymbol{S}$ in (3) that

$$\hat{d}(\boldsymbol{Y}, \boldsymbol{MY}) = (n - m)\mathrm{tr}(\boldsymbol{SS}^{-1}) = (n - m)p. \tag{A.7}$$

Substituting (A.7) into (A.6) yields

$$\hat{d}(\boldsymbol{Y}, \hat{\boldsymbol{Y}}) = (n - m)p + \hat{d}(\boldsymbol{MY}, \hat{\boldsymbol{Y}}). \tag{A.8}$$

By (7), (A.8), and the definition of the bias, Lemma 2 is proved.

A.3 The Proof of Lemma 3

Let $\boldsymbol{W}$ be a $p \times p$ random matrix defined by

$$\boldsymbol{W} = (n - m)\boldsymbol{\Omega S \Omega}, \tag{A.9}$$

where $\boldsymbol{\Omega}$ is given by (A.1). Then, $\boldsymbol{W} \sim \mathcal{W}_p(n - m, \boldsymbol{I}_p)$. Let $\boldsymbol{Q}_2$ be a $p \times (p - q)$ matrix satisfying $\boldsymbol{Q}_2'\boldsymbol{Q}_2 = \boldsymbol{I}_{p-q}$ and $\boldsymbol{Q}_1'\boldsymbol{Q}_2 = \boldsymbol{O}_{q,p-q}$, where $\boldsymbol{Q}_1$ is given by (10), and let $\boldsymbol{Q}$ be the $p \times p$ orthogonal matrix defined by $\boldsymbol{Q} = (\boldsymbol{Q}_1, \boldsymbol{Q}_2)$. Then, the three matrices $\boldsymbol{C}_1$, $\boldsymbol{C}_2$, and $\boldsymbol{C}_3$ given in (9) can be rewritten as

$$\begin{aligned}
\boldsymbol{C}_1 &= \boldsymbol{Q}_1 \left(\boldsymbol{Q}_1'\boldsymbol{W}^{-1}\boldsymbol{Q}_1\right)^{-1} \boldsymbol{Q}_1'\boldsymbol{W}^{-1}(\boldsymbol{Q}_1, \boldsymbol{Q}_2)\boldsymbol{Q}', \\
\boldsymbol{C}_2 &= (n - m)\boldsymbol{Q} \begin{pmatrix} \boldsymbol{Q}_1' \\ \boldsymbol{Q}_2' \end{pmatrix} \boldsymbol{W}^{-1}\boldsymbol{Q}_1 \left(\boldsymbol{Q}_1'\boldsymbol{W}^{-1}\boldsymbol{Q}_1\right)^{-1} \boldsymbol{Q}_1'\boldsymbol{W}^{-1}(\boldsymbol{Q}_1, \boldsymbol{Q}_2)\boldsymbol{Q}', \\
\boldsymbol{C}_3 &= \boldsymbol{Q} \begin{pmatrix} \boldsymbol{Q}_1' \\ \boldsymbol{Q}_2' \end{pmatrix} \boldsymbol{W}^{-1}\boldsymbol{Q}_1 \left(\boldsymbol{Q}_1'\boldsymbol{W}^{-1}\boldsymbol{Q}_1\right)^{-2} \boldsymbol{Q}_1'\boldsymbol{W}^{-1}(\boldsymbol{Q}_1, \boldsymbol{Q}_2)\boldsymbol{Q}'.
\end{aligned} \tag{A.10}$$

Let $\boldsymbol{V}$ be a $p \times p$ random matrix defined by $\boldsymbol{V} = \boldsymbol{Q}'\boldsymbol{WQ}$, and let $\boldsymbol{Z}_1$ and $\boldsymbol{Z}_2$ be respectively $(n - m) \times q$ and $(n - m) \times (p - q)$ independent random matrices distributed as $(\boldsymbol{Z}_1, \boldsymbol{Z}_2) \sim \mathcal{N}_{(n-m)\times p}(\boldsymbol{O}_{n-m,p}, \boldsymbol{I}_{(n-m)p})$. By using $\boldsymbol{V} \sim \mathcal{W}_p(n - m, \boldsymbol{I}_p)$, the partitioned $\boldsymbol{V}$ can be rewritten in $\boldsymbol{Z}_1$ and $\boldsymbol{Z}_2$ as

$$\boldsymbol{V} = \begin{pmatrix} \boldsymbol{V}_{11} & \boldsymbol{V}_{12} \\ \boldsymbol{V}_{12}' & \boldsymbol{V}_{22} \end{pmatrix} = \begin{pmatrix} \boldsymbol{Z}_1'\boldsymbol{Z}_1 & \boldsymbol{Z}_1'\boldsymbol{Z}_2 \\ \boldsymbol{Z}_2'\boldsymbol{Z}_1 & \boldsymbol{Z}_2'\boldsymbol{Z}_2 \end{pmatrix}.$$

Hence, it follows from $\boldsymbol{V}^{-1} = \boldsymbol{Q}'\boldsymbol{W}^{-1}\boldsymbol{Q}$ and the general formula for the inverse of a partitioned matrix, e.g., th. 8.5.11 in [2], that

$$\begin{aligned}
\boldsymbol{V}^{-1} &= \begin{pmatrix} \boldsymbol{Q}_1'\boldsymbol{W}^{-1}\boldsymbol{Q}_1 & \boldsymbol{Q}_1'\boldsymbol{W}^{-1}\boldsymbol{Q}_2 \\ \boldsymbol{Q}_2'\boldsymbol{W}^{-1}\boldsymbol{Q}_1 & \boldsymbol{Q}_2'\boldsymbol{W}^{-1}\boldsymbol{Q}_2 \end{pmatrix} \\
&= \begin{pmatrix} \boldsymbol{V}_{11\cdot2}^{-1} & -\boldsymbol{V}_{11\cdot2}^{-1}\boldsymbol{V}_{12}\boldsymbol{V}_{22}^{-1} \\ -\boldsymbol{V}_{22}^{-1}\boldsymbol{V}_{12}'\boldsymbol{V}_{11\cdot2}^{-1} & \boldsymbol{V}_{22}^{-1} + \boldsymbol{V}_{22}^{-1}\boldsymbol{V}_{12}'\boldsymbol{V}_{11\cdot2}^{-1}\boldsymbol{V}_{12}\boldsymbol{V}_{22}^{-1} \end{pmatrix},
\end{aligned} \tag{A.11}$$

where $\boldsymbol{V}_{11\cdot 2} = \boldsymbol{V}_{11} - \boldsymbol{V}_{12}\boldsymbol{V}_{22}^{-1}\boldsymbol{V}_{12}'$. Substituting (A.11) into (A.10) yields

$$\begin{aligned} \boldsymbol{C}_1 &= \boldsymbol{Q}_1 \left(\boldsymbol{I}_q, -\boldsymbol{V}_{12}\boldsymbol{V}_{22}^{-1}\right) \boldsymbol{Q}', \\ \boldsymbol{C}_2 &= (n-m)\boldsymbol{Q} \left\{ \boldsymbol{V}^{-1} - \begin{pmatrix} \boldsymbol{O}_{q,q} & \boldsymbol{O}_{q,p-q} \\ \boldsymbol{O}_{p-q,q} & \boldsymbol{V}_{22}^{-1} \end{pmatrix} \right\} \boldsymbol{Q}', \\ \boldsymbol{C}_3 &= \boldsymbol{Q} \begin{pmatrix} \boldsymbol{I}_q & -\boldsymbol{V}_{12}\boldsymbol{V}_{22}^{-1} \\ -\boldsymbol{V}_{22}^{-1}\boldsymbol{V}_{12}' & \boldsymbol{V}_{22}^{-1}\boldsymbol{V}_{12}'\boldsymbol{V}_{12}\boldsymbol{V}_{22}^{-1} \end{pmatrix} \boldsymbol{Q}'. \end{aligned} \tag{A.12}$$

By using the independence of $\boldsymbol{Z}_1$ and $\boldsymbol{Z}_2$, and formulas of expectations of the matrix normal distribution and Wishart distribution, we have

$$\begin{aligned} &E\left[\boldsymbol{V}^{-1}\right] = \frac{c_1}{n-m}\boldsymbol{I}_p, \quad E\left[\boldsymbol{V}_{22}^{-1}\right] = \frac{c_2}{n-m}\boldsymbol{I}_{p-q}, \\ &E\left[\boldsymbol{V}_{12}\boldsymbol{V}_{22}^{-1}\right] = E\left[\boldsymbol{Z}_1'\right] E\left[\boldsymbol{Z}_2\left(\boldsymbol{Z}_2'\boldsymbol{Z}_2\right)^{-1}\right] = \boldsymbol{O}_{q,p-q}, \end{aligned}$$

where c_1 and c_2 are given by (11). It follows from the result $E[\boldsymbol{Z}_1\boldsymbol{Z}_1'] = q\boldsymbol{I}_{n-m}$ and the independence of $\boldsymbol{Z}_1$ and $\boldsymbol{Z}_2$ that

$$\begin{aligned} E\left[\boldsymbol{V}_{22}^{-1}\boldsymbol{V}_{12}'\boldsymbol{V}_{12}\boldsymbol{V}_{22}^{-1}\right] &= E\left[\left(\boldsymbol{Z}_2'\boldsymbol{Z}_2\right)^{-1}\boldsymbol{Z}_2'\boldsymbol{Z}_1\boldsymbol{Z}_1'\boldsymbol{Z}_2\left(\boldsymbol{Z}_2'\boldsymbol{Z}_2\right)^{-1}\right] \\ &= qE\left[\left(\boldsymbol{Z}_2'\boldsymbol{Z}_2\right)^{-1}\boldsymbol{Z}_2'\boldsymbol{Z}_2\left(\boldsymbol{Z}_2'\boldsymbol{Z}_2\right)^{-1}\right] \\ &= qE\left[\boldsymbol{V}_{22}^{-1}\right] = c_3\boldsymbol{I}_{p-q}, \end{aligned}$$

where c_3 is given by (11). By using the above expectations, (A.12), and $\boldsymbol{Q}_2\boldsymbol{Q}_2' = \boldsymbol{I}_p - \boldsymbol{Q}_1\boldsymbol{Q}_1'$, we have

$$\begin{aligned} E\left[\boldsymbol{C}_1\right] &= \boldsymbol{Q}_1\left(\boldsymbol{I}_q, \boldsymbol{O}_{q,p-q}\right)' \begin{pmatrix} \boldsymbol{Q}_1' \\ \boldsymbol{Q}_2' \end{pmatrix} = \boldsymbol{Q}_1\boldsymbol{Q}_1', \\ E\left[\boldsymbol{C}_2\right] &= \boldsymbol{Q}\left\{ c_1\boldsymbol{I}_p - \begin{pmatrix} \boldsymbol{O}_{q,q} & \boldsymbol{O}_{q,p-q} \\ \boldsymbol{O}_{p-q,q} & c_2\boldsymbol{I}_{p-q} \end{pmatrix} \right\} \boldsymbol{Q}' = c_2\boldsymbol{Q}_1\boldsymbol{Q}_1' + c_1c_3\boldsymbol{I}_p, \\ E\left[\boldsymbol{C}_3\right] &= \boldsymbol{Q}\begin{pmatrix} \boldsymbol{I}_q & \boldsymbol{O}_{p-q,q} \\ \boldsymbol{O}_{q,p-q} & c_3\boldsymbol{I}_{p-q} \end{pmatrix} \boldsymbol{Q}' = \frac{1}{c_1}E\left[\boldsymbol{C}_2\right]. \end{aligned}$$

Therefore, Lemma 3 is proved.

A.4 The Proof of Lemma 4

Let $\boldsymbol{U}_0$, $\boldsymbol{U}_1$ and $\boldsymbol{U}_2$ be $p \times p$ symmetric random matrices defined by

$$\boldsymbol{U}_0 = \boldsymbol{\varOmega}\boldsymbol{Y}'\boldsymbol{M}\boldsymbol{Y}\boldsymbol{\varOmega}, \quad \boldsymbol{U}_1 = \boldsymbol{\varOmega}\boldsymbol{Y}'\boldsymbol{H}\boldsymbol{Y}\boldsymbol{\varOmega}, \quad \boldsymbol{U}_2 = \boldsymbol{\varOmega}\boldsymbol{Y}'\boldsymbol{H}^2\boldsymbol{Y}\boldsymbol{\varOmega},$$

where $\boldsymbol{\varOmega}$ is given by (A.1). It follows from the definitions of $\boldsymbol{C}_1$, $\boldsymbol{C}_2$, and $\boldsymbol{C}_3$ in (9) that

$$\begin{aligned} &\boldsymbol{\varOmega}^{-1}\boldsymbol{G}\boldsymbol{\varOmega} = \boldsymbol{C}_1', \quad \boldsymbol{C}_1'\boldsymbol{C}_1 = \boldsymbol{C}_3, \quad \boldsymbol{\varOmega}^{-1}\boldsymbol{S}^{-1}\boldsymbol{\varOmega}^{-1}\boldsymbol{C}_1 = \boldsymbol{C}_2, \\ &\boldsymbol{C}_1'\boldsymbol{\varOmega}^{-1}\boldsymbol{S}^{-1}\boldsymbol{\varOmega}^{-1}\boldsymbol{C}_1 = \boldsymbol{C}_2. \end{aligned}$$

By using these results, the definitions of d and $\hat{d}$ in (5), and the assumptions of $\boldsymbol{M}$ and $\boldsymbol{H}$ in (2), $d(\boldsymbol{MY}, \hat{\boldsymbol{Y}})$ and $\hat{d}(\boldsymbol{MY}, \hat{\boldsymbol{Y}})$ in (8) can be rewritten as

$$d(\boldsymbol{MY}, \hat{\boldsymbol{Y}}) = \mathrm{tr}(\boldsymbol{U}_0) - 2\mathrm{tr}(\boldsymbol{U}_1\boldsymbol{C}_1) + \mathrm{tr}(\boldsymbol{U}_2\boldsymbol{C}_3),$$
$$\hat{d}(\boldsymbol{MY}, \hat{\boldsymbol{Y}}) = (n-m)\mathrm{tr}(\boldsymbol{U}_0\boldsymbol{W}^{-1}) - 2\mathrm{tr}(\boldsymbol{U}_1\boldsymbol{C}_2) + \mathrm{tr}(\boldsymbol{U}_2\boldsymbol{C}_2),$$

where $\boldsymbol{W}$ is given by (A.9). Notice that $\boldsymbol{U}_0$ and $\boldsymbol{S}$, $\boldsymbol{U}_1$ and $\boldsymbol{S}$, and $\boldsymbol{U}_2$ and $\boldsymbol{S}$ are independent because of $\boldsymbol{M}(\boldsymbol{I}_n - \boldsymbol{M}) = \boldsymbol{O}_{n,n}$ and $\boldsymbol{H}(\boldsymbol{I}_n - \boldsymbol{M}) = \boldsymbol{O}_{n,n}$, and $\boldsymbol{C}_1$, $\boldsymbol{C}_2$, and $\boldsymbol{C}_3$ are random matrices in which $\boldsymbol{S}$ is the only random variable. These imply that

$$E\left[d(\boldsymbol{MY}, \hat{\boldsymbol{Y}})\right] = \mathrm{tr}\left(\boldsymbol{\Delta}_0\right) - 2\mathrm{tr}\left(\boldsymbol{\Delta}_1 E\left[\boldsymbol{C}_1\right]\right) + \mathrm{tr}\left(\boldsymbol{\Delta}_2 E\left[\boldsymbol{C}_3\right]\right),$$
$$E\left[\hat{d}(\boldsymbol{MY}, \hat{\boldsymbol{Y}})\right] = (n-m)\mathrm{tr}\left(\boldsymbol{\Delta}_0 E\left[\boldsymbol{W}^{-1}\right]\right) - 2\mathrm{tr}\left(\boldsymbol{\Delta}_1 E\left[\boldsymbol{C}_2\right]\right) + \mathrm{tr}\left(\boldsymbol{\Delta}_2 E\left[\boldsymbol{C}_2\right]\right),$$

where $\boldsymbol{\Delta}_0 = E[\boldsymbol{U}_0]$, $\boldsymbol{\Delta}_1 = E[\boldsymbol{U}_1]$, and $\boldsymbol{\Delta}_2 = E[\boldsymbol{U}_2]$. Notice that $(n-m)E[\boldsymbol{W}^{-1}] = c_1\boldsymbol{I}_p$, where c_1 is given by (11). By using this result and Lemma 3, we have

$$\begin{aligned} E\left[d(\boldsymbol{MY}, \hat{\boldsymbol{Y}})\right] &= \mathrm{tr}\left(\boldsymbol{\Delta}_0\right) - 2\mathrm{tr}\left(\boldsymbol{\Delta}_1\boldsymbol{Q}_1\boldsymbol{Q}_1'\right) + \frac{1}{c_1}\mathrm{tr}\left(\boldsymbol{\Delta}_2 E\left[\boldsymbol{C}_2\right]\right), \\ E\left[\hat{d}(\boldsymbol{MY}, \hat{\boldsymbol{Y}})\right] &= c_1\mathrm{tr}\left(\boldsymbol{\Delta}_0\right) - 2\mathrm{tr}\left(\boldsymbol{\Delta}_1 E\left[\boldsymbol{C}_2\right]\right) + \mathrm{tr}\left(\boldsymbol{\Delta}_2 E\left[\boldsymbol{C}_2\right]\right), \end{aligned} \quad \text{(A.13)}$$

where c_2 and c_3 are given by (11). Let $L = \mathrm{tr}\{\boldsymbol{U}_1((n-m)\boldsymbol{W}^{-1} - \boldsymbol{C}_2)\}$. It is easy to see that $c_2^{-1} = c_1^{-1} + c_2^{-1}c_3$ and $E[L] = c_1\mathrm{tr}(\boldsymbol{\Delta}_1) - \mathrm{tr}(\boldsymbol{\Delta}_1 E[\boldsymbol{C}_2])$. Hence, we can derive

$$\begin{aligned} \mathrm{tr}\left(\boldsymbol{\Delta}_1\boldsymbol{Q}_1\boldsymbol{Q}_1'\right) &= \frac{1}{c_2}\mathrm{tr}\left(\boldsymbol{\Delta}_1 E\left[\boldsymbol{C}_2\right]\right) - \frac{c_1c_3}{c_2}\mathrm{tr}\left(\boldsymbol{\Delta}_1\right) \\ &= \frac{1}{c_1}\mathrm{tr}\left(\boldsymbol{\Delta}_1 E\left[\boldsymbol{C}_2\right]\right) - \frac{c_3}{c_2}E\left[L\right]. \end{aligned} \quad \text{(A.14)}$$

A simple calculation shows that $c_2^{-1}c_3 = q/(n-m)$ and $L = \mathrm{tr}(\boldsymbol{R})$, where $\boldsymbol{R}$ is given by (12). Using these results, (A.13), and (A.14) yields

$$E\left[d(\boldsymbol{MY}, \hat{\boldsymbol{Y}})\right] = \frac{1}{c_1}E\left[\hat{d}(\boldsymbol{MY}, \hat{\boldsymbol{Y}})\right] + \frac{2q}{n-m}E\left[\mathrm{tr}(\boldsymbol{R})\right].$$

Consequently, Lemma 4 is proved.

A.5 The Proof of Theorem 1

From Lemmas 1 and 4, an unbiased estimator of R_p in (6) can be given by

$$\hat{R}_p = (n-m)p + 2q\mathrm{tr}(\boldsymbol{H}) + \hat{D}, \quad \text{(A.15)}$$

where $\hat{D}$ is given by (13). Notice that $\{1 - (p+1)/(n-m)\}(n-m)p = (n-m)p - p(p+1)$. Hence, the result and (A.8) imply that

$$\hat{D} = \left(1 - \frac{p+1}{n-m}\right)\hat{d}(\boldsymbol{Y}, \hat{\boldsymbol{Y}}) - (n-m)p + p(p+1) + \frac{2q}{n-m}\mathrm{tr}(\boldsymbol{R}). \qquad \text{(A.16)}$$

Substituting (A.16) into (A.15) yields that an unbiased estimator $\hat{R}_p$ coincides with MC_p in (14).

A.6 The Proof of Equation in (15)

The $\hat{d}(\boldsymbol{MY}, \hat{\boldsymbol{Y}})$ when $\boldsymbol{H} = \boldsymbol{M}$ can be rewritten as

$$\hat{d}(\boldsymbol{MY}, \hat{\boldsymbol{Y}}) = \mathrm{tr}\left\{\boldsymbol{Y}'\boldsymbol{MY}(\boldsymbol{I}_p - \boldsymbol{G})\boldsymbol{S}^{-1}(\boldsymbol{I}_p - \boldsymbol{G})'\right\}.$$

Since $\boldsymbol{P} = \boldsymbol{S}^{1/2}(\boldsymbol{I}_p - \boldsymbol{G})\boldsymbol{S}^{-1/2}$ is a symmetric idempotent matrix, we have

$$(\boldsymbol{I}_p - \boldsymbol{G})\boldsymbol{S}^{-1}(\boldsymbol{I}_p - \boldsymbol{G})' = \boldsymbol{S}^{-1/2}\boldsymbol{P}^2\boldsymbol{S}^{-1/2} = (\boldsymbol{I}_p - \boldsymbol{G})\boldsymbol{S}^{-1}.$$

This implies that $\hat{d}(\boldsymbol{MY}, \hat{\boldsymbol{Y}}) = \mathrm{tr}\{\boldsymbol{Y}'\boldsymbol{MY}(\boldsymbol{I}_p - \boldsymbol{G})\boldsymbol{S}^{-1}\} = \mathrm{tr}(\boldsymbol{R})$. From this result and (A.8), $\hat{d}(\boldsymbol{Y}, \hat{\boldsymbol{Y}}) = (n-m)p + \mathrm{tr}(\boldsymbol{R})$ is derived.

References

1. Fujikoshi, Y., Satoh, K.: Modified AIC and C_p in multivariate linear regression. Biometrika **84**, 707–716 (1997). https://doi.org/10.1093/biomet/84.3.707
2. Harville, D.A.: Matrix Algebra from a Statistician's Perspective. Springer, New York (1997). https://doi.org/10.1007/b98818
3. Mallows, C.L.: Some comments on C_p. Technometrics **15**, 661–675 (1973). https://doi.org/10.2307/1267380
4. Mallows, C.L.: More comments on C_p. Technometrics **37**, 362–372 (1995). https://doi.org/10.2307/1269729
5. Potthoff, R.F., Roy, S.N.: A generalized multivariate analysis of variance model useful especially for growth curve problems. Biometrika **51**, 313–326 (1964). https://doi.org/10.2307/2334137
6. Satoh, K., Kobayashi, M., Fujikoshi, Y.: Variable selection for the growth curve model. J. Multivar. Anal. **60**, 277–292 (1997). https://doi.org/10.1006/jmva.1996.1658
7. Sparks, R.S., Coutsourides, D., Troskie, L.: The multivariate C_p. Comm. Stat. Theor. Meth. **12**, 1775–1793 (1983). https://doi.org/10.1080/03610928308828569
8. Yanagihara, H., Satoh, K.: An unbiased C_p criterion for multivariate ridge regression. J. Multivar. Anal. **101**, 1226–1238 (2010). https://doi.org/10.1016/j.jmva.2009.09.017
9. Yanagihara, H., Nagai, I., Satoh, K.: A bias-corrected C_p criterion for optimizing ridge parameters in multivariate generalized ridge regression. Jpn. J. Appl. Stat. **38**, 151–172 (2009) (in Japanese). https://doi.org/10.5023/jappstat.38.151

Geographically Weighted Sparse Group Lasso: Local and Global Variable Selections for GWR

Mineaki Ohishi[1(✉)], Koki Kirishima[2], Kensuke Okamura[3], Yoshimichi Itoh[3], and Hirokazu Yanagihara[2]

[1] Tohoku University, Sendai 980-8576, Japan
mineaki.ohishi.a4@tohoku.ac.jp
[2] Hiroshima University, Higashi-Hiroshima 739-8526, Japan
[3] Tokyo Kantei Co., Ltd., Shinagawa 141-0021, Japan

Abstract. This paper deals with the variable selection problem in geographically weighted regression (GWR). GWR is a local estimation method that continuously evaluates geographical effects in regression involving spatial data. Specifically, the method estimates regression coefficients for each observed point using a varying coefficient model. With such a model, variable selection has two aspects: local selection, which applies to each observed point, and global selection, which is common for all observed points. We approach both variable selections simultaneously via sparse group Lasso. To illustrate the proposed method, we apply it to apartment rent data in Tokyo.

Keywords: geographically weighted regression · sparse group Lasso · variable selection

1 Introduction

Consider a spatial data sample of size n represented by $\{y_i, \boldsymbol{x}_i, \boldsymbol{s}_i\}$ $(i = 1, \ldots, n)$, where y_i is a response variable, $\boldsymbol{x}_i$ is a $(k+1)$-dimensional vector of explanatory variables of which the first element is 1, and $\boldsymbol{s}_i$ is a vector representing an observed point. The $\boldsymbol{s}_i$ could, for example, be a two-dimensional vector of longitude and latitude. This paper deals with data that include not only y_i and $\boldsymbol{x}_i$ but also $\boldsymbol{s}_i$. That is, we assume that y_i depends on not only $\boldsymbol{x}_i$ but also $\boldsymbol{s}_i$ (for example, the prices of real estate and land, or meteorological data such as temperature and weather).

When modeling with spatial data, the way in which geographical effects are described is important. Since it is generally considered that neighbor data have similar trends, a local estimation method is often adopted. For regression analyses, geographically weighted regression (GWR), as proposed by [3], has been a popular approach. The GWR model can be described as follows:

$$y_i = \boldsymbol{x}_i' \boldsymbol{\theta}(\boldsymbol{s}_i) + \varepsilon_i, \tag{1}$$

where $\boldsymbol{\theta}(\boldsymbol{s})$ is a $(k+1)$-dimensional unknown vector of varying coefficients for an observed point $\boldsymbol{s}$, and ε_i is an error variable. By varying regression coefficients

I. Czarnowski et al. (Eds.): KESIDT 2023, SIST 352, pp. 183–192, 2023.
https://doi.org/10.1007/978-981-99-2969-6_16

by observed point, GWR is able to flexibly capture geographical effects. The varying coefficients in the GWR model are estimated for each observed point by a weighted least square method. Note that if there are duplicate observed points, estimates for n points are unnecessary since $\boldsymbol{\theta}(\boldsymbol{s}_i) = \boldsymbol{\theta}(\boldsymbol{s}_j)$ holds if $\boldsymbol{s}_i = \boldsymbol{s}_j$ $(i \neq j)$. For example, suppose we are dealing with real estate data; if some rooms in the same building are included in the observations, then duplication of an observed point occurs. In considering such a case, we can show unique observed points as $\boldsymbol{s}_1^*, \ldots, \boldsymbol{s}_m^*$ $(m \leq n)$ and express a vector of varying coefficients for $\boldsymbol{s}_\ell^*$ $(\ell \in \{1, \ldots, m\})$ as $\boldsymbol{\theta}(\boldsymbol{s}_\ell^*) = \boldsymbol{\theta}_\ell = (\theta_{\ell 0}, \theta_{\ell 1}, \ldots, \theta_{\ell k})'$. Let $\boldsymbol{y}$ and $\boldsymbol{X}$ be an n-dimensional vector and an $n \times (k+1)$ matrix defined by $\boldsymbol{y} = (y_1, \ldots, y_n)'$ and $\boldsymbol{X} = (\boldsymbol{x}_1, \ldots, \boldsymbol{x}_n)'$ with $\mathrm{rank}(\boldsymbol{X}) = k+1$, respectively. Moreover, with the geographical weight matrix $\boldsymbol{W}_\ell = \mathrm{diag}(w_{\ell 1}, \ldots, w_{\ell n})$ $(\ell \in \{1, \ldots, m\})$, we transform $\boldsymbol{y}$ and $\boldsymbol{X}$ as $\boldsymbol{u}_\ell = \boldsymbol{W}_\ell^{1/2}\boldsymbol{y}$ and $\boldsymbol{Z}_\ell = \boldsymbol{W}_\ell^{1/2}\boldsymbol{X}$ with $\mathrm{rank}(\boldsymbol{Z}_\ell) = k+1$, respectively, where $w_{\ell i}$ is a non-negative geographical weight between the two points $\boldsymbol{s}_\ell^*$ and $\boldsymbol{s}_i$. Then, the GWR estimator of $\boldsymbol{\theta}_\ell$ is given by

$$\tilde{\boldsymbol{\theta}}_\ell = \arg\min_{\boldsymbol{\theta} \in \mathbb{R}^{k+1}} r_\ell(\boldsymbol{\theta}) = (\boldsymbol{Z}_\ell' \boldsymbol{Z}_\ell)^{-1} \boldsymbol{Z}_\ell' \boldsymbol{u}_\ell, \quad r_\ell(\boldsymbol{\theta}) = \frac{1}{2} \| \boldsymbol{u}_\ell - \boldsymbol{Z}_\ell \boldsymbol{\theta} \|^2, \tag{2}$$

where $\| \cdot \|$ is the ℓ_2 norm. The weight $w_{\ell i}$ is generally defined by a kernel, which is monotonically decreasing for the distance $\| \boldsymbol{s}_\ell^* - \boldsymbol{s}_i \|$, based on the knowledge that neighboring data have similar trends; here, the Gaussian kernel is often used. The package `spgwr` (e.g., ver. 0.6–35 [1]) for implementing GWR is part of the statistical software R (e.g., ver. 4.2.2 [13]) and adapts the Gaussian kernel as the default setting, with bisquare and tricube kernels as other options. The fitted value of GWR is given by $\hat{y}_i = \boldsymbol{x}_i' \tilde{\boldsymbol{\theta}}_{\ell_i}$ $(\ell_i \in \{1, \ldots, m\}$ *s.t.* $\boldsymbol{s}_i = \boldsymbol{s}_{\ell_i}^*)$, and the ith row vector of a hat matrix $\boldsymbol{H}$ satisfying $\hat{\boldsymbol{y}} = (\hat{y}_1, \ldots, \hat{y}_n)' = \boldsymbol{H}\boldsymbol{y}$ is given by $\boldsymbol{x}_i' (\boldsymbol{Z}_{\ell_i}' \boldsymbol{Z}_{\ell_i})^{-1} \boldsymbol{X}' \boldsymbol{W}_{\ell_i}$.

In this paper, we focus on the variable selection problem for GWR. In general, the variable selection problem involves searching for the best combination of variables. Conceptually, the problem can be solved via an all-possible-combinations search (APCS), which compares all possible combinations of variables. However, APCS is impractical when the number of variables is large because of exponentially increasing the number of combinations. For such large-scale cases, a number of algorithms have been proposed, e.g., forward or backward selection methods and the kick-one-out method (see e.g., [10,20]). These algorithms compare some (but not all) possible combinations of variables in a way similar to APCS. Another approach using sparse estimation is also popular. The sparse estimation method simultaneously implements the estimation and variable selection by estimating several parameters to exactly zero. Lasso, as proposed by [15], is one of the basic methods for sparse estimation. It estimates parameters using a penalized estimation method based on the ℓ_1 norm. Unfortunately, the Lasso estimator cannot be obtained in closed form. Still, Lasso has become widely used as a standard method based on several algorithms for efficiently obtaining estimates, e.g., LARS [6], coordinate descent algorithm [8], and ADMM [2]. Moreover, Lasso is scalable for the number of variables. While Lasso estimates several individual parameters to be exactly zero, [19] proposed group Lasso, which is an extension of Lasso that estimates a sub-vector of parameters as a vector of exact zeros. Group Lasso allows variable selection for grouped variables within the framework of penalized regression. In addition to group Lasso, there are a variety of Lasso extensions, e.g., fused Lasso [16], which penalizes differences in parameters, and sparse group Lasso [14], which is a hybrid of Lasso and group Lasso. Like group Lasso, Elastic net [22] is a hybrid method. One of merits of a penalized estimation method is that we can implement various estimation methods by combining multiple penalties as needed

In variable selection for GWR, [18] proposed geographically weighted Lasso (GW Lasso), which applies Lasso to GWR locally. That is, in the estimation (2) for the model (1), GW Lasso estimates several elements of $\boldsymbol{\theta}_\ell$ to exactly zero and implements local variable selection. Such local variable selection reveals geographical differences in the explanatory variables affecting the response variable. We are also interested in global variable selection, which is common variable selection for all the observed points. Let $\boldsymbol{\Theta}$ be an $m \times (k+1)$ matrix of varying coefficients written as

$$\boldsymbol{\Theta} = (\boldsymbol{\theta}_1, \ldots, \boldsymbol{\theta}_m)' = (\boldsymbol{\theta}_{(0)}, \boldsymbol{\theta}_{(1)}, \ldots, \boldsymbol{\theta}_{(k)}).$$

Then, global variable selection corresponds to whether $\boldsymbol{\theta}_{(j)}$ ($j \in \{1, \ldots, k\}$) is $\mathbf{0}_m$ (it does not consider $j = 0$ since $\boldsymbol{\theta}_{(0)}$ is the parameter vector for the intercept), where $\mathbf{0}_m$ is an m-dimensional vector of zeros. Hence, applying group Lasso to $\boldsymbol{\theta}_{(j)}$ allows us to perform global variable selection. Our proposed approach applies sparse group Lasso, a hybrid of Lasso and group Lasso, to achieve both local and variable selections simultaneously.

We call our approach geographically weighted sparse group Lasso (GWSGL) and define its estimator as

$$\hat{\boldsymbol{\Theta}}_{\lambda,\alpha} = \arg \min_{\boldsymbol{\Theta} \in \mathbb{R}^{m \times n}} f(\boldsymbol{\Theta} \mid \lambda, \alpha),$$

$$f(\boldsymbol{\Theta} \mid \lambda, \alpha) = \sum_{\ell=1}^{m} r_\ell(\boldsymbol{\theta}_\ell) + \lambda \sum_{j=1}^{k} \left\{ \alpha \sum_{\ell=1}^{m} v_{1,\ell j} |\theta_{\ell j}| + (1-\alpha) v_{2,j} \|\boldsymbol{\theta}_{(j)}\| \right\}, \tag{3}$$

where $\lambda \in [0, \infty)$ and $\alpha \in [0, 1]$ are tuning parameters, and $v_{1,\ell j}$ and $v_{2,j}$ are non-negative penalty weights inspired by adaptive Lasso [21]. We call a weighted penalty, like the second and third terms in (3), an adaptive-type penalty. The λ value adjusts the strength of the hybrid penalty against the model fit corresponding to the first term in (3); when $\lambda = 0$, GWSGL coincides with ordinary GWR. The α value adjusts the balance between the two penalties. The first penalty is for each element of $\boldsymbol{\Theta}$ and implements local variable selection. The second penalty is a group-type penalty for each column vector of $\boldsymbol{\Theta}$ and implements global variable selection. Hence, the estimation based on minimizing (3) allows us to implement local and global variable selections simultaneously, where α determines the weight for the two variable selections. In particular, only local variable selection is implemented when $\alpha = 1$ and only global variable selection is implemented when $\alpha = 0$. Regarding the penalty weights, ordinary sparse group Lasso requires $v_{1,\ell j} = 1$ and $v_{2,j} = \sqrt{m}$, while for adaptive-type penalties, $v_{1,\ell j} = 1/|\theta^{\dagger}_{\ell j}|$ and $v_{2,j} = m/\|\boldsymbol{\theta}^{\dagger}_{(j)}\|$ are often adopted, where $\theta^{\dagger}_{\ell j}$ and $\boldsymbol{\theta}^{\dagger}_{(j)}$ are some type of estimators for $\theta_{\ell j}$ and $\boldsymbol{\theta}_{(j)}$, respectively.

Since the GWSGL estimator $\hat{\boldsymbol{\Theta}}_{\lambda,\alpha}$ cannot be obtained in closed form, it is necessary to search for the minimizer of (3) using a numerical search algorithm. For sparse group Lasso, [14] solved the optimization problem by combining the block-wise coordinate descent algorithm (BCDA) and the iterative thresholding algorithm (ITA; e.g., [7]). Specifically, a sub-problem for each block is solved by ITA and the procedure is repeated until solution convergence. In our problem, BCDA can reduce the optimization problem for GWSGL to that for ordinary sparse group Lasso. Hence, in this paper, we derive an algorithm consisting of BCDA and ITA to minimize (3). To demonstrate its performance, we apply the proposed method to apartment rent data for Tokyo's 23 wards.

The remainder of the paper is organized as follows: In Sect. 2, we describe our algorithm for solving the GWSGL optimization problem. In Sect. 3, we show an example using real data.

2 Optimization Algorithm for GWSGL

2.1 Preliminaries

We begin by specifying the proposed algorithm for minimizing the objective function $f(\boldsymbol{\Theta} \mid \lambda, \alpha)$ in (3). Since, as mentioned earlier, it is difficult to directly minimize the objective function, we apply BCDA in a way similar to [14]. Note that "block" indicates each column vector $\boldsymbol{\theta}_{(j)}$ $(j \in \{0, 1, \dots, k\})$ of $\boldsymbol{\Theta}$. We search for the minimizer $\hat{\boldsymbol{\Theta}}_{\lambda,\alpha}$ by repeatedly minimizing the objective function with respect to $\boldsymbol{\theta}_{(j)}$. Since the penalty term consisting of the second and third terms is already expanded about $\boldsymbol{\theta}_{(j)}$, we give a decomposition of the weighted residual sum of squares $r_\ell(\boldsymbol{\theta}_\ell)$ in the first term with respect to $\boldsymbol{\theta}_{(j)}$.

We write the column vectors of $\boldsymbol{Z}_\ell$ as $\boldsymbol{Z}_\ell = (\boldsymbol{z}_{\ell,(0)}, \boldsymbol{z}_{\ell,(1)}, \dots, \boldsymbol{z}_{\ell,(k)})$. Then, $r_\ell(\boldsymbol{\theta}_\ell)$ is rewritten with respect to $\theta_{\ell j}$ $(j \in \{0, 1, \dots, k\})$ as

$$r_\ell(\boldsymbol{\theta}_\ell) = \frac{1}{2}\|\boldsymbol{u}_{\ell,j} - \boldsymbol{z}_{\ell,(j)}\theta_{\ell j}\|^2 = \frac{1}{2}\|\boldsymbol{z}_{\ell,(j)}\|^2\theta_{\ell j}^2 - c_{\ell j}\theta_{\ell j} + \frac{1}{2}\|\boldsymbol{u}_{\ell,j}\|^2,$$

where $\boldsymbol{u}_{\ell,j} = \boldsymbol{u}_\ell - \sum_{i \neq j}^{k} \boldsymbol{z}_{\ell,(i)}\theta_{\ell i}$ and $c_{\ell j} = \boldsymbol{u}'_{\ell,j}\boldsymbol{z}_{\ell,(j)}$. Hence, the first term in the objective function is decomposed with respect to $\boldsymbol{\theta}_{(j)}$ as

$$\sum_{\ell=1}^{m} r_\ell(\boldsymbol{\theta}_\ell) = \frac{1}{2}\boldsymbol{\theta}'_{(j)}\boldsymbol{D}_j\boldsymbol{\theta}_{(j)} - \boldsymbol{c}'_{(j)}\boldsymbol{\theta}_{(j)} + \frac{1}{2}\sum_{\ell=1}^{m}\|\boldsymbol{u}_{\ell,j}\|^2,$$

where $\boldsymbol{D}_j = \mathrm{diag}(\|\boldsymbol{z}_{1,(j)}\|^2, \ldots, \|\boldsymbol{z}_{m,(j)}\|^2)$ and $\boldsymbol{c}_{(j)} = (c_{1j}, \ldots, c_{mj})'$. Ignoring terms, which do not depend on $\boldsymbol{\theta}_{(j)}$, gives the block-wise objective function as

$$f_j(\boldsymbol{\theta}_{(j)} \mid \lambda, \alpha) = \frac{1}{2}\boldsymbol{\theta}'_{(j)}\boldsymbol{D}_j\boldsymbol{\theta}_{(j)} - \boldsymbol{c}'_{(j)}\boldsymbol{\theta}_{(j)} + p_j(\boldsymbol{\theta}_{(j)} \mid \lambda, \alpha), \tag{4}$$

$$p_j(\boldsymbol{\theta}_{(j)} \mid \lambda, \alpha) = \begin{cases} 0 & (j = 0) \\ \lambda \left[\alpha \sum_{\ell=1}^{m} v_{1,\ell j}|\theta_{\ell j}| + (1-\alpha) v_{2,j} \|\boldsymbol{\theta}_{(j)}\| \right] & (j = 1, \ldots, k) \end{cases}.$$

BCDA searches for the minimizer $\hat{\boldsymbol{\Theta}}_{\lambda,\alpha}$ by repeatedly minimizing $f_j(\boldsymbol{\theta}_{(j)} \mid \lambda, \alpha)$ for $j = 0, 1, \ldots, k$ until solution convergence.

2.2 Solution Update and Main Algorithm

In the previous section, we described the use of BCDA to minimize the objective function (3) and derived the block-wise objective function (4). Here, we give the solution update of BCDA by minimizing (4).

The block-wise objective function for $j = 0$ does not include the penalty. Hence, the solution update for $j = 0$ is given by

$$\hat{\boldsymbol{\theta}}_{(0)} = \boldsymbol{D}_0^{-1}\boldsymbol{c}_{(0)}. \tag{5}$$

Since the block-wise objective function for $j \in \{1, \ldots, k\}$ includes the penalty, direct minimization is difficult. However, $f_j(\boldsymbol{\theta}_{(j)} \mid \lambda, \alpha)$ is essentially equivalent to the block-wise objective function for ordinary sparse group Lasso. Hence, we can minimize $f_j(\boldsymbol{\theta}_{(j)} \mid \lambda, \alpha)$ via ITA in a similar way to [14]. We use the following notation:

$$\boldsymbol{g}_j(\boldsymbol{\theta}_{(j)}) = \frac{\partial}{\partial \boldsymbol{\theta}_{(j)}} f_j(\boldsymbol{\theta}_{(j)} \mid 0, \alpha) = \boldsymbol{D}_j\boldsymbol{\theta}_{(j)} - \boldsymbol{c}_{(j)}, \quad L_j = \max_{\ell \in \{1,\ldots,m\}} \|\boldsymbol{z}_{\ell,(j)}\|^2,$$

$$\boldsymbol{b}_j(\boldsymbol{\theta}_{(j)}) = \boldsymbol{\theta}_{(j)} - \boldsymbol{g}(\boldsymbol{\theta}_{(j)})/L_j, \quad \boldsymbol{v}_{1,j} = (v_{1,1j}, \ldots, v_{1,mj})'.$$

With the current solution $\boldsymbol{\theta}^{\mathrm{c}}_{(j)}$ for $\boldsymbol{\theta}_{(j)}$, the solution for $j \in \{1, \ldots, k\}$ is updated by

$$\hat{\boldsymbol{\theta}}_{(j)} = \left(1 - \frac{\lambda(1-\alpha)v_{2,j}}{L_j\|\boldsymbol{s}(\boldsymbol{\theta}^{\mathrm{c}}_{(j)})\|}\right)_+ \boldsymbol{s}(\boldsymbol{\theta}^{\mathrm{c}}_{(j)}), \ \boldsymbol{s}(\boldsymbol{\theta}_{(j)}) = \mathbf{soft}\left(\boldsymbol{b}_j(\boldsymbol{\theta}_{(j)}), \frac{\lambda\alpha}{L_j}\boldsymbol{v}_{1,j}\right), \tag{6}$$

where $(x)_+ = \max\{x, 0\}$ and $\mathbf{soft}(\boldsymbol{x}, \boldsymbol{a})$ is a vector element-wise-applying soft-thresholding operator [5] defined by $\mathrm{soft}(x, a) = \mathrm{sign}(x)\ (|x| - a)_+$. If $L_j\|\boldsymbol{s}(\boldsymbol{\theta}^{\mathrm{c}}_{(j)})\| \leq \lambda(1-\alpha)v_{2j}$, $\boldsymbol{\theta}_{(j)}$ is updated as $\mathbf{0}_m$. If not, whether each element of $\boldsymbol{\theta}_{(j)}$ is zero is determined by the soft-thresholding operator.

From the results, the algorithm for solving the GWSGL optimization problem is summarized in Algorithm 1. After the algorithm terminated, if $\hat{\boldsymbol{\theta}}_{(j)} = \mathbf{0}_m$, the jth explanatory variable is not globally selected. If not, the jth explanatory variable is globally selected. Furthermore, if the ℓth element of $\hat{\boldsymbol{\theta}}_{(j)}$ is zero, the jth explanatory variable is not locally selected at the ℓth observed point.

Algorithm 1. Main algorithm to minimize (3)

Require: λ, α, and initial matrix for $\boldsymbol{\Theta}$
  repeat
    update $\boldsymbol{\theta}_{(0)}$ by (5)
    for $j = 1, \ldots, k$ **do**
      update $\boldsymbol{\theta}_{(j)}$ by (6)
    end for
  until solution converges

3 Real Data Example

3.1 Data and Method

In this section, we present an illustrative application of GWSGL to an actual dataset. The dataset consists of studio apartment rental data for Tokyo's 23 wards as reported by Tokyo Kantei Co., Ltd. [17] between April 2014 and April 2015. This is the same data used in [11]. The sample size is n = 61,999; the number of unique observed data points is m = 25,516. The data items are listed in Table 1. For the GWR model where the response variable is Rent and the explanatory variables are otherwise, we estimate the varying coefficients for each observed point and select the explanatory variables locally and globally by applying GWSGL. Note that although the sample size is 61,999, the varying coefficients are estimated for only 25,516 points.

Table 1. Data items

Rent	Monthly rent for an apartment (yen)
Area	Floor area of an apartment (m^2)
Age	Building age (years)
Floor	Interaction of logarithmic transformations of the top floor and a room floor
Walk.	Walking time (min) to the nearest station
Park.	With a parking lot or not
Condm.	Condominium or not
Corner	Corner apartment or not
Fixed	Fixed-term tenancy agreement or not
South	Facing south or not
R.C.	Reinforced concrete or not

To execute GWSGL, the geographical weight $w_{\ell i}$, the two tuning parameters λ and α, and penalty weights $v_{1,\ell j}$ and $v_{2,j}$ must be determined. We define $w_{\ell i}$ by the Gaussian kernel as $w_{\ell i} = \kappa(\|\boldsymbol{s}_\ell^* - \boldsymbol{s}_i\| \mid \delta)$ and $\kappa(d \mid \delta) = \exp\{-(d/\delta)^2\}^{1/2}$, where $\delta > 0$ is a bandwidth. In this application, the value of δ is optimized under the ordinary GWR in advance, and GWSGL is implemented for the optimal

δ afterwards. The δ value is typically selected based on minimizing a model selection criterion; CV and AIC$_c$ [9] are available in R package spgwr. Here, we select δ by using AIC$_c$ as

$$\hat{\delta} = \arg\min_{\delta>0} \mathrm{AIC}_c(\delta, 0, \alpha),$$

$$\mathrm{AIC}_c(\delta, \lambda, \alpha) = \log \frac{1}{n}\sum_{i=1}^{n}\{y_i - \boldsymbol{x}_i'\hat{\boldsymbol{\theta}}(\boldsymbol{s}_i \mid \delta, \lambda, \alpha)\}^2 + \frac{2\{\mathrm{df}(\delta, \lambda, \alpha) + 1\}}{n - \mathrm{df}(\delta, \lambda, \alpha) - 2},$$

where $\hat{\boldsymbol{\theta}}(\boldsymbol{s} \mid \delta, \lambda, \alpha) = (\hat{\theta}_0(\boldsymbol{s} \mid \delta, \lambda, \alpha), \hat{\theta}_1(\boldsymbol{s} \mid \delta, \lambda, \alpha), \ldots, \hat{\theta}_k(\boldsymbol{s} \mid \delta, \lambda, \alpha))'$ is the GWSGL estimator for varying coefficients at an observed point $\boldsymbol{s}$, and $\mathrm{df}(\delta, \lambda, \alpha)$ is the degrees of freedom in the model. When $\lambda = 0$, $\hat{\boldsymbol{\theta}}(\boldsymbol{s} \mid \delta, \lambda, \alpha)$ coincides with the ordinary GWR estimator (2). With the local active set $\mathcal{A}_i = \{j \in \{1, \ldots, k\} \mid \hat{\theta}_j(\boldsymbol{s}_i \mid \delta, \lambda, \alpha) \neq 0\}$, $\mathrm{df}(\delta, \lambda, \alpha)$ is defined by

$$\mathrm{df}(\delta, \lambda, \alpha) = \sum_{i=1}^{n} \boldsymbol{x}_{\mathcal{A}_i,i}'(\boldsymbol{X}_{\mathcal{A}_i}'\boldsymbol{W}_{\ell_i}\boldsymbol{X}_{\mathcal{A}_i})^{-1}\boldsymbol{X}_{\mathcal{A}_i}'\boldsymbol{W}_{\ell_i}\boldsymbol{e}_i, \ \ell_i \in \{1, \ldots, m\} \ s.t. \ \boldsymbol{s}_i = \boldsymbol{s}_{\ell_i}^*,$$

where $\boldsymbol{X}_{\mathcal{A}_i}$ is a matrix consisting of $\boldsymbol{1}_n$ and $\boldsymbol{x}_{(j)}$ $(j \in \mathcal{A}_i)$ from the column vectors of $\boldsymbol{X} = (\boldsymbol{1}_n, \boldsymbol{x}_{(1)}, \ldots, \boldsymbol{x}_{(k)})$, $\boldsymbol{x}_{\mathcal{A}_i,l}$ is the lth row vector of $\boldsymbol{X}_{\mathcal{A}_i}$ $(l \in \{1, \ldots, n\})$, $\boldsymbol{e}_i$ is an n-dimensional unit vector of which the first element is one, and $\boldsymbol{1}_n$ is an n-dimensional vector of ones. In addition to AIC$_c$, various model selection criteria have been proposed, e.g., the extended GCV (EGCV) criterion [12] defined by

$$\mathrm{EGCV}(\delta, \lambda, \alpha \mid \gamma) = \frac{\sum_{i=1}^{n}\{y_i - \boldsymbol{x}_i'\hat{\boldsymbol{\theta}}(\boldsymbol{s}_i \mid \delta, \lambda, \alpha)\}^2}{\{1 - \mathrm{df}(\delta, \lambda, \alpha)/n\}^{\gamma}},$$

where γ is a non-negative parameter adjusting the strength of the penalty. The EGCV criterion coincides with the GCV criterion [4] when $\gamma = 2$. Regarding the two tuning parameters, Algorithm 1 is executed for each candidate pair of λ and α, and the optimal pair is selected based on minimizing AIC$_c$ as

$$(\hat{\lambda}, \hat{\alpha}) = \arg\min_{\lambda \geq 0, \ \alpha \in [0,1]} \mathrm{AIC}_c(\hat{\delta}, \lambda, \alpha).$$

The penalty weights are defined as adaptive-type with the ordinary GWR estimator.

3.2 Results

The application results are given below. For comparison, we also show the results of the ordinary GWR along with the results of GWSGL tuned by the EGCV criterion with $\gamma = \log n$. We denote GWSGL tuned by AIC$_c$ as GWSGL1 and GWSGL tuned by the EGCV criterion as GWSGL2.

Table 2 provides a summary of the application results in which R^2 represents the coefficient of determination and MER represents the median error rate. Note that AIC$_c$ and the EGCV criterion selected the same bandwidth. Since the $\hat{\alpha}$

Table 2. Summary

	$\hat{\delta}$	$\hat{\lambda}$	$\hat{\alpha}$	R^2	MER (%)	runtime (min.)
GWR	0.014	–	–	0.8728	6.21	66.0
GWSGL1	0.014	1,109,026,762	0.9	0.8727	6.23	143.1
GWSGL2	0.014	8,308,319,375	0.9	0.8717	6.28	144.0

values are both 0.9, we found that the weight for the local variable selection is very large. Although the GWR fit is very good from the values of R^2 and MER, the GWSGL fits are both comparable to the GWR fit despite the fact that their estimates have shrunk. In particular, we can see that $\hat{\lambda}$ for GWSGL2 is much larger than is the case for GWSGL1. This indicates that the estimates of GWSGL2 have more shrunk. Nevertheless, the GWSGL2 fit is not inferior.

Table 3 shows the number of active observed points for each explanatory variable (a variable is considered active when the corresponding estimate is not zero). For both GWSGL1 and 2, since the numbers for Area and Age are equal to the number of observed points m, these variables are active for all observed points. Park. of GWSGL1 is active for 7,861 points, meaning that Park. is not selected for 17,655 points. Similarly, we can obtain the local variable selection results from the number of active points. For all variables, the values for GWSGL2 are less than or equal to that for GWSGL1. Moreover, regarding the global variable selection here, while the values for GWSGL1 are all non-zero, only the value for South for GWSGL2 is zero. This means that GWSGL1 indicates that all variables are globally active, while GWSGL2 indicates that South is globally non-active. These differences are due to the fact that the amount of shrinkage with the EGCV criterion is larger than that for AIC_c.

Table 3. Number of local active variables

	Area	Age	Floor	Walk.	Park.	Condm.	Corner	Fixed	South	R.C.
GWSGL1	25,516	25,516	25,514	25,412	7,861	18,001	18,197	16,225	12,174	23,002
GWSGL2	25,516	25,516	25,254	23,602	3,518	8,304	11,603	8,511	0	21,082

Figures 1 and 2 give estimates of Area and R.C. for each observed point by color; a blue point indicates that the estimate is zero. Since Area is active for all points for both GWSGL1 and 2, there are no blue points. We can see that the estimates of GWSGL1 and 2 have the same trends as the estimates of GWR. By comparison, since R.C. has non-active points, the figure shows many blue points. In particular, the number of non-active points for GWSGL2 is larger than that for GWSGL1, and the blue area for GWSGL2 looks as though it extends from the blue area for GWSGL1.

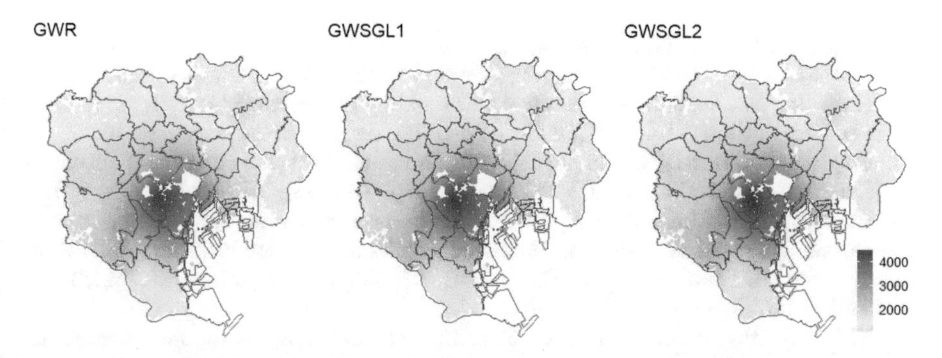

Fig. 1. GWR and GWSGL estimates for Area

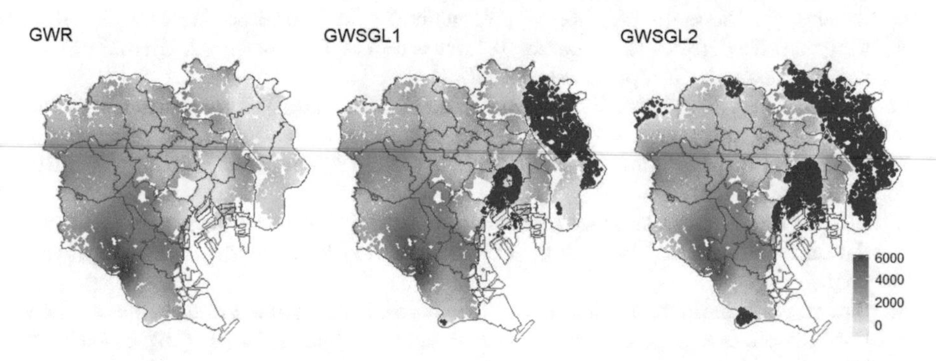

Fig. 2. GWR and GWSGL estimates for R.C.

Acknowledgments. This work was partially supported by JSPS KAKENHI Grant Numbers JP20H04151 and JP21K13834. The authors thank the associate editor and the two reviewers for their valuable comments.

References

1. Bivand, R., Yu, D.: spgwr: Geographically Weighted Regression (2022). https://CRAN.R-project.org/package=spgwr. R package version 0.6-35
2. Boyd, S., Parikh, N., Chu, E., Peleato, B., Eckstein, J.: Distributed optimization and statistical learning via the alternating direction method of multipliers. Found. Trends Mach. Learn. **3**, 1–122 (2011). https://doi.org/10.1561/2200000016
3. Brunsdon, C., Fotheringham, S., Charlton, M.: Geographically weighted regression: a method for exploring spatial nonstationarity. Geogr. Anal. **28**, 281–298 (1996). https://doi.org/10.1111/j.1538-4632.1996.tb00936.x
4. Craven, P., Wahba, G.: Smoothing noisy data with spline functions: estimating the correct degree of smoothing by the method of generalized cross-validation. Numer. Math. **31**, 377–403 (1979). https://doi.org/10.1007/BF01404567
5. Donoho, D.L., Johnstone, I.M.: Ideal spatial adaptation by wavelet shrinkage. Biometrika **81**, 425–455 (1994). https://doi.org/10.1093/biomet/81.3.425

6. Efron, B., Hastie, T., Johnstone, I., Tibshirani, R.: Least angle regression. Ann. Statist. **32**, 407–499 (2004). https://doi.org/10.1214/009053604000000067
7. Fornasier, M., Rauhut, H.: Iterative thresholding algorithms. Appl. Comput. Harmon. Anal. **25**(2), 187–208 (2008). https://doi.org/10.1016/j.acha.2007.10.005
8. Friedman, J., Hastie, T., Höfling, H., Tibshirani, R.: Pathwise coordinate optimization. Ann. Appl. Stat. **1**, 302–332 (2007). https://doi.org/10.1214/07-AOAS131
9. Hurvich, C.M., Simonoff, J.S., Tsai, C.-L.: Smoothing parameter selection in nonparametric regression using an improved Akaike information criterion. J. R. Stat. Soc. Ser. B. Stat. Methodol. **60**, 271–293 (1998). https://doi.org/10.1111/1467-9868.00125
10. Nishii, R., Bai, Z.D., Krishnaiah, P.R.: Strong consistency of the information criterion for model selection in multivariate analysis. Hiroshima Math. J. 18, 451–462 (1988). https://doi.org/10.32917/hmj/1206129611
11. Ohishi, M., Okamura, K., Itoh, Y., Yanagihara, H.: Coordinate descent algorithm for generalized group fused Lasso. Technical report TR-No. 21–02, Hiroshima Statistical Research Group, Hiroshima (2021)
12. Ohishi, M., Yanagihara, H., Fujikoshi, Y.: A fast algorithm for optimizing ridge parameters in a generalized ridge regression by minimizing a model selection criterion. J. Statist. Plann. Inference **204**, 187–205 (2020). https://doi.org/10.1016/j.jspi.2019.04.010
13. R Core Team: R: A Language and Environment for Statistical Computing. R Foundation for Statistical Computing, Vienna, Austria (2022). https://www.R-project.org/
14. Simon, N., Friedman, J., Hastie, T., Tibshirani, R.: A sparse-group Lasso. J. Comput. Graph. Statist. **22**, 231–245 (2013). https://doi.org/10.1080/10618600.2012.681250
15. Tibshirani, R.: Regression shrinkage and selection via the Lasso. J. R. Stat. Soc. Ser. B Stat Methodol. **58**, 267–288 (1996). https://doi.org/10.1111/j.2517-6161.1996.tb02080.x
16. Tibshirani, R., Saunders, M., Rosset, S., Zhu, J., Knight, K.: Sparsity and smoothness via the fused Lasso. J. R. Stat. Soc. Ser. B Stat Methodol. **67**, 91–108 (2005). https://doi.org/10.1111/j.1467-9868.2005.00490.x
17. Tokyo Kantei Co., Ltd. https://www.kantei.ne.jp
18. Wheeler, D.: Simultaneous coefficient penalization and model selection in geographically weighted regression: the geographically weighted Lasso. Environ. Plann. A **41**, 722–742 (2009). https://doi.org/10.1068/a40256
19. Yuan, M., Lin, Y.: Model selection and estimation in regression with grouped variables. J. R. Stat. Soc. Ser. B Stat Methodol. **68**, 49–67 (2006). https://doi.org/10.1111/j.1467-9868.2005.00532.x
20. Zhao, L.C., Krishnaiah, P.R., Bai, Z.D.: On detection of the number of signals in presence of white noise. J. Multivar. Anal. **20**, 1–25 (1986). https://doi.org/10.1016/0047-259X(86)90017-5
21. Zou, H.: The adaptive Lasso and its oracle properties. J. Am. Stat. Assoc. **101**, 1418–1429 (2006). https://doi.org/10.1198/016214506000000735
22. Zou, H., Hastie, T.: Regularization and variable selection via the elastic net. J. R. Stat. Soc. Ser. B Stat Methodol. **67**, 301–320 (2005). https://doi.org/10.1111/j.1467-9868.2005.00503.x

Kick-One-Out-Based Variable Selection Method Using Ridge-Type C_p Criterion in High-Dimensional Multi-response Linear Regression Models

Ryoya Oda(✉)

Graduate School of Advanced Science and Engineering, Hiroshima University, 1-3-1 Kagamiyama, Higashi-Hiroshima, Hiroshima, Japan
ryoya-oda@hiroshima-u.ac.jp

Abstract. In this paper, the kick-one-out method using a ridge-type C_p criterion is proposed for variable selection in multi-response linear regression models. Sufficient conditions for the consistency of this method are obtained under a high-dimensional asymptotic framework such that the number of explanatory variables and response variables, k and p, may go to infinity with the sample size n, and p may exceed n but k is less than n. It is expected that the method satisfying these sufficient conditions has a high probability of selecting the true model, even when $p > n$.

Keywords: consistency · high-dimension · multi-response linear regression · variable selection

1 Introduction

The multi-response linear regression model is one of the basic models in statistical analysis and appears in many textbooks on statistics (e.g., [5,11]). The model can be described as follows. Suppose $\boldsymbol{Y}$ denotes an $n \times p$ response matrix and $\boldsymbol{X}$ denotes an $n \times k$ explanatory matrix with $\text{rank}(\boldsymbol{X}) = k$, where n is the sample size, and p and k are the numbers of response variables and explanatory variables, respectively. If we let j denote a subset of $\omega = \{1, \ldots, k\}$ containing k_j elements and $\boldsymbol{X}_j$ denote an $n \times k_j$ matrix consisting of $\boldsymbol{X}$ columns indexed by elements of j, where k_A denotes the number of elements in a set A, i.e., $k_A = \#(A)$, then the multi-response linear regression model with $\boldsymbol{Y}$ and $\boldsymbol{X}_j$ can be expressed as

$$j: \ \boldsymbol{Y} = \boldsymbol{X}_j \boldsymbol{\Theta}_j + \boldsymbol{\mathcal{E}}, \tag{1}$$

where $\boldsymbol{\Theta}_j$ is a $k_j \times p$ unknown matrix of regression coefficients and $\boldsymbol{\mathcal{E}}$ is an $n \times p$ error matrix. One of the key variable selection problems is selecting the true explanatory variables affecting the response variables. This is regarded as

I. Czarnowski et al. (Eds.): KESIDT 2023, SIST 352, pp. 193–202, 2023.
https://doi.org/10.1007/978-981-99-2969-6_17

the problem of finding the true model j_* ($\subset \omega$) among candidate models in (1). When the number of candidate models is not too large, the usual method to solve this problem is to select the model $\arg\min_{j\in\mathcal{J}} \mathrm{SC}(j)$, where $\mathcal{J}$ is a set of candidate models and $\mathrm{SC}(\cdot)$ is a variable selection criterion that have consistency. Consistency here means that the equation $P(\hat{j} = j_*) \to 1$ holds under an asymptotic framework. Under this condition, it can be expected that the criterion has consistency and a high probability of selecting the true model j_* under appropriate circumstances for the asymptotic framework being used. For example, [4,14] obtained sufficient conditions for the consistency of the C_p-type criterion when p goes to infinity with n, but p/n is less than 1 and k is fixed.

In this paper, we consider a high-dimensional case such that k and p are large although p is allowed to exceed n. In this case, there are two problems with the usual method of selecting models. When p exceeds n, the C_p-type criterion does not work, because the inverse matrix of the estimator for the covariance matrix used in the criterion does not exist. [6] proposed an AIC-type criterion by estimating the inverse of the covariance matrix based on a Lasso-type penalized likelihood function. [13] proposed a Moore-Penrose-type C_p criterion defined by replacing the inverse matrix of the estimator of the covariance matrix with the Moore-Penrose inverse matrix (e.g., [12]). [7] proposed a ridge-type C_p criterion using a ridge-type estimator of the covariance matrix. Let $\boldsymbol{P}_j = \boldsymbol{X}_j(\boldsymbol{X}_j'\boldsymbol{X}_j)^{-1}\boldsymbol{X}_j'$ and $\boldsymbol{S}_j = (n-k_j)^{-1}\boldsymbol{Y}'(\boldsymbol{I}_n - \boldsymbol{P}_j)\boldsymbol{Y}$ for a subset j ($\subset \omega$). [9] has defined the ridge-type generalized C_p (RGC_p) criterion as follows:

$$RGC_p(j|\alpha,\lambda) = (n-k_j)\mathrm{tr}(\boldsymbol{S}_j\boldsymbol{V}_\lambda^{-1}) + pk_j\alpha,$$

where $\boldsymbol{V}_\lambda = \boldsymbol{S}_\omega + \lambda^{-1}\mathrm{tr}(\boldsymbol{S}_\omega)\boldsymbol{I}_p$, λ is a positive ridge parameter, and α is a positive constant. In [9], the sufficient conditions for consistency of the RGC_p criterion are obtained when n goes to infinity and p may go to infinity and is allowed to exceed n, but k is fixed. On the other hand, it is difficult to calculate variable selection criteria for all candidate models when k is large and the number of candidate models is also large. [8,15] proposed a practicable selection method even when k is large, a method that [1] named the kick-one-out (KOO) method. [1,10] obtained the sufficient conditions for consistency of the KOO method using the generalized C_p criterion when p and k may tend to infinity with n but $n > p$. The contents of the KOO method are summarized in [3].

In order to propose a variable selection method that has consistency when k and p are large although p is allowed to exceed n, we treat the best model by the KOO method using the RGC_p criterion as follows:

$$\hat{j}_{\alpha,\lambda} = \{\ell \in \omega \mid RGC_p(\omega_\ell|\alpha,\lambda) > RGC_p(\omega|\alpha,\lambda)\}, \tag{2}$$

where $\omega_\ell = \omega\backslash\{\ell\}$. The KOO method (2) can be regarded as a kind of variable reduction methods. The aim of this paper is to obtain sufficient conditions for consistency of (2), i.e., $P(\hat{j}_{\alpha,\lambda} = j_*) \to 1$ under the following high-dimensional (HD) asymptotic framework:

$$\mathrm{HD}: n \to \infty,\ p/n \to r_1 \in [0,\infty],\ k/n \to r_2 \in [0,1). \tag{3}$$

Under the HD asymptotic framework, n always goes to infinity, but p, k, and k_{j_*} do not necessarily have to go to infinity; moreover, $r_1 = \infty$ means that p/n goes to ∞. Using the obtained sufficient conditions, we have a consistent variable selection method under the HD asymptotic framework.

The remainder of the paper is organized as follows. In Sect. 2, we establish sufficient conditions for the consistency of (2) and show a consistent variable selection method by the obtained conditions under the HD asymptotic framework. In Sect. 3, we conduct numerical experiments for verification purposes. Technical details are given in the Appendix.

2 Sufficient Conditions for Consistency

We begin by defining our notation and the assumptions necessary to ensure the consistency of (2), i.e., $P(\hat{j}_{\alpha,\lambda} = j_*) \to 1$ under the HD asymptotic framework (3). Let a $p \times p$ non-centrality matrix be denoted by

$$\boldsymbol{\Delta}_\ell = (np)^{-1}\boldsymbol{\Theta}_{j_*}'\boldsymbol{X}_{j_*}'(\boldsymbol{I}_n - \boldsymbol{P}_{\omega_\ell})\boldsymbol{X}_{j_*}\boldsymbol{\Theta}_{j_*}.$$

Note that $\boldsymbol{\Delta}_\ell = \boldsymbol{O}_{p,p}$ when $\ell \notin j_*$. Suppose that the rows of the true error matrix $\boldsymbol{\mathcal{E}}$ are mutually independent error vectors from a distribution of $\boldsymbol{\varepsilon}$ with the expectation $\mathbf{0}_p$ and positive-definite covariance matrix $\boldsymbol{\Sigma}_*$, where $\mathbf{0}_p$ is a p-dimensional vector of zeros. Then, the following assumptions apply:

Assumption A1. $j_* \subset \omega$.

Assumption A2. *There exists* $c > 0$ *such that for all* $\ell \in j_*$, $\mathrm{tr}(\boldsymbol{\Delta}_\ell) > c$.

Assumption A3. $\limsup_{p\to\infty} p^{-1}\mathrm{tr}(\boldsymbol{\Sigma}_*) < \infty$.

Assumption A4. $\limsup_{p\to\infty} \kappa_4/\mathrm{tr}(\boldsymbol{\Sigma}_*)^2 < \infty$, *where* $\kappa_4 = E[||\boldsymbol{\varepsilon}||^4] - \mathrm{tr}(\boldsymbol{\Sigma}_*)^2 - 2\mathrm{tr}(\boldsymbol{\Sigma}_*^2)$.

From the definition of consistency, Assumption A1 is needed. Assumption A2 ensures that the non-centrality matrix does not vanish for $\ell \in j_*$. This assumption is similar to the assumption used by [4,9]. Assumption A3 is a regularity assumption for the true covariance matrix. Assumption A4 restricts a class of distributions of the true error vector; for example, Assumption A4 holds when the distribution of $\boldsymbol{\varepsilon}$ is one of the elliptical distributions and the 4-th moment of $\boldsymbol{\varepsilon}$ exists.

Next, we obtain the conditions for the consistency of (2) under the HD asymptotic framework. Let $\mathcal{D}_\ell = RGC_p(\omega_\ell|\alpha,\lambda) - RGC_p(\omega|\alpha,\lambda)$. The probability $P(\hat{j}_{\alpha,\lambda} = j_*)$ is expressed by

$$\begin{aligned} P(\hat{j}_{\alpha,\lambda} = j_*) &= P\left(\{\cap_{\ell\notin j_*}\{\mathcal{D}_\ell < 0\}\} \cap \{\cap_{\ell\in j_*}\{\mathcal{D}_\ell > 0\}\}\right) \\ &\geq 1 - P\left(\cup_{\ell\notin j_*}\{\mathcal{D}_\ell \geq 0\}\right) - P\left(\cup_{\ell\in j_*}\{\mathcal{D}_\ell \leq 0\}\right). \end{aligned} \quad (4)$$

We then obtain the orders of the two probabilities on the right-hand side of (4) (the proof is given in Appendix 1).

Theorem 1. *Suppose that Assumptions A1, A2, and A3 hold. In addition, we assume that for some constant* τ *satisfying* $0 < \tau < 1$,

$$\lim_{n\to\infty,p/n\to r_1,k/n\to r_2} \lambda^{-1}p\alpha\tau > 1, \quad \lim_{n\to\infty,p/n\to r_1,k/n\to r_2} n^{-1}(1+\lambda^{-1})p\alpha = 0. \quad (5)$$

Then, we have

$$\begin{aligned} &P(\hat{j}_{\alpha,\lambda} = j_*) \\ &\geq 1 - O\left(k\xi^2\mathrm{tr}(\boldsymbol{\Sigma}_*)^{-2}(\lambda^{-1}\tau p\alpha - 1)^{-2}\right) - O\left(\xi^2\mathrm{tr}(\boldsymbol{\Sigma}_*)^{-2}n^{-1}\right) \\ &\quad - O\left(k_{j_*}\xi^2 n^{-2}p^{-2}\right) - O\left(k_{j_*}E[||\boldsymbol{\varepsilon}||^4]n^{-2}p^{-2}\right) - O\left(k_{j_*}\lambda_{\max}(\boldsymbol{\Sigma}_*)^2 n^{-2}p^{-2}\right), \end{aligned} \quad (6)$$

where $\xi^2 = \max\{\kappa_4, \mathrm{tr}(\boldsymbol{\Sigma}_*^2)\}$ *and* $\lambda_{\max}(\boldsymbol{A})$ *is the maximum eigenvalue of a square matrix* $\boldsymbol{A}$.

Since selection method (2) has consistency if the right-hand side of (6) goes to 1, we can directly obtain the conditions for consistency from Theorem 1 by adding Assumption A4.

Theorem 2. *Suppose that Assumptions A1, A2, A3, and A4 hold. Then, the selection method* (2) *has consistency under the HD asymptotic framework if the following conditions are satisfied:*

$$\lim_{n\to\infty,p/n\to r_1,k/n\to r_2} k^{-1/2}\lambda^{-1}p\alpha = \infty, \quad \lim_{n\to\infty,p/n\to r_1,k/n\to r_2} n^{-1}(1+\lambda^{-1})p\alpha = 0. \quad (7)$$

We used Assumption A4 regarding the 4-th moment of the true error vector to derive Theorem 2. We now relax the consistency conditions (7) by replacing Assumption A4 with the following assumption:

Assumption A5. $\limsup_{p\to\infty} E[||\boldsymbol{\varepsilon}||^8]/\mathrm{tr}(\boldsymbol{\Sigma}_*)^4 < \infty$.

Since $(E[||\boldsymbol{\varepsilon}||^4])^2 \leq E[||\boldsymbol{\varepsilon}||^8]$ holds, Assumption A5 is stronger than Assumption A4. However, as with Assumption A4, Assumption A5 holds when the distribution of $\boldsymbol{\varepsilon}$ is one of the elliptical distributions and the 8-th moment of $\boldsymbol{\varepsilon}$ exists. Using Assumption A5, we can derive the result of the relaxed consistency conditions (2) (the proof is given in Appendix 2).

Theorem 3. *Suppose that Assumptions A1, A2, A3, and A5 hold. If equation* (5) *holds, then we have*

$$P(\hat{j}_{\alpha,\lambda} = j_*) \geq 1 - O\left(k\lambda^4 p^{-4}\alpha^{-4}\right) - O\left(n^{-1}\right).$$

Moreover, the selection method (2) *has consistency under the HD asymptotic framework if the following conditions are satisfied:*

$$\lim_{n\to\infty,p/n\to r_1,k/n\to r_2} k^{-1/4}\lambda^{-1}p\alpha = \infty, \quad \lim_{n\to\infty,p/n\to r_1,k/n\to r_2} n^{-1}(1+\lambda^{-1})p\alpha = 0. \quad (8)$$

We observe that conditions (8) are more relaxed than (7) by the amount $k^{1/4}$ since the conditions (8) were obtained by using the higher moment of $\boldsymbol{\varepsilon}$. We can show an example. Let the values of λ and α be given by

$$\lambda = n^{1/2}, \ \alpha = p^{-1}n^{1/2}(k \log n / \log\log p)^{1/4}. \tag{9}$$

From Theorem 3, the method (2) with (9) has consistency when $\log\log p / \log n \to 0$ under the HD asymptotic framework.

3 Numerical Studies

Various numerical results were used to examine the probabilities of selecting the true model j_* by the selection method (2) with (9). Here, we designate the RGC_p criterion with (9) as the RGC_p^* criterion. The probabilities were calculated by Monte Carlo simulations with 10,000 iterations. For comparison, we also considered applying the KOO method to three criteria, HGIC$_i$ $(i = 1, 2, 3)$, in [6] given by

$$\text{HGIC}_i(j) = p + \log\left|(1 - k_j/n)\boldsymbol{D}_{\boldsymbol{S}_j}\right| + \beta_i p k_j, \ \beta_i = n^{-1}(\log p)(\log\log p)^{i/2}$$

where $\boldsymbol{D}_{\boldsymbol{S}_j} = \text{diag}\{(\boldsymbol{S}_j)_{11}, \ldots, (\boldsymbol{S}_j)_{pp}\}$, $(\boldsymbol{A})_{ij}$ is the (i,j)-th element of a matrix $\boldsymbol{A}$ and $\text{diag}\{(\boldsymbol{A})_{11}, \ldots, (\boldsymbol{A})_{pp}\}$ is the diagonal matrix with diagonal elements corresponding to those of a $p \times p$ matrix $\boldsymbol{A}$. The HGIC$_i$ $(i = 1, 2, 3)$ have consistency under the high-dimensional settings in [6]; however, whether the KOO method using the HGIC$_i$ $(i = 1, 2, 3)$ has consistency under our high-dimensional settings is unclear. We set the true model as $j_* = \{1, \ldots, k_{j_*}\}$. The data $\boldsymbol{X}$ and the true parameters were determined as follows:

$$\boldsymbol{X} = (\boldsymbol{1}_n, \tilde{\boldsymbol{X}}), \ \tilde{\boldsymbol{X}} \sim \mathcal{N}_{n\times(k-1)}(\boldsymbol{O}_{n,k-1}, \boldsymbol{\Psi} \otimes \boldsymbol{I}_n),$$
$$\boldsymbol{\Theta}_{j_*} \sim \mathcal{N}_{k_{j_*}\times p}(\boldsymbol{O}_{k_{j_*},p}, \boldsymbol{I}_p \otimes \boldsymbol{I}_{k_{j_*}}), \ \boldsymbol{\Sigma}_* = (1 - 0.8)\boldsymbol{I}_p + 0.8\boldsymbol{1}_p\boldsymbol{1}_p',$$

where $(\boldsymbol{\Psi})_{ab} = (0.5)^{|a-b|}$, the notation $\mathcal{N}_{n\times p}(\boldsymbol{A}, \boldsymbol{B} \otimes \boldsymbol{I}_n)$ expresses the $n \times p$ matrix normal distribution with mean matrix $\boldsymbol{A}$ and covariance matrix $\boldsymbol{B} \otimes \boldsymbol{I}_n$, the notation $\otimes$ denotes the Kronecker product and $\boldsymbol{1}_n$ is an n-dimensional vector of ones. Moreover, we express the true error vector as $\boldsymbol{\varepsilon} = \boldsymbol{\Sigma}_*^{1/2}\boldsymbol{z}$, where $\boldsymbol{z} = (z_1, \ldots, z_p)'$. Let $\boldsymbol{\nu} = (\nu_1, \ldots, \nu_p)' \sim \mathcal{N}_{p\times 1}(\boldsymbol{0}_p, \boldsymbol{I}_p)$. Then, $\boldsymbol{z}$ was generated from the following distributions:

Case 1. Multivariate normal distribution: $\boldsymbol{z} = \boldsymbol{\nu}$.
Case 2. Element-wise independent and standardized log-normal distribution:

$$z_a = (e^{\nu_a} - e^{1/2})/\{e(e-1)\}^{1/2}.$$

Table 1 shows the probabilities of selecting the true model j_* by the KOO methods using the RGC_p^* criterion and HGIC$_i$ $(i = 1, 2, 3)$. From Table 1, we observe that the KOO method using the RGC_p^* criterion appears to have consistency for all the combinations of n, p, k, and k_{j_*}. This is thought to be because the λ and α in the RGC_p^* criterion satisfy the consistency conditions under the HD asymptotic framework. The probabilities by the KOO methods using the HGIC$_i$ $(i = 1, 2, 3)$ seem to be low for some combinations of n, p, k, and k_{j_*}.

Table 1. Probabilities (%) of selecting the true model j_* by the KOO methods using the RGC_p^* criterion and HGIC_i $(i = 1, 2, 3)$.

n	p	k	k_{j_*}	Case 1				Case 2			
				RGC_p^*	HGIC_1	HGIC_2	HGIC_3	RGC_p^*	HGIC_1	HGIC_2	HGIC_3
50	10	10	5	99.41	42.77	37.86	33.44	99.01	42.02	37.05	32.63
100	10	10	5	99.99	49.41	44.19	39.52	99.98	48.21	43.01	38.27
500	10	10	5	100.00	53.49	48.07	42.93	100.00	53.10	48.13	43.10
50	25	10	5	99.88	70.91	75.16	78.81	99.91	69.70	73.70	77.20
100	50	10	5	100.00	87.53	91.94	95.17	100.00	87.17	92.14	95.45
500	250	10	5	100.00	98.54	99.67	99.95	100.00	98.26	99.63	99.96
50	10	25	5	91.84	0.74	0.41	0.24	90.12	0.95	0.55	0.35
100	10	50	5	98.46	0.00	0.00	0.00	97.97	0.00	0.00	0.00
500	10	250	5	99.96	0.00	0.00	0.00	99.98	0.00	0.00	0.00
50	10	25	12	77.80	3.22	2.17	1.49	78.32	3.33	2.48	1.61
100	10	50	25	90.69	0.16	0.08	0.04	88.83	0.20	0.13	0.08
500	10	250	125	99.10	0.00	0.00	0.00	99.12	0.00	0.00	0.00
50	25	25	5	99.47	8.96	11.68	14.88	99.17	8.12	10.71	13.87
100	50	50	5	100.00	3.71	8.29	16.63	100.00	3.43	7.98	15.26
500	250	250	5	100.00	0.10	3.91	28.29	100.00	0.11	3.35	27.56
50	25	25	12	99.16	17.58	21.30	25.52	97.68	16.83	20.50	24.45
100	50	50	25	100.00	12.15	21.33	32.38	100.00	11.59	19.83	30.22
500	250	250	125	100.00	2.15	16.09	49.94	100.00	2.05	15.36	49.84
50	500	10	5	99.77	97.87	97.83	73.80	99.77	97.83	97.93	76.33
100	1000	10	5	100.00	99.44	99.90	100.00	100.00	99.35	99.91	100.00
500	5000	10	5	100.00	99.93	100.00	100.00	100.00	99.95	100.00	100.00
50	500	25	5	99.56	68.06	82.56	65.92	99.70	65.58	80.74	67.88
100	1000	50	5	100.00	55.96	84.77	96.71	99.99	55.20	84.01	96.63
500	5000	250	5	100.00	29.21	82.65	98.98	100.00	28.68	81.96	98.70
50	500	25	12	99.79	74.81	78.82	43.94	99.78	74.27	78.85	47.53
100	1000	50	25	99.99	71.02	90.11	97.76	100.00	69.06	89.66	97.90
500	5000	250	125	100.00	51.82	90.55	99.43	100.00	50.77	90.54	99.33

Acknowledgments. The author would like to thank two reviewers for valuable comments. This work was supported by funding from JSPS KAKENHI grant numbers JP20K14363, JP20H04151, and JP19K21672.

Appendix 1: Proof of Theorem 1

First, we consider the case of $\ell \notin j_*$. Note that $(\boldsymbol{P}_\omega - \boldsymbol{P}_{\omega_\ell})\boldsymbol{X}_{j_*} = \boldsymbol{O}_{n,k_{j_*}}$ holds. Let $\boldsymbol{W} = \boldsymbol{\mathcal{E}}'(\boldsymbol{I}_n - \boldsymbol{P}_\omega)\boldsymbol{\mathcal{E}}$ and $\boldsymbol{V}_{\omega,\omega_\ell} = \boldsymbol{\mathcal{E}}'(\boldsymbol{P}_\omega - \boldsymbol{P}_{\omega_\ell})\boldsymbol{\mathcal{E}}$. Then, the upper bound of $\mathcal{D}_\ell$ can be written as

$$\mathcal{D}_\ell = \mathrm{tr}(\boldsymbol{V}_{\omega,\omega_\ell}\boldsymbol{S}_\lambda^{-1}) - p\alpha \le \lambda(n-k)\frac{\mathrm{tr}(\boldsymbol{V}_{\omega,\omega_\ell})}{\mathrm{tr}(\boldsymbol{W})} - p\alpha. \tag{10}$$

Let E_1 be the event defined by $E_1 = \{(n-k)^{-1}\mathrm{tr}(\boldsymbol{W}) \ge \tau\mathrm{tr}(\boldsymbol{\Sigma}_*)\}$. Using (10) and the event E_1, we have

$$P\left(\cup_{\ell\notin j_*}\{\mathcal{D}_\ell \ge 0\}\right) \le \sum_{\ell\in j_*^c} P\left(\mathrm{tr}(\boldsymbol{V}_{\omega,\omega_\ell}) \ge \lambda^{-1}\tau p\alpha\mathrm{tr}(\boldsymbol{\Sigma}_*)\right) + P(E_1^c). \tag{11}$$

Applying (i) and (iii) of Lemma 1 in [9] to (11), the following equation can be derived:

$$\begin{aligned}&P\left(\cup_{\ell\notin j_*}\{\mathcal{D}_\ell \ge 0\}\right)\\ &\le O\left(k\xi^2\mathrm{tr}(\boldsymbol{\Sigma}_*)^{-2}(\lambda^{-1}\tau p\alpha - 1)^{-2}\right) + O\left(\xi^2\mathrm{tr}(\boldsymbol{\Sigma}_*)^{-2}n^{-1}(\tau-1)^{-2}\right).\end{aligned} \tag{12}$$

Next, we consider the case of $\ell \in j_*$. Since $(\boldsymbol{I}_n - \boldsymbol{P}_\omega)\boldsymbol{X}_{j_*} = \boldsymbol{O}_{n,k_{j_*}}$ holds, notice that

$$\mathrm{tr}\{\boldsymbol{Y}'(\boldsymbol{P}_\omega - \boldsymbol{P}_{\omega_\ell})\boldsymbol{Y}\} = \mathrm{tr}(\boldsymbol{V}_{\omega,\omega_\ell}) + 2\mathrm{tr}(\boldsymbol{U}_{\omega_\ell}) + np\delta_\ell^2,$$

where $\delta_\ell^2 = \mathrm{tr}(\boldsymbol{\Delta}_\ell)$ and $\boldsymbol{U}_{\omega_\ell} = \boldsymbol{\Theta}_{j_*}'\boldsymbol{X}_{j_*}'(\boldsymbol{I}_n - \boldsymbol{P}_{\omega_\ell})\boldsymbol{\mathcal{E}}$. Using this notation, the lower bound of $\mathcal{D}_\ell$ can be written as

$$\mathcal{D}_\ell \ge (1+\lambda^{-1})^{-1}(n-k)\mathrm{tr}(\boldsymbol{W})^{-1}\left\{\mathrm{tr}(\boldsymbol{V}_{\omega,\omega_\ell}) + 2\mathrm{tr}(\boldsymbol{U}_{\omega_\ell}) + np\delta_\ell^2\right\} - p\alpha. \tag{13}$$

Let E_2 and $E_{3,\ell}$ be the events defined by $E_2 = \{(n-k)^{-1}\mathrm{tr}(\boldsymbol{W}) \le 3\mathrm{tr}(\boldsymbol{\Sigma}_*)/2\}$ and $E_{3,\ell} = \{\mathrm{tr}(\boldsymbol{U}_{\omega_\ell}) \ge -np\delta_\ell^2/4\}$. Using this notation and (13), we have

$$\begin{aligned}&P\left(\cup_{\ell\in j_*}\{\mathcal{D}_\ell \le 0\}\right)\\ &\le \sum_{\ell\in j_*} P\left(\mathrm{tr}(\boldsymbol{V}_{\omega,\omega_\ell}) + np\min_{\ell\in j_*}\delta_\ell^2/2 \le 3(1+\lambda^{-1})p\alpha\mathrm{tr}(\boldsymbol{\Sigma}_*)/2\right)\\ &\quad + P(E_2^c) + \sum_{\ell\in j_*} P(E_{3,\ell}^c)\\ &\le \sum_{\ell\in j_*} P\left(\mathrm{tr}(\boldsymbol{V}_{\omega,\omega_\ell}) - \mathrm{tr}(\boldsymbol{\Sigma}_*) + npc/2 \le \mathrm{tr}(\boldsymbol{\Sigma}_*)\left\{3(1+\lambda^{-1})p\alpha/2 - 1\right\}\right)\\ &\quad + P(E_2^c) + \sum_{\ell\in j_*} P(E_{3,\ell}^c).\end{aligned} \tag{14}$$

Applying (i), (ii), and (iv) of Lemma 1 in [9] to (14), the following equation can be derived:

$$\begin{aligned}P\left(\cup_{\ell\in j_*}\{\mathcal{D}_\ell \le 0\}\right) &\le O\left(k_{j_*}\xi^2 n^{-2}p^{-2}\right) + O\left(\xi^2\mathrm{tr}(\boldsymbol{\Sigma}_*)^{-2}n^{-1}\right)\\ &\quad + O\left(k_{j_*}E[||\boldsymbol{\varepsilon}||^4]n^{-2}p^{-2}\right) + O\left(k_{j_*}\lambda_{\max}(\boldsymbol{\Sigma}_*)^2 n^{-2}p^{-2}\right).\end{aligned} \tag{15}$$

Therefore, (12) and (15) complete the proof of Theorem 1. □

Appendix 2: Proof of Theorem 2

To prove Theorem 2, we need the following lemma concerning the 8-th moment of ε (the proof is given in Appendix 3):

Lemma 1. *Let* $\boldsymbol{A}$ *be an* $n \times n$ *symmetric idempotent matrix satisfying* $\mathrm{rank}(\boldsymbol{A}) = m < n$. *Then,* $E[\mathrm{tr}(\boldsymbol{\mathcal{E}}'\boldsymbol{A}\boldsymbol{\mathcal{E}})^4] \le \phi m^3 E[||\varepsilon||^8]$ *holds, where* ϕ *is a constant not depending on* n, p, *and* m.

Using Lemma 1, for $\ell \notin j_*$ we have

$$\begin{aligned} P\left(\mathrm{tr}(\boldsymbol{V}_{\omega,\omega_\ell}) \ge \lambda^{-1}\tau p\alpha \mathrm{tr}(\boldsymbol{\Sigma}_*)\right) &\le E[\mathrm{tr}(\boldsymbol{V}_{\omega,\omega_\ell})^4]/\{\lambda^{-1}\tau p\alpha \mathrm{tr}(\boldsymbol{\Sigma}_*)\}^4 \\ &\le \phi E[||\varepsilon||^8]/\{\lambda^{-1}\tau p\alpha \mathrm{tr}(\boldsymbol{\Sigma}_*)\}^4 \\ &= O\left(\lambda^4 p^{-4}\alpha^{-4}\right). \end{aligned} \tag{16}$$

Notice that $\kappa_4 \le E[||\varepsilon||^4] \le E[||\varepsilon||^8]^{1/2}$ holds. Therefore, (11), (12), (15), and (16) complete the proof of Theorem 2. □

Appendix 3: Proof of Lemma 1

Denote the i-th row vector of $\boldsymbol{\mathcal{E}}$ as ε_i. The expectation $E[\mathrm{tr}(\boldsymbol{\mathcal{E}}'\boldsymbol{A}\boldsymbol{\mathcal{E}})^4]$ can be expressed as follows:

$$\begin{aligned} &E[\mathrm{tr}(\boldsymbol{\mathcal{E}}'\boldsymbol{A}\boldsymbol{\mathcal{E}})^4] \\ &= \sum_{i=1}^n \{(\boldsymbol{A})_{ii}\}^4 E[||\varepsilon_i||^8] + \phi_1 \sum_{i\ne j}^n \{(\boldsymbol{A})_{ii}\}^3 (\boldsymbol{A})_{jj} E[||\varepsilon_i||^6 ||\varepsilon_j||^2] \\ &\quad + \phi_2 \sum_{i\ne j}^n \{(\boldsymbol{A})_{ii}\}^2 \{(\boldsymbol{A})_{ij}\}^2 E[||\varepsilon_i||^4 (\varepsilon_i'\varepsilon_j)^2] \\ &\quad + \phi_3 \sum_{i\ne j}^n \{(\boldsymbol{A})_{ii}\}^2 \{(\boldsymbol{A})_{jj}\}^2 E[||\varepsilon_i||^4 ||\varepsilon_j||^4] + \phi_4 \sum_{i\ne j}^n \{(\boldsymbol{A})_{ij}\}^4 E[(\varepsilon_i'\varepsilon_j)^4] \\ &\quad + \phi_5 \sum_{i\ne j}^n (\boldsymbol{A})_{ii} (\boldsymbol{A})_{jj} \{(\boldsymbol{A})_{ij}\}^2 E[||\varepsilon_i||^2 ||\varepsilon_j||^2 (\varepsilon_i'\varepsilon_j)^2] \\ &\quad + \phi_6 \sum_{i\ne j\ne k}^n (\boldsymbol{A})_{ii} (\boldsymbol{A})_{jj} (\boldsymbol{A})_{kk} (\boldsymbol{A})_{ij} E[||\varepsilon_i||^2 ||\varepsilon_j||^2 ||\varepsilon_k||^2 (\varepsilon_i'\varepsilon_j)] \\ &\quad + \phi_7 \sum_{i\ne j\ne k}^n (\boldsymbol{A})_{ii} \{(\boldsymbol{A})_{jk}\}^3 E[||\varepsilon_i||^2 (\varepsilon_j'\varepsilon_k)^3] \\ &\quad + \phi_8 \sum_{i\ne j\ne k}^n (\boldsymbol{A})_{ii} (\boldsymbol{A})_{jj} (\boldsymbol{A})_{ik} (\boldsymbol{A})_{jk} E[||\varepsilon_i||^2 ||\varepsilon_j||^2 (\varepsilon_i'\varepsilon_k)(\varepsilon_j'\varepsilon_k)] \\ &\quad + \phi_9 \sum_{i\ne j\ne k}^n \{(\boldsymbol{A})_{ij}\}^2 (\boldsymbol{A})_{jk} (\boldsymbol{A})_{ki} E[(\varepsilon_i'\varepsilon_j)^2 (\varepsilon_j'\varepsilon_k)(\varepsilon_k'\varepsilon_i)] \end{aligned}$$

$$+ \phi_{10} \sum_{i \neq j \neq k}^{n} (\boldsymbol{A})_{ii}(\boldsymbol{A})_{ij}\{(\boldsymbol{A})_{jk}\}^2 E[||\boldsymbol{\varepsilon}_i||^2(\boldsymbol{\varepsilon}_i'\boldsymbol{\varepsilon}_j)(\boldsymbol{\varepsilon}_j'\boldsymbol{\varepsilon}_k)^2],$$

where the summation $\sum_{i \neq j \neq k}^{n}$ is defined by $\sum_{i \neq j}^{n} \sum_{k:k \neq i,j}^{n}$ and $\phi_1, \ldots, \phi_{10}$ are natural numbers not depending on n, p, and m. Since $\boldsymbol{A}$ is positive semi-definite and is also symmetric idempotent, $0 \leq (\boldsymbol{A})_{ii} \leq 1$, $\{(\boldsymbol{A})_{ij}\}^2 \leq (\boldsymbol{A})_{ii}(\boldsymbol{A})_{jj}$ (e.g., [2], Fact 8.9.9), $\boldsymbol{1}_n'\boldsymbol{D_A}\boldsymbol{A}\boldsymbol{D_A}\boldsymbol{1}_n \leq \mathrm{tr}(\boldsymbol{A})$ and $\boldsymbol{1}_n'\boldsymbol{D_A}^2\boldsymbol{A}\boldsymbol{D_A}\boldsymbol{1}_n \leq \mathrm{tr}(\boldsymbol{A})$ hold. Recalling $\boldsymbol{D_A} = \mathrm{diag}\{(\boldsymbol{A})_{11}, \ldots, (\boldsymbol{A})_{nn}\}$ and using these facts, we have

$$\sum_{i=1}^{n}\{(\boldsymbol{A})_{ii}\}^4 \leq m, \ \sum_{i \neq j}^{n}\{(\boldsymbol{A})_{ii}\}^3(\boldsymbol{A})_{jj} \leq m^2, \ \sum_{i \neq j}^{n}\{(\boldsymbol{A})_{ii}\}^2\{(\boldsymbol{A})_{ij}\}^2 \leq m^2,$$

$$\sum_{i \neq j}^{n}\{(\boldsymbol{A})_{ii}\}^2\{(\boldsymbol{A})_{jj}\}^2 \leq m^2, \ \sum_{i \neq j}^{n}\{(\boldsymbol{A})_{ij}\}^4 \leq m^2, \ \sum_{i \neq j}^{n}(\boldsymbol{A})_{ii}(\boldsymbol{A})_{jj}\{(\boldsymbol{A})_{ij}\}^2 \leq m^2,$$

$$\sum_{i \neq j \neq k}^{n} (\boldsymbol{A})_{ii}(\boldsymbol{A})_{jj}(\boldsymbol{A})_{kk}(\boldsymbol{A})_{ij} \leq m(m+4), \ \sum_{i \neq j \neq k}^{n} (\boldsymbol{A})_{ii}\{(\boldsymbol{A})_{jk}\}^3 \leq m(m^2+m+1),$$

$$\sum_{i \neq j \neq k}^{n} (\boldsymbol{A})_{ii}(\boldsymbol{A})_{jj}(\boldsymbol{A})_{ik}(\boldsymbol{A})_{jk} \leq 5m, \ \sum_{i \neq j \neq k}^{n} \{(\boldsymbol{A})_{ij}\}^2(\boldsymbol{A})_{jk}(\boldsymbol{A})_{ki} \leq m(3m+2),$$

$$\sum_{i \neq j \neq k}^{n} (\boldsymbol{A})_{ii}(\boldsymbol{A})_{ij}\{(\boldsymbol{A})_{jk}\}^2 \leq m(m+6).$$

Moreover, for any $a, b, c, d, e, f, g, h \in \mathbb{N}$, $E[(\boldsymbol{\varepsilon}_a'\boldsymbol{\varepsilon}_b)(\boldsymbol{\varepsilon}_c'\boldsymbol{\varepsilon}_d)(\boldsymbol{\varepsilon}_e'\boldsymbol{\varepsilon}_f)(\boldsymbol{\varepsilon}_g'\boldsymbol{\varepsilon}_h)] \leq E[||\boldsymbol{\varepsilon}_a||^8]$ holds. Therefore, the proof of Lemma 1 is completed. □

References

1. Bai, Z.D., Fujikoshi, Y., Hu, J.: Strong consistency of the AIC, BIC, C_p and KOO methods in high-dimensional multivariate linear regression. TR No. 18-9, Statistical Research Group, Hiroshima University (2018)
2. Bernstein, D.H.: Matrix Mathematics: Theory, Facts, and Formulas. Princeton University Press, Princeton (2009)
3. Fujikoshi, Y.: High-dimensional consistencies of KOO methods in multivariate regression model and discriminant analysis. J. Multivariate Anal. **188**, 104860 (2022). https://doi.org/10.1016/j.jmva.2021.104860
4. Fujikoshi, Y., Sakurai, T., Yanagihara, H.: Consistency of high-dimensional AIC-type and C_p-type criteria in multivariate linear regression. J. Multivariate Anal. **123**, 184–200 (2014). https://doi.org/10.1016/j.jmva.2013.09.006
5. Fujikoshi, Y., Ulyanov, V.V., Shimizu, R.: Multivariate Statistics: High-Dimensional and Large-Sample Approximations. Wiley, Hoboken (2010)
6. Katayama, S., Imori, S.: Lasso penalized model selection criteria for high-dimensional multivariate linear regression analysis. J. Multivariate Anal. **132**, 138–150 (2014). https://doi.org/10.1016/j.jmva.2014.08.002
7. Kubokawa, T., Srivastava, M.S.: Selection of variables in multivariate regression models for large dimensions. Comm. Statist. A-Theory Methods **41**, 2465–2489 (2012). https://doi.org/10.1080/03610926.2011.624242

8. Nishii, R., Bai, Z.D., Krishnaiah, P.R.: Strong consistency of the information criterion for model selection in multivariate analysis. Hiroshima Math. J. **18**, 451–462 (1988). https://doi.org/10.32917/hmj/1206129611
9. Oda, R.: Consistent variable selection criteria in multivariate linear regression even when dimension exceeds sample size. Hiroshima Math. J. **50**, 339–374 (2020). https://doi.org/10.32917/hmj/1607396493
10. Oda, R., Yanagihara, H.: A fast and consistent variable selection method for high-dimensional multivariate linear regression with a large number of explanatory variables. Electron. J. Statist. **14**, 1386–1412 (2020). https://doi.org/10.1214/20-EJS1701
11. Srivastava, M.S.: Methods of Multivariate Statistics. Wiley, New York (2002)
12. Srivastava, M.S.: Multivariate theory for analyzing high dimensional data. J. Japan Statist. Soc. **37**, 53–86 (2007). https://doi.org/10.14490/jjss.37.53
13. Yamamura, M., Yanagihara, H., Srivastava, M. S.: Variable selection by C_p statistic in multiple responses regression with fewer sample size than the dimension. In: Setchi, R., et al. (eds.) Knowledge-Based and Intelligent Information and Engineering Systems, vol. 6278, pp. 7–14 (2010). https://doi.org/10.1007/978-3-642-15393-8_2
14. Yanagihara, H.: A high-dimensionality-adjusted consistent C_p-type statistic for selecting variables in a normality-assumed linear regression with multiple responses. Procedia Comput. Sci. **96**, 1096–1105 (2016). https://doi.org/10.1016/j.procs.2016.08.151
15. Zhao, L.C., Krishnaiah, P.R., Bai, Z.D.: On detection of the number of signals in presence of white noise. J. Multivariate Anal. **20**, 1–25 (1986). https://doi.org/10.1016/0047-259X(86)90017-5

Estimation Algorithms for MLE of Three-Mode GMANOVA Model with Kronecker Product Covariance Matrix

Keito Horikawa[1(✉)], Isamu Nagai[2], Rei Monden[1], and Hirokazu Yanagihara[3]

[1] Informatics and Data Science Program, Graduate School of Advanced Science and Engineering, Hiroshima University, Higashi-Hiroshima, Hiroshima 739-8521, Japan
m221271@hiroshima-u.ac.jp

[2] Faculty of Liberal Arts and Science, Chukyo University, Nagoya, Aichi 466-8666, Japan

[3] Mathematics Program, Graduate School of Advanced Science and Engineering, Hiroshima University, Higashi-Hiroshima, Hiroshima 739-8526, Japan

Abstract. Currently, there are a lot of measurement data on different items collected over time. The GMANOVA model is appropriate for analyzing the trends in such data, in order to analyze some longitudinal data collected on different items simultaneously, we need to create separate GMANOVA models, but separate GMANOVA models cannot reveal the relationships between items. Also, if we try to apply one model to all of these data, the problem arises that there are too many parameters and the covariance structure is too large. Therefore, we propose an extended GMANOVA model and two estimation algorithms. The proposed model's covariance matrix has a Kronecker structure. Therefore, because the estimator cannot be uniquely determined using the log-likelihood, we propose two parameter estimation algorithms.

Keywords: GMANOVA · Kronecker product covariance structure · Longitudinal data · Three-mode principal component analysis

1 Introduction

In this paper, we consider longitudinal data on each of multiple items measured over time for each explanatory variable, none of which depend on time. Such data are obtained by simultaneously measuring items along with explanatory variables, for example, the heights and weights of individuals measured every year. When the number of items is one, we can estimate the longitudinal trend by using the generalized multivariate analysis of variance (GMANOVA) model proposed by [8] for when measurements are taken with the same timing from all individuals (see, e.g., [9]). When the number of items is one, we consider $\boldsymbol{y}_{i1} = \boldsymbol{X}\boldsymbol{\Xi}_1'\boldsymbol{a}_i + \boldsymbol{\nu}_{i1} (i = 1, \cdots, n)$, where n means the number of individuals,

I. Czarnowski et al. (Eds.): KESIDT 2023, SIST 352, pp. 203–213, 2023.
https://doi.org/10.1007/978-981-99-2969-6_18

$\boldsymbol{y}_{i1}$ is a p-dimensional vector of ith individual's longitudinal data, $\boldsymbol{a}_i$ is a k-dimensional explanatory vector of ith individual, $\boldsymbol{X}$ indicates what functions are used for estimating the longitudinal trend which is $p \times q$ matrix, $\boldsymbol{\Xi}_1$ is a $k \times q$ unknown matrix, and $\boldsymbol{\nu}_{ij}$ is a p-dimensional error vector. Here, $\boldsymbol{\Xi}_1$ are the coefficients for the functions, so estimating the longitudinal trend coincides with estimating this matrix.

When the number of items is more than one, we can express the jth item's model as $\boldsymbol{y}_{ij} = \boldsymbol{X}\boldsymbol{\Xi}_j'\boldsymbol{a}_i + \boldsymbol{\nu}_{ij} (j = 1, \cdots, m)$, where $\boldsymbol{\Xi}_j$ is a $k \times q$ unknown matrix. Note that $\boldsymbol{\Xi}_j$ means that the coefficients for each function are different for each item. Thus, we can estimate the longitudinal trend for each item by using the ordinary GMANOVA model. However, we cannot see the relationships between items in this model since we estimate $\boldsymbol{\Xi}_j$ separately for each item.

In this paper, we consider that the coefficients are different for each item and they show the relation between items. In order to consider these different coefficients, we add an explanatory variable about jth item as $c_{je}(e = 1, \cdots, l)$. Then, we consider that the longitudinal trend for the jth item can be obtained from a weighted sum of the longitudinal trends. Hence, we consider $\boldsymbol{y}_{ij} = \sum_{e=1}^{l} c_{je}\boldsymbol{X}\boldsymbol{\Xi}_e'\boldsymbol{a}_i + \boldsymbol{\nu}_{ij}$, where $\boldsymbol{\Xi}_e$ is a $k \times q$ unknown matrix and l is the number of explanatory variables for the item. Based on this expression, and letting $\boldsymbol{\Theta} = (\boldsymbol{\Xi}_1, \ldots, \boldsymbol{\Xi}_l)$, $\boldsymbol{Y}_j = (\boldsymbol{y}_{1j}, \ldots, \boldsymbol{y}_{nj})'$ and $\boldsymbol{\mathcal{E}}_j = (\boldsymbol{\nu}_{1j}, \ldots, \boldsymbol{\nu}_{nj})$ be an $n \times p$ matrix of n individuals measured p times longitudinally for the jth item, we can rewrite the previous model as the following GMANOVA model:

$$\boldsymbol{Y} = \boldsymbol{A}\boldsymbol{\Theta}(\boldsymbol{C} \otimes \boldsymbol{X})' + \boldsymbol{\mathcal{E}}, \tag{1}$$

where $\boldsymbol{Y} = (\boldsymbol{Y}_1, \ldots, \boldsymbol{Y}_m)$, $\boldsymbol{A}$, $\boldsymbol{C}$, and $\boldsymbol{X}$ are known appropriate matrices, $\boldsymbol{\Theta}$ is an unknown matrix, and $\boldsymbol{\mathcal{E}} = (\boldsymbol{\mathcal{E}}_1, ..., \boldsymbol{\mathcal{E}}_m)$ is an error matrix. Here, $\otimes$ denotes the Kronecker product of matrices (see, e.g., [4, chap. 16]). In particular, the Kronecker product covariance structure (see e.g., [2,5,10]), which will be described later, will be considered for the covariance matrix. Further details of the above representation are given in Sect. 2.

The data dealt with in this paper are also called three-mode data. As the principal component analysis for three-mode data called three-mode principal component analysis, as proposed by [3], we will name the model (1) the three-mode GMANOVA model. In this model, the relationship between items is expressed by adding $\boldsymbol{C}$ representing items to the normal GMANOVA model. In this paper, we try to obtain maximum likelihood estimators (MLEs) of the three-mode GMANOVA model when the matrix normal distribution is assumed for the model (1). Unfortunately, it is not possible to obtain explicit forms for the MLEs, so the purpose of this paper is to propose algorithms for obtaining the MLEs.

In Sect. 2, we summarize our model and assumptions, and show the probability density and log-likelihood functions. We propose two types of estimation methods by using maximum likelihood estimation in Sect. 3. Additionally, we show algorithms based on the proposed estimation methods. In Sect. 4, we compare the proposed estimation algorithms and another algorithm through some numerical experiments. Technical details are in the Appendix 1.

2 Model and Log-Likelihood Function

In this section, we first give the model description, as follows:

$$\boldsymbol{Y} = \begin{pmatrix} \boldsymbol{y}_1' \\ \vdots \\ \boldsymbol{y}_n' \end{pmatrix} = \begin{pmatrix} \boldsymbol{a}_1'\boldsymbol{\Theta}(\boldsymbol{C}\otimes\boldsymbol{X})' + \boldsymbol{\varepsilon}_1' \\ \vdots \\ \boldsymbol{a}_n'\boldsymbol{\Theta}(\boldsymbol{C}\otimes\boldsymbol{X})' + \boldsymbol{\varepsilon}_n' \end{pmatrix} = \boldsymbol{A\Theta}(\boldsymbol{C}\otimes\boldsymbol{X})' + \boldsymbol{\mathcal{E}},$$

where $\boldsymbol{Y}$ is an $n \times mp$ matrix having the structure $\boldsymbol{Y} = (\boldsymbol{Y}_1, \ldots, \boldsymbol{Y}_m)$, in which $\boldsymbol{Y}_j$ is an $n \times p$ matrix collected for the jth item for each individual measured p times, $\boldsymbol{A} = (\boldsymbol{a}_1, \ldots, \boldsymbol{a}_n)'$ is an $n \times k$ between-individual explanatory matrix with $\mathrm{rank}(\boldsymbol{A}) = k\ (< n)$, $\boldsymbol{\Theta}$ is a $k \times lq$ unknown matrix, $\boldsymbol{C}$ is an $m \times l$ between-item explanatory matrix with $\mathrm{rank}(\boldsymbol{C}) = l\ (< m)$, $\boldsymbol{X}$ is a $p \times q$ matrix with $\mathrm{rank}(\boldsymbol{X}) = q\ (< p)$ that represents the function express longitudinal trend, $\boldsymbol{\mathcal{E}} = (\boldsymbol{\varepsilon}_1, \ldots, \boldsymbol{\varepsilon}_n)'$ is an $n \times mp$ error matrix with $\boldsymbol{\varepsilon}_1, \ldots, \boldsymbol{\varepsilon}_n \overset{\text{i.i.d.}}{\sim} \mathcal{N}_{mp}(\boldsymbol{0}_{mp}, \boldsymbol{\Sigma}^*)$, $\boldsymbol{0}_a$ indicates the a-dimensional zero vector, and $\boldsymbol{\Sigma}^*$ is an $mp \times mp$ unknown positive definite matrix. Here, $\mathcal{N}_{mp}(\boldsymbol{x}, \boldsymbol{H})$ means the mp-dimensional normal distribution with expected value $\boldsymbol{x}$ and covariance matrix $\boldsymbol{H}$.

The number of unknown parameters in $\boldsymbol{\Sigma}^*$ becomes very large when the number of items m or the number of measurements p becomes large. In the case of the number of unknown parameters in $\boldsymbol{\Sigma}^*$ being large, we need a large n if we use one of the usual estimation methods. Thus, in this paper, we assume $\boldsymbol{\Sigma}^*$ as $\boldsymbol{\Psi}\otimes\boldsymbol{\Sigma}$, where $\boldsymbol{\Psi}$ and $\boldsymbol{\Sigma}$ are respectively $m \times m$ and $p \times p$ unknown positive definite matrices. Also, we assume that the $(1,1)$th element of $\boldsymbol{\Psi}$ is 1 as in [10] in order to ensure that it is possible to estimate $\boldsymbol{\Psi}$ and $\boldsymbol{\Sigma}$. This assumption is called assuming a Kronecker product covariance structure (see, e.g., [2,5,10]). For an example of applying this structure to data analysis, we refer the reader to [6].

In order to obtain the MLEs for $\boldsymbol{\Theta}$, $\boldsymbol{\Psi}$, and $\boldsymbol{\Sigma}$, we consider the log-likelihood function. First, from the assumptions on $\boldsymbol{\varepsilon}_i$ $(i = 1, \ldots, n)$, we obtain

$$\boldsymbol{y}_i \sim \mathcal{N}_{mp}((\boldsymbol{C}\otimes\boldsymbol{X})\boldsymbol{\Theta}'\boldsymbol{a}_i, \boldsymbol{\Psi}\otimes\boldsymbol{\Sigma}),$$

and $\boldsymbol{y}_1, \ldots, \boldsymbol{y}_n$ are mutually independent. Thus, the probability density function of $\boldsymbol{y}_i$ is

$$f(\boldsymbol{y}_i) = \left(\frac{1}{2\pi}\right)^{mp/2} |\boldsymbol{\Psi}\otimes\boldsymbol{\Sigma}|^{-1/2} \times \exp\left\{-\frac{1}{2}\left(\boldsymbol{y}_i - (\boldsymbol{C}\otimes\boldsymbol{X})\boldsymbol{\Theta}'\boldsymbol{a}_i\right)'\left(\boldsymbol{\Psi}\otimes\boldsymbol{\Sigma}\right)^{-1}\left(\boldsymbol{y}_i - (\boldsymbol{C}\otimes\boldsymbol{X})\boldsymbol{\Theta}'\boldsymbol{a}_i\right)\right\},$$

where $|\boldsymbol{\Psi}\otimes\boldsymbol{\Sigma}|$ means the determinant of $\boldsymbol{\Psi}\otimes\boldsymbol{\Sigma}$. Here, we note that $|\boldsymbol{\Psi}\otimes\boldsymbol{\Sigma}| = |\boldsymbol{\Psi}|^p|\boldsymbol{\Sigma}|^m$. Since $\boldsymbol{y}_1, \cdots, \boldsymbol{y}_n$ are independent, using the above probability density function of $\boldsymbol{y}_i$ $(i = 1, \ldots, n)$, the log-likelihood function for $\boldsymbol{y}_i$ can be expressed as follows:

$$
\begin{aligned}
\ell(\boldsymbol{\Sigma}, \boldsymbol{\Psi}, \boldsymbol{\Theta}) &\stackrel{\text{def.}}{=} \log \prod_{i=1}^{n} f(\boldsymbol{y}_i) \\
&= -\frac{1}{2}\{nmp \log(2\pi) + np \log|\boldsymbol{\Psi}| + nm \log|\boldsymbol{\Sigma}| \\
&\qquad + d(\boldsymbol{Y}, \boldsymbol{A\Theta}(\boldsymbol{C} \otimes \boldsymbol{X})|\boldsymbol{\Psi} \otimes \boldsymbol{\Sigma})\},
\end{aligned} \tag{2}
$$

where $d(\boldsymbol{D}, \boldsymbol{E}|\boldsymbol{F}) \stackrel{\text{def.}}{=} \text{tr}\{\boldsymbol{V}(\boldsymbol{D}, \boldsymbol{E}|\boldsymbol{F})\}$ and $\boldsymbol{V}(\boldsymbol{D}, \boldsymbol{E}|\boldsymbol{F}) \stackrel{\text{def.}}{=} (\boldsymbol{D} - \boldsymbol{E})\boldsymbol{F}^{-1}(\boldsymbol{D} - \boldsymbol{E})'$ with appropriate matrices $\boldsymbol{D}$, $\boldsymbol{E}$, and $\boldsymbol{F}$. By maximizing the log-likelihood function, we can obtain MLEs for $\boldsymbol{\Theta}$, $\boldsymbol{\Psi}$, and $\boldsymbol{\Sigma}$, since maximizing likelihood is the same thing as maximizing log-likelihood that taken the log of likelihood. We are now ready to propose estimation algorithms for the MLEs in the next section.

3 Proposed Algorithms

3.1 Block-Wise Coordinate Descent Algorithm

The log-likelihood function (2) with two of the matrices $\boldsymbol{\Theta}$, $\boldsymbol{\Sigma}$, and $\boldsymbol{\Psi}$ fixed can be expressed as a function of the unfixed matrix, and maximization of this function is easy. This is the same as maximization in the unfixed matrix direction, and for this reason, we will refer to such a maximization method as the block-wise coordinate descent algorithms. In particular, when $q = p$ and $l = m$, our proposed algorithm is equivalent to the "flip-flop" algorithm proposed in [5].

MLE for $\boldsymbol{\Theta}$ given $\boldsymbol{\Sigma}$ and $\boldsymbol{\Psi}$: First, we will calculate the MLE for $\boldsymbol{\Theta}$ given $\boldsymbol{\Psi}$ and $\boldsymbol{\Sigma}$. As in the estimation method of the GMANOVA model, by differentiating the log-likelihood function with respect to $\boldsymbol{\Theta}$, the MLE of $\boldsymbol{\Theta}$ is given by

$$
\hat{\boldsymbol{\Theta}}(\boldsymbol{\Sigma}, \boldsymbol{\Psi}) = (\boldsymbol{A}'\boldsymbol{A})^{-1}\boldsymbol{A}'\boldsymbol{Y}\left\{\boldsymbol{\Psi}^{-1}\boldsymbol{C}(\boldsymbol{C}'\boldsymbol{\Psi}^{-1}\boldsymbol{C})^{-1} \otimes \boldsymbol{\Sigma}^{-1}\boldsymbol{X}(\boldsymbol{X}'\boldsymbol{\Sigma}^{-1}\boldsymbol{X})^{-1}\right\}, \tag{3}
$$

since $(\boldsymbol{\Psi} \otimes \boldsymbol{\Sigma})^{-1} = \boldsymbol{\Psi}^{-1} \otimes \boldsymbol{\Sigma}^{-1}$.

MLE for $\boldsymbol{\Sigma}$ given $\boldsymbol{\Theta}$ and $\boldsymbol{\Psi}$: To obtain the MLE for $\boldsymbol{\Sigma}$, we first obtain

$$
\frac{\partial}{\partial \boldsymbol{\Sigma}}\ell(\boldsymbol{\Sigma}, \boldsymbol{\Psi}, \boldsymbol{\Theta}) = -\frac{1}{2}\frac{\partial}{\partial \boldsymbol{\Sigma}}\{np \log|\boldsymbol{\Psi}| + nm \log|\boldsymbol{\Sigma}| + d(\boldsymbol{Y}, \boldsymbol{A\Theta}(\boldsymbol{C} \otimes \boldsymbol{X})|\boldsymbol{\Psi} \otimes \boldsymbol{\Sigma})\}.
$$

Here, putting $\boldsymbol{G}_i(\boldsymbol{\Theta}) = (\boldsymbol{e}_{i1}, \ldots, \boldsymbol{e}_{im})'$, where $(\boldsymbol{e}'_{i1}, \ldots, \boldsymbol{e}'_{im})' = \boldsymbol{y}_i - (\boldsymbol{C} \otimes \boldsymbol{X})\boldsymbol{\Theta}'\boldsymbol{a}_i$ and $\boldsymbol{e}_{ij}$ is a p-dimensional vector, after some calculations (see the Appendix 1), we obtain

$$
d(\boldsymbol{Y}, \boldsymbol{A\Theta}(\boldsymbol{C} \otimes \boldsymbol{X})|\boldsymbol{\Psi} \otimes \boldsymbol{\Sigma}) = \sum_{i=1}^{n} \text{tr}\left\{\boldsymbol{G}_i(\boldsymbol{\Theta})\boldsymbol{\Sigma}^{-1}\boldsymbol{G}'_i(\boldsymbol{\Theta})\boldsymbol{\Psi}^{-1}\right\}. \tag{4}
$$

Using the result (4) and a formula for the differential, $\partial\ell(\boldsymbol{\Sigma}, \boldsymbol{\Psi}, \boldsymbol{\Theta})/(\partial\boldsymbol{\Sigma})$ can be rewritten as follows:

$$
\frac{\partial}{\partial \boldsymbol{\Sigma}}\ell(\boldsymbol{\Sigma}, \boldsymbol{\Psi}, \boldsymbol{\Theta}) = -\frac{1}{2}\left\{nm\boldsymbol{\Sigma}^{-1} - \sum_{i=1}^{n} \boldsymbol{\Sigma}^{-1}\boldsymbol{G}'_i(\boldsymbol{\Theta})\boldsymbol{\Psi}^{-1}\boldsymbol{G}_i(\boldsymbol{\Theta})\boldsymbol{\Sigma}^{-1}\right\}.
$$

In order to derive the MLE for $\boldsymbol{\Sigma}$, we solve $\partial\, \ell(\boldsymbol{\Sigma}, \boldsymbol{\Psi}, \boldsymbol{\Theta})/(\partial\, \boldsymbol{\Sigma})|_{\boldsymbol{\Sigma}=\hat{\boldsymbol{\Sigma}}} = \mathbf{0}_p\mathbf{0}_p'$. Then, we obtain the following MLE for $\boldsymbol{\Sigma}$ given $\boldsymbol{\Psi}$ and $\boldsymbol{\Theta}$;

$$\hat{\boldsymbol{\Sigma}}(\boldsymbol{\Psi}, \boldsymbol{\Theta}) = \frac{1}{nm}\sum_{i=1}^{n} \boldsymbol{G}_i'(\boldsymbol{\Theta})\boldsymbol{\Psi}^{-1}\boldsymbol{G}_i(\boldsymbol{\Theta}). \tag{5}$$

MLE for $\boldsymbol{\Psi}$ given $\boldsymbol{\Theta}$ and $\boldsymbol{\Sigma}$: From a similar calculation of the differential and using the same $\boldsymbol{G}_i(\boldsymbol{\Theta})$ as in the previous section, we also obtain the following result:

$$\frac{\partial}{\partial \boldsymbol{\Psi}}\ell(\boldsymbol{\Sigma}, \boldsymbol{\Psi}, \boldsymbol{\Theta}) = -\frac{1}{2}\left\{ np\boldsymbol{\Psi}^{-1} - \sum_{i=1}^{n} \boldsymbol{\Psi}^{-1}\boldsymbol{G}_i(\boldsymbol{\Theta})\boldsymbol{\Sigma}^{-1}\boldsymbol{G}_i'(\boldsymbol{\Theta})\boldsymbol{\Psi}^{-1} \right\}.$$

Thus, we obtain the MLE for $\boldsymbol{\Psi}$ given $\boldsymbol{\Sigma}$ and $\boldsymbol{\Theta}$ as follows:

$$\hat{\boldsymbol{\Psi}}(\boldsymbol{\Sigma}, \boldsymbol{\Theta}) = \frac{1}{np}\sum_{i=1}^{n} \boldsymbol{G}_i(\boldsymbol{\Theta})\boldsymbol{\Sigma}^{-1}\boldsymbol{G}_i'(\boldsymbol{\Theta}). \tag{6}$$

Estimating Algorithm 1: Since $\hat{\boldsymbol{\Theta}}(\boldsymbol{\Sigma}, \boldsymbol{\Psi})$ in (3) needs $\boldsymbol{\Sigma}$ and $\boldsymbol{\Psi}$, $\hat{\boldsymbol{\Sigma}}(\boldsymbol{\Psi}, \boldsymbol{\Theta})$ in (5) needs $\boldsymbol{\Psi}$ and $\boldsymbol{\Theta}$, and $\hat{\boldsymbol{\Psi}}(\boldsymbol{\Sigma}, \boldsymbol{\Theta})$ in (6) needs $\boldsymbol{\Sigma}$ and $\boldsymbol{\Theta}$, we use the following renewal algorithm to obtain them. In Step 1 of this algorithm, r is used to check if the value is small enough to converge.

Algorithm 1

Step 1 Set $s = 1$, a small value r, and initial matrices for $\boldsymbol{\Sigma}$ and $\boldsymbol{\Psi}$ as $\hat{\boldsymbol{\Sigma}}^{(0)}$ and $\hat{\boldsymbol{\Psi}}^{(0)}$. Here $\hat{\boldsymbol{\Sigma}}^{(0)}$ and $\hat{\boldsymbol{\Psi}}^{(0)}$ are assumed to be positive definite matrices, and the $(1, 1)$th element of $\hat{\boldsymbol{\Psi}}^{(0)}$ is assumed to be 1. If there is no particular problem, $\hat{\boldsymbol{\Sigma}}^{(0)}$ and $\hat{\boldsymbol{\Psi}}^{(0)}$ are assumed to be unit matrices.
Step 2 $\hat{\boldsymbol{\Theta}}^{(s)}$ as $\hat{\boldsymbol{\Theta}}(\hat{\boldsymbol{\Sigma}}^{(s-1)}, \hat{\boldsymbol{\Psi}}^{(s-1)})$ is derived from (3) by substituting $\hat{\boldsymbol{\Psi}}^{(s-1)}$ in for $\boldsymbol{\Psi}$ and $\hat{\boldsymbol{\Sigma}}^{(s-1)}$ in for $\boldsymbol{\Sigma}$.
Step 3 $\hat{\boldsymbol{\Sigma}}^{(s)}$ as $\hat{\boldsymbol{\Sigma}}(\hat{\boldsymbol{\Psi}}^{(s-1)}, \hat{\boldsymbol{\Theta}}^{(s)})$ is derived from (5) by substituting $\hat{\boldsymbol{\Psi}}^{(s-1)}$ in for $\boldsymbol{\Psi}$ and $\hat{\boldsymbol{\Theta}}^{(s)}$ in for $\boldsymbol{\Theta}$.
Step 4 $\hat{\boldsymbol{\Psi}}^{(s)}$ as $\hat{\boldsymbol{\Psi}}(\hat{\boldsymbol{\Sigma}}^{(s)}, \hat{\boldsymbol{\Theta}}^{(s)})$ is derived from (6) with substituting $\hat{\boldsymbol{\Sigma}}^{(s)}$ in for $\boldsymbol{\Sigma}$ and $\hat{\boldsymbol{\Theta}}^{(s)}$ in for $\boldsymbol{\Theta}$.
Step 5 Divide $\hat{\boldsymbol{\Psi}}^{(s)}$ by the $(1, 1)$th component of $\hat{\boldsymbol{\Psi}}^{(s)}$ and multiply $\hat{\boldsymbol{\Sigma}}^{(s)}$ by the $(1, 1)$th component of $\hat{\boldsymbol{\Psi}}^{(s)}$.
Step 6 If $|\ell(\hat{\boldsymbol{\Sigma}}^{(s)}, \hat{\boldsymbol{\Psi}}^{(s)}, \hat{\boldsymbol{\Theta}}^{(s)}) - \ell(\hat{\boldsymbol{\Sigma}}^{(s-1)}, \hat{\boldsymbol{\Psi}}^{(s-1)}, \hat{\boldsymbol{\Theta}}^{(s-1)})| < r|\ell(\hat{\boldsymbol{\Sigma}}^{(s-1)}, \hat{\boldsymbol{\Psi}}^{(s-1)}, \hat{\boldsymbol{\Theta}}^{(s-1)})|$, we obtain the estimators for the MLEs as $\hat{\boldsymbol{\Theta}}^{(s)}$, $\hat{\boldsymbol{\Sigma}}^{(s)}$, and $\hat{\boldsymbol{\Psi}}^{(s)}$. Otherwise, we set s as $s + 1$ and return to Step 2.

In this algorithm, the condition in Step 6 determines how many updates of the log-likelihood are performed using the renewal matrices.

In these estimation algorithms, two unknown matrices are required to obtain one matrix. Therefore, we need a large number of parameters in order to obtain a small number of parameters in many situations, for example, when we estimate kl parameters in $\boldsymbol{\Theta}$, we have to fix $\boldsymbol{\Sigma}$ and $\boldsymbol{\Psi}$. As a result, the number of parameters to be estimated is kl, while the number of parameters to be fixed is $mp(m+1)(p+1)/4$. When $kl < mp(m+1)(p+1)/4$, the number of parameters to be fixed is more than to be estimated. Thus, we propose an alternative estimation algorithm that solves this problem.

3.2 Dual-Block-Wise Coordinate Descent Algorithm

The log-likelihood function (2) with $\boldsymbol{\Psi}$ or $\boldsymbol{\Sigma}$ fixed can be regarded as a function of $(\boldsymbol{\Sigma}, \boldsymbol{\Theta})$ or $(\boldsymbol{\Psi}, \boldsymbol{\Theta})$, and the maximization of the functions can be easily achieved using the maximization of the log-likelihood of the usual GMANOVA model. This is the same as maximization in direction of two matrices, and for this reason we will refer to this maximization method as the dual-block-wise coordinate descent algorithm.

This algorithm requires matrices with different arrangements of $\boldsymbol{Y}$ and $\boldsymbol{\Theta}$. Let $\boldsymbol{Y}_{\star,i} = (\boldsymbol{y}_{i1}, \ldots, \boldsymbol{y}_{im})$ and $\boldsymbol{\Theta}_{\star,j} = (\boldsymbol{\theta}_{j1}, \ldots, \boldsymbol{\theta}_{jl})$, where $(\boldsymbol{y}'_{i1}, \ldots, \boldsymbol{y}'_{im})$ is the ith row of $\boldsymbol{Y}$ and $(\boldsymbol{\theta}'_{j1}, \ldots, \boldsymbol{\theta}'_{jl})$ is the jth row of $\boldsymbol{\Theta}$. For our algorithm, we prepare matrices with different arrangements of $\boldsymbol{Y}$ and $\boldsymbol{\Theta}$ as follows:

$$\boldsymbol{Y}_{(1)} = \begin{pmatrix} \boldsymbol{Y}'_{\star,1} \\ \vdots \\ \boldsymbol{Y}'_{\star,n} \end{pmatrix}, \ \boldsymbol{Y}_{(2)} = \begin{pmatrix} \boldsymbol{Y}_{\star,1} \\ \vdots \\ \boldsymbol{Y}_{\star,n} \end{pmatrix}, \ \boldsymbol{\Theta}_{(1)} = \begin{pmatrix} \boldsymbol{\Theta}'_{\star,1} \\ \vdots \\ \boldsymbol{\Theta}'_{\star,k} \end{pmatrix}, \ \boldsymbol{\Theta}_{(2)} = \begin{pmatrix} \boldsymbol{\Theta}_{\star,1} \\ \vdots \\ \boldsymbol{\Theta}_{\star,k} \end{pmatrix}.$$

MLEs for $\boldsymbol{\Sigma}$ and $\boldsymbol{\Theta}_{(1)}$ given $\boldsymbol{\Psi}$: Letting $\boldsymbol{Z} = (\boldsymbol{I}_n \otimes \boldsymbol{\Psi}^{-1/2})\boldsymbol{Y}_{(1)}$, and $\boldsymbol{B} = \boldsymbol{A} \otimes \boldsymbol{\Psi}^{-1/2}\boldsymbol{C}$, under $\boldsymbol{\Psi} = \hat{\boldsymbol{\Psi}}$, the log-likelihood function can be rewritten as

$$\ell\left(\boldsymbol{\Sigma}, \hat{\boldsymbol{\Psi}}, \boldsymbol{\Theta}_{(1)}\right) = -\frac{1}{2}\left\{nmp\log(2\pi) + nm\log|\boldsymbol{\Sigma}| + np\log|\hat{\boldsymbol{\Psi}}| + d\left(\boldsymbol{Z}, \boldsymbol{B}\boldsymbol{\Theta}_{(1)}\boldsymbol{X}'|\boldsymbol{\Sigma}\right)\right\},$$

from $\boldsymbol{Y}_{(1)}$ has similar distribution as $\boldsymbol{Y}$.

Since this log-likelihood function is the same as that in the GMANOVA model under the normal distribution assumption with $\mathrm{rank}(\boldsymbol{B}) = kl$ (from the Kronecker product's properties), we can obtain the MLE for $\boldsymbol{\Theta}_{(1)}$ given $\hat{\boldsymbol{\Psi}}$ as

$$\begin{aligned} \hat{\boldsymbol{\Theta}}_{(1)}(\hat{\boldsymbol{\Psi}}) &= (\boldsymbol{B}'\boldsymbol{B})^{-1}\boldsymbol{B}'\boldsymbol{Z}\boldsymbol{W}_{(1)}^{-1}\boldsymbol{X}\left(\boldsymbol{X}'\boldsymbol{W}_{(1)}^{-1}\boldsymbol{X}\right)^{-1} \\ &= \left\{(\boldsymbol{A}'\boldsymbol{A})^{-1}\boldsymbol{A}' \otimes (\boldsymbol{C}'\hat{\boldsymbol{\Psi}}^{-1}\boldsymbol{C})^{-1}\boldsymbol{C}'\hat{\boldsymbol{\Psi}}^{-1}\right\}\boldsymbol{Y}_{(1)}\boldsymbol{W}_{(1)}^{-1}\boldsymbol{X}\left(\boldsymbol{X}'\boldsymbol{W}_{(1)}^{-1}\boldsymbol{X}\right)^{-1}, \end{aligned} \tag{7}$$

where

$$\begin{aligned} \boldsymbol{W}_{(1)} &= \frac{1}{nm-kl}\boldsymbol{Z}'\{\boldsymbol{I}_{nm} - \boldsymbol{P}(\boldsymbol{B})\}\boldsymbol{Z} \\ &= \frac{1}{nm-kl}\boldsymbol{Y}'_{(1)}(\boldsymbol{I}_n \otimes \hat{\boldsymbol{\Psi}}^{-1/2})\left\{\boldsymbol{I}_{nm} - \boldsymbol{P}(\boldsymbol{A} \otimes \hat{\boldsymbol{\Psi}}^{-1/2}\boldsymbol{C})\right\}(\boldsymbol{I}_n \otimes \hat{\boldsymbol{\Psi}}^{-1/2})\boldsymbol{Y}_{(1)}, \end{aligned}$$

and $\boldsymbol{P}(\boldsymbol{K}) = \boldsymbol{K}(\boldsymbol{K}'\boldsymbol{K})^{-1}\boldsymbol{K}'$ by using the properties of the Kronecker product, and assuming $\boldsymbol{W}_{(1)}^{-1}$ is exist. Here, we note that $nm > kl$ since $n > k$ and $m > l$.

In addition, we obtain the MLE for $\boldsymbol{\Sigma}$ as follows:

$$\hat{\boldsymbol{\Sigma}}(\hat{\boldsymbol{\Psi}}, \hat{\boldsymbol{\Theta}}_{(1)}(\hat{\boldsymbol{\Psi}})) = \frac{1}{nm}\boldsymbol{U}(\boldsymbol{Y}_{(1)}, (\boldsymbol{A} \otimes \boldsymbol{C})\hat{\boldsymbol{\Theta}}_{(1)}(\hat{\boldsymbol{\Psi}})\boldsymbol{X}' | \boldsymbol{I}_n \otimes \hat{\boldsymbol{\Psi}}), \tag{8}$$

where $\boldsymbol{U}(\boldsymbol{L}, \boldsymbol{M}|\boldsymbol{N}) \overset{\text{def.}}{=} (\boldsymbol{L} - \boldsymbol{M})'\boldsymbol{N}^{-1}(\boldsymbol{L} - \boldsymbol{M}) = \boldsymbol{V}(\boldsymbol{L}', \boldsymbol{M}'|\boldsymbol{N})$, and $\boldsymbol{L}$, $\boldsymbol{M}$, and $\boldsymbol{N}$ are appropriate matrices.

MLEs for $\boldsymbol{\Psi}$ and $\boldsymbol{\Theta}_{(2)}$ Under Given $\boldsymbol{\Sigma}$: Following a similar argument, using $\boldsymbol{Y}_{(2)}$, the log-likelihood function given $\boldsymbol{\Sigma} = \hat{\boldsymbol{\Sigma}}$ has a similar form to that in the previous section.In particular, we can obtain the MLE for $\boldsymbol{\Theta}_{(2)}$ given $\hat{\boldsymbol{\Sigma}}$ as follows:

$$\hat{\boldsymbol{\Theta}}_{(2)}(\hat{\boldsymbol{\Sigma}}) = \left\{(\boldsymbol{A}'\boldsymbol{A})^{-1}\boldsymbol{A}' \otimes (\boldsymbol{X}'\hat{\boldsymbol{\Sigma}}^{-1}\boldsymbol{X})^{-1}\boldsymbol{X}'\hat{\boldsymbol{\Sigma}}^{-1}\right\}\boldsymbol{Y}_{(2)}\boldsymbol{W}_{(2)}^{-1}\boldsymbol{C}\left(\boldsymbol{C}'\boldsymbol{W}_{(2)}^{-1}\boldsymbol{C}\right)^{-1}, \tag{9}$$

where

$$\boldsymbol{W}_{(2)} = \frac{1}{np - kq}\boldsymbol{Y}_{(2)}'(\boldsymbol{I}_n \otimes \hat{\boldsymbol{\Sigma}}^{-1/2})\left\{\boldsymbol{I}_{np} - \boldsymbol{P}(\boldsymbol{A} \otimes \hat{\boldsymbol{\Sigma}}^{-1/2}\boldsymbol{X})\right\}(\boldsymbol{I}_n \otimes \hat{\boldsymbol{\Sigma}}^{-1/2})\boldsymbol{Y}_{(2)},$$

when $\boldsymbol{W}_{(2)}^{-1}$ exists. We also note that $np > kq$ from $n > k$ and $p > q$.

Finally, the MLE for $\boldsymbol{\Psi}$ is derived as follows:

$$\hat{\boldsymbol{\Psi}}(\hat{\boldsymbol{\Sigma}}, \hat{\boldsymbol{\Theta}}_{(2)}(\hat{\boldsymbol{\Sigma}})) = \frac{1}{np}\boldsymbol{U}(\boldsymbol{Y}_{(2)}, (\boldsymbol{A} \otimes \boldsymbol{X})\hat{\boldsymbol{\Theta}}_{(2)}(\hat{\boldsymbol{\Sigma}})\boldsymbol{C}' | \boldsymbol{I}_n \otimes \hat{\boldsymbol{\Sigma}}). \tag{10}$$

Estimating Algorithm 2: We estimate $\boldsymbol{\Sigma}$, $\boldsymbol{\Psi}$, and $\boldsymbol{\Theta}$ by using the following algorithm based on the above estimation methods:

Algorithm 2

Step 1 Set $s = 1$, a small value r, and initial matrices for $\boldsymbol{\Sigma}$ and $\boldsymbol{\Psi}$ as $\hat{\boldsymbol{\Sigma}}^{(0)}$ and $\hat{\boldsymbol{\Psi}}^{(0)}$. Here $\hat{\boldsymbol{\Sigma}}^{(0)}$ and $\hat{\boldsymbol{\Psi}}^{(0)}$ are assumed to be positive definite matrices, and the $(1,1)$th element of $\hat{\boldsymbol{\Psi}}^{(0)}$ is assumed to be 1. If there is no particular problem, $\hat{\boldsymbol{\Sigma}}^{(0)}$ and $\hat{\boldsymbol{\Psi}}^{(0)}$ are assumed to be unit matrices.

Step 2 Given $\hat{\boldsymbol{\Psi}} = \hat{\boldsymbol{\Psi}}^{(s-1)}$, $\hat{\boldsymbol{\Theta}}_{(1)}^{(s)}$ as $\hat{\boldsymbol{\Theta}}_{(1)}(\hat{\boldsymbol{\Psi}})$ is derived from (7) and $\hat{\boldsymbol{\Sigma}}^{(s)}$ as $\hat{\boldsymbol{\Sigma}}(\hat{\boldsymbol{\Psi}}, \hat{\boldsymbol{\Theta}}_{(1)}^{(s)})$ is derived from (8).

Step 3 Given $\hat{\boldsymbol{\Sigma}} = \hat{\boldsymbol{\Sigma}}^{(s)}$, $\hat{\boldsymbol{\Theta}}_{(2)}^{(s)}$ as $\hat{\boldsymbol{\Theta}}_{(2)}(\hat{\boldsymbol{\Sigma}})$ is derived from (9) and $\hat{\boldsymbol{\Psi}}^{(s)}$ as $\hat{\boldsymbol{\Psi}}(\hat{\boldsymbol{\Sigma}}, \hat{\boldsymbol{\Theta}}_{(2)}^{(s)})$ is derived from (10).

Step 4 Divide $\hat{\boldsymbol{\Psi}}^{(s)}$ by the $(1,1)$th component of $\hat{\boldsymbol{\Psi}}^{(s)}$ and multiply $\hat{\boldsymbol{\Sigma}}^{(s)}$ by the $(1,1)$th component of $\hat{\boldsymbol{\Psi}}^{(s)}$.

Step 5 If $|\ell(\hat{\boldsymbol{\Sigma}}^{(s)}, \hat{\boldsymbol{\Psi}}^{(s)}, \hat{\boldsymbol{\Theta}}^{(s)}) - \ell(\hat{\boldsymbol{\Sigma}}^{(s-1)}, \hat{\boldsymbol{\Psi}}^{(s-1)}, \hat{\boldsymbol{\Theta}}^{(s-1)})| < r|\ell(\hat{\boldsymbol{\Sigma}}^{(s-1)}, \hat{\boldsymbol{\Psi}}^{(s-1)}, \hat{\boldsymbol{\Theta}}^{(s-1)})|$, we obtain the estimators for the MLEs as $\hat{\boldsymbol{\Sigma}}^{(s)}$, $\hat{\boldsymbol{\Psi}}^{(s)}$, and $\hat{\boldsymbol{\Theta}}^{(s)}$ by rearranging $\hat{\boldsymbol{\Theta}}^{(s)}_{(1)}$ or $\hat{\boldsymbol{\Theta}}^{(s)}_{(2)}$. Otherwise, we set s as $s+1$ and return to Step 2.

We note that $d(\boldsymbol{Z}, \boldsymbol{B}\hat{\boldsymbol{\Theta}}_{(1)}\boldsymbol{X}'|\hat{\boldsymbol{\Sigma}})$ or the corresponding value in the log-likelihood may become constant.

4 Numerical Experiments

In order to verify the estimation results when the dimensions of each explanatory matrix and the estimation algorithms are changed, we examined five cases with different (n, p, q, m, l) with the two proposed algorithms and a commonly used optimization algorithm. All elements of $\boldsymbol{A}$ and $\boldsymbol{C}$ were generated independently from the uniform distribution between -1 to 1, denoted $\mathcal{U}(-1, 1)$. We take $\boldsymbol{\Theta} = \boldsymbol{T} - \mathbf{1}_{k,lq}$, where all elements of $\boldsymbol{T}$ were generated independently from the $\mathcal{U}(-2, 2)$ and all elements of $k \times lq$ matrix $\mathbf{1}_{k,lq}$ were 1. The (a, b)th element of $p \times q$ matrix $\boldsymbol{X}$ is equal to $a^{(b-1)}$ $(a = 1, \dots, p;\ b = 1, \dots, q)$. The data vector $\boldsymbol{Y}$ used for the simulation is generated from (1) with $k = 5$. Here, we generated $\boldsymbol{\Sigma}$ and $\boldsymbol{\Psi}$ by using the `datasets.make_spd_matrix` function [11] in sklearn library of Python package that generates a positive definite matrix, and the mp-dimensional vectors $\boldsymbol{\varepsilon}_1, \dots, \boldsymbol{\varepsilon}_n$ were generated independently from $\mathcal{N}_{mp}(\mathbf{0}_{mp}, \boldsymbol{\Psi} \otimes \boldsymbol{\Sigma})$.

In numerical experiments, the BFGS (Broyden-Fletcher-Goldfarb-Shanno) algorithm was used as the commonly used algorithm for comparison with the two proposed algorithms. This algorithm is used to solve nonlinear optimization problems (see, e.g., [1]). It is known as the quasi-Newton method because it optimizes by approximating of the Hessian matrix instead of the exact matrix, and then uses the approximation to update the unknown parameters in the Newton method. The BFGS algorithm can be performed using the `optimize.minimize` function [12] in scipy library of Python. Further details of this method can be found in, e.g., [7, chap. 6.1]. In the BFGS algorithm, we set $\boldsymbol{\Sigma} = \boldsymbol{\Lambda}_1'\boldsymbol{\Lambda}_2$, $\boldsymbol{\Psi} = \boldsymbol{\Lambda}_2'\boldsymbol{\Lambda}_2$, where $\boldsymbol{\Lambda}_1$ and $\boldsymbol{\Lambda}_2$ are $p \times p$ and $q \times q$ upper triangular matrices, respectively, and the $(1, 1)$th element of $\boldsymbol{\Lambda}_2$ is assumed to be 1, and execute the algorithm with $\boldsymbol{\Lambda}_1$ and $\boldsymbol{\Lambda}_2$ as the unknown parameter matrices instead of $\boldsymbol{\Sigma}$ and $\boldsymbol{\Psi}$.

Given these data, we set fixed parameters and explanatory matrices and estimated 100 times with different $\boldsymbol{\mathcal{E}}$. Table 1 shows the averages of the number of iterations required for convergence (s), the processing time (t) until convergence, and the $(-1)\times$ likelihood value divided by nmp (L) at convergence. In the proposed algorithms, we set $r = 10^{-8}$ for use in judging convergence. Note that smaller s, t, and L represent a better algorithm. In the table, algorithms 1, 2, and 3 are the proposed algorithms 1 and 2, and the commonly used algorithm, respectively. Moreover, for algorithm 3, s is not shown since their values are extremely large (about 6,000 to 40,000 iterations) and there is no point to compare.

Table 1. Results of numerical experiments

n	p	q	m	l	Algorithm	s (time)	t (sec.)	L
100	4	3	3	2	1	4.900	0.067	1.168
					2	4.060	0.051	1.170
					3	—	4.642	1.167
300	4	3	3	2	1	4.000	0.196	1.184
					2	3.930	0.246	1.184
					3	—	3.997	1.185
300	6	3	3	2	1	4.000	0.185	1.015
					2	3.990	0.336	1.015
					3	—	19.579	1.042
300	4	3	6	4	1	4.000	0.220	0.886
					2	4.000	0.405	0.887
					3	—	27.600	0.886
300	6	3	6	4	1	4.000	0.220	0.718
					2	4.000	0.482	0.719
					3	—	63.395	0.719

Regarding these results, we note that both of our proposed methods optimized the unknown matrices in 4 or 5 iterations. Also, the processing times of our proposed algorithms were much shorter than that of the commonly used optimization method. Moreover, when p, m, or/and l become large, the proposed method's processing time becomes still shorter. On the other hand, the values of the negative log-likelihood were similar. This means that our proposed algorithms converge to matrices similar to those of the commonly used optimize method. Thus, we think the proposed algorithms converge to appropriate matrices. From these results, the proposed methods are very useful for estimating and optimizing the three-mode GMANOVA model.

Acknowledgments. The second and last authors' research was partially supported by JSPS Bilateral Program Grant Number JPJSBP 120219927, and the last author's research was also partially supported by JSPS KAKENHI Grant Number 20H04151.

Appendix 1: Calculation for (4)

In this section, we derive the formula in (4). Recalling that $(\boldsymbol{e}'_{i1}, \ldots, \boldsymbol{e}'_{im})' = \boldsymbol{y}_i - (\boldsymbol{C} \otimes \boldsymbol{X})\boldsymbol{\Theta}'\boldsymbol{a}_i$ and letting $\boldsymbol{e}_i = (\boldsymbol{e}'_{i1}, \ldots, \boldsymbol{e}'_{im})'$, we calculate $d(\boldsymbol{Y}, \boldsymbol{A\Theta}(\boldsymbol{C} \otimes \boldsymbol{X})|\boldsymbol{\Psi} \otimes \boldsymbol{\Sigma})$ as follows:

$$d(\boldsymbol{Y}, \boldsymbol{A\Theta}(\boldsymbol{C} \otimes \boldsymbol{X})|\boldsymbol{\Psi} \otimes \boldsymbol{\Sigma}) = \mathrm{tr}\left[\begin{pmatrix} \boldsymbol{e}_1' \\ \vdots \\ \boldsymbol{e}_n' \end{pmatrix} (\boldsymbol{\Psi} \otimes \boldsymbol{\Sigma})^{-1}(\boldsymbol{e}_1, \cdots \boldsymbol{e}_n)\right]$$
$$= \sum_{i=1}^{n} \boldsymbol{e}_i'(\boldsymbol{\Psi} \otimes \boldsymbol{\Sigma})^{-1}\boldsymbol{e}_i.$$

Using the p-dimensional vector $\boldsymbol{e}_{ij}$ and $(\boldsymbol{\Psi} \otimes \boldsymbol{\Sigma})^{-1} = \boldsymbol{\Psi}^{-1} \otimes \boldsymbol{\Sigma}^{-1}$, and letting $(\boldsymbol{Q})_{uv}$ be the (u, v)th element of some matrix $\boldsymbol{Q}$, we obtain the following:

$$\begin{aligned}
&\boldsymbol{e}_i'(\boldsymbol{\Psi} \otimes \boldsymbol{\Sigma})^{-1}\boldsymbol{e}_i \\
&= (\boldsymbol{e}_{i1}', \cdots, \boldsymbol{e}_{im}') \begin{pmatrix} (\boldsymbol{\Psi}^{-1})_{11}\boldsymbol{\Sigma}^{-1} & \cdots & (\boldsymbol{\Psi}^{-1})_{1m}\boldsymbol{\Sigma}^{-1} \\ \vdots & \ddots & \vdots \\ (\boldsymbol{\Psi}^{-1})_{m1}\boldsymbol{\Sigma}^{-1} & \cdots & (\boldsymbol{\Psi}^{-1})_{mm}\boldsymbol{\Sigma}^{-1} \end{pmatrix} \begin{pmatrix} \boldsymbol{e}_{i1} \\ \vdots \\ \boldsymbol{e}_{im} \end{pmatrix} \\
&= \sum_{u=1}^{m}\sum_{v=1}^{m} (\boldsymbol{\Psi}^{-1})_{vu}\boldsymbol{e}_{iv}'\boldsymbol{\Sigma}^{-1}\boldsymbol{e}_{iu}.
\end{aligned}$$

On the other hand, recalling $\boldsymbol{G}_i(\boldsymbol{\Theta}) = (\boldsymbol{e}_{i1}, \ldots, \boldsymbol{e}_{im})'$, we obtain the following result:

$$\begin{aligned}
&\mathrm{tr}\{\boldsymbol{G}_i(\boldsymbol{\Theta})\boldsymbol{\Sigma}^{-1}\boldsymbol{G}_i'(\boldsymbol{\Theta})\boldsymbol{\Psi}^{-1}\} \\
&= \mathrm{tr}\left\{\begin{pmatrix} \boldsymbol{e}_{i1}'\boldsymbol{\Sigma}^{-1}\boldsymbol{e}_{i1} & \cdots & \boldsymbol{e}_{i1}'\boldsymbol{\Sigma}^{-1}\boldsymbol{e}_{im} \\ \vdots & \ddots & \vdots \\ \boldsymbol{e}_{im}'\boldsymbol{\Sigma}^{-1}\boldsymbol{e}_{i1} & \cdots & \boldsymbol{e}_{im}'\boldsymbol{\Sigma}^{-1}\boldsymbol{e}_{im} \end{pmatrix} \boldsymbol{\Psi}\right\} \\
&= \mathrm{tr}\begin{pmatrix} \sum_{u=1}^{m} \boldsymbol{e}_{i1}'\boldsymbol{\Sigma}^{-1}\boldsymbol{e}_{iu}(\boldsymbol{\Psi}^{-1})_{u1} & & * \\ & \ddots & \\ * & & \sum_{u=1}^{m} \boldsymbol{e}_{im}'\boldsymbol{\Sigma}^{-1}\boldsymbol{e}_{iu}(\boldsymbol{\Psi}^{-1})_{um} \end{pmatrix} \\
&= \sum_{v=1}^{m}\sum_{u=1}^{m} (\boldsymbol{\Psi}^{-1})_{uv}\boldsymbol{e}_{iu}'\boldsymbol{\Sigma}^{-1}\boldsymbol{e}_{iv},
\end{aligned}$$

where $*$ means some negligible values. Thus, since $\boldsymbol{\Psi}$ and $\boldsymbol{\Sigma}$ are symmetric matrices, we obtain $\boldsymbol{e}_i'(\boldsymbol{\Psi} \otimes \boldsymbol{\Sigma})^{-1}\boldsymbol{e}_i = \mathrm{tr}\{\boldsymbol{G}_i(\boldsymbol{\Theta})\boldsymbol{\Sigma}^{-1}\boldsymbol{G}_i'(\boldsymbol{\Theta})\boldsymbol{\Psi}^{-1}\}$. Hence, we obtain the formula in (4).

References

1. Fletcher, R.: Practical Methods of Optimization. Wiley, Hoboken (2013)
2. Guggenberger, P., Kleibergen, F., Mavroeidis, S.: A test for Kronecker product structure covariance matrix. J. Econom. **233**(1), 88–112 (2022). https://doi.org/10.1016/j.jeconom.2022.01.005
3. Kroonenberg, P.M., de Leeuw, J.: Principal component analysis of three-mode data by means of alternating least squares algorithms. Psychometrika **45**, 69–97 (1980). https://doi.org/10.1007/BF02293599

4. Harville, D.A.: Matrix Algebra from a Statistician's Perspective. Springer, New York (1997). https://doi.org/10.1007/b98818
5. Lu, H., Zimmerman, D.L.: The likelihood ratio test for a separable covariance matrix. Stat. Probabil. Lett. **73**, 449–457 (2005). https://doi.org/10.1016/j.spl.2005.04.020
6. Martini, J.W.R., Crossa, J., Toledo, F.H., Cuevas, J.: On Hadamard and Kronecker products in covariance structures for genotype × environment interaction. Plant Genome **13**, e20033 1–12 (2020). https://doi.org/10.1002/tpg2.20033
7. Nocedal, J., Wright, S.J.: Numerical Optimization, 2nd edn. Springer, New York (2006). https://doi.org/10.1007/978-0-387-40065-5
8. Potthoff, R.F., Roy, S.N.: A generalized multivariate analysis of variance model useful especially for growth curve problems. Biometrika **51**, 313–326 (1964). https://doi.org/10.2307/2334137
9. Satoh, K., Yanagihara, H.: Estimation of varying coefficients for a growth curve model. Am. J. Math. Manag. Sci. **30**, 243–256 (2010). https://doi.org/10.1080/01966324.2010.10737787
10. Srivastava, M.S., von Rosen, T., von Rosen, D.: Models with a Kronecker product covariance structure: estimation and testing. Math. Methods Stat. **17**, 357–370 (2008). https://doi.org/10.3103/S1066530708040066
11. Pedregosa, F., et al.: Scikit-learn: machine learning in python. J. Mach. Learn. Res. **12**, 2825–2830 (2011)
12. Virtanen, P., et al.: SciPy 1.0 contributors: SciPy 1.0: fundamental algorithms for scientific computing in python. Nat. Methods **7**, 261–272 (2020). https://rdcu.be/b08Wh. https://doi.org/10.1038/s41592-019-0686-2

Implications of the Usage of Three-Mode Principal Component Analysis with a Fixed Polynomial Basis

Rei Monden[1(✉)], Isamu Nagai[2], and Hirokazu Yanagihara[3]

[1] Informatics and Data Science Program, Graduate School of Advanced Science and Engineering, Hiroshima University, Higashi-Hiroshima, Hiroshima 739-8521, Japan
mondenr@hiroshima-u.ac.jp

[2] Faculty of Liberal Arts and Science, Chukyo University, Nagoya, Aichi 466-8666, Japan

[3] Mathematics Program, Graduate School of Advanced Science and Engineering, Hiroshima University, Higashi-Hiroshima, Hiroshima 739-8526, Japan

Abstract. When I subjects answer questions regarding J variables K times, the data can be stored in a three-mode data set of size $I \times J \times K$. Among the various component analysis approaches to summarize such data, Three-mode Principal Component Analysis (3MPCA) is often used. 3MPCA expresses a three-mode data set by means of A-, B-, and C-mode components and a core array, but this approach is not suitable for predicting data at unobserved time points. To handle this issue, the current study suggests a fixed orthonormal polynomial basis for the 3MPCA model. Briefly, we fix the C-mode component matrix with an orthonormal polynomial basis. More specifically, i) measurement time points, and ii) log transformation of the measurement time points were used as an orthonormal polynomial basis. The use of our proposed methods was demonstrated by applying them to a longitudinal data set collected from patients with depression.

Keywords: Array data · Longitudinal data · Polynomial basis · Three-mode principal component analysis

1 Three-Mode Principal Component Analysis: 3MPCA

Three-mode data are observed in various research fields, including engineering, psychology, and chemistry. For instance, data collected from I subjects for J variables at K time points are referred to as "three-mode/three-way data" or a "three-way array". You may imagine a certain number of patients with depression who are giving answers to questions regarding the severity of their depressive symptoms over time, or monthly measurements of a lagoon from multiple sampling sites. Various three-mode analysis techniques are available to explore the structure of such three-mode data. One of the most commonly used approaches

I. Czarnowski et al. (Eds.): KESIDT 2023, SIST 352, pp. 214–224, 2023.
https://doi.org/10.1007/978-981-99-2969-6_19

is Three-mode Principal Component Analysis (3MPCA; [5,6]), also known as Tucker 3 [10] or Three-way Principal Component Analysis. This is a generalization of the standard principal component analysis approach, which can describe three-mode data by means of a small number of components for each entity as well as their interactions.

Suppose that we denote $I \times J \times K$ three-mode data as $\underline{\mathbf{X}}$, as shown in Fig. 1, and fit a three-mode principal component model by setting the numbers of components as P, Q, and R for the three respective modes.

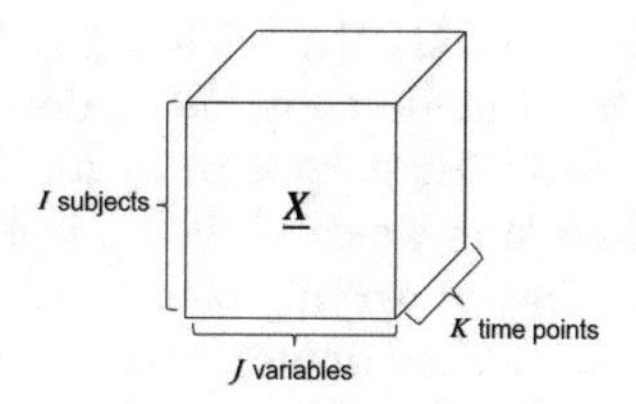

Fig. 1. Three-mode data of $I \times J \times K$

Then, the three-mode principal component analysis can be expressed as follows:

$$x_{ijk} = \sum_{p=1}^{P}\sum_{q=1}^{Q}\sum_{r=1}^{R} g_{pqr}a_{ip}b_{jq}c_{kr} + e_{ijk}, \tag{1}$$

where a_{ip}, b_{jq}, and c_{kr} are elements of the component matrices $\boldsymbol{A}$ $(I \times P)$, $\boldsymbol{B}$ $(J \times Q)$, and $\boldsymbol{C}$ $(K \times R)$, which are typically called the A-mode, B-mode and C-mode components, respectively. These component matrices are often constrained to be orthogonal, since the resulting core is easier to interpret and the model computation is much faster. Further, g_{pqr} denotes the (p, q, r)th element of the core array $\underline{\mathbf{G}}$ $(P \times Q \times R)$, and e_{ijk} denotes the (i, j, k)th element of the error $\underline{\mathbf{E}}$ $(I \times J \times K)$ for element x_{ijk}. Equivalently, by rearranging the $I \times J \times K$ three-mode data into an $I \times JK$ matrix, $\boldsymbol{X}$, the model can be re-written as follows:

$$\boldsymbol{X} = \boldsymbol{A}\boldsymbol{G}(\boldsymbol{C} \otimes \boldsymbol{B})' + \boldsymbol{E}, \tag{2}$$

where $\boldsymbol{G}$ $(P \times QR)$ and $\boldsymbol{E}$ $(I \times JK)$ denote the unfolded two-way matrices of the three-way arrays $\underline{\mathbf{G}}$ and $\underline{\mathbf{E}}$, respectively, and $\otimes$ denotes the Kronecker product.

Although this 3MPCA model has been shown to be useful in practice for summarizing three-mode data [7], it is not suitable when i) the variance of the three-way interaction of interest does not include the majority of the total variance, or ii) the interest of the analysis is to predict data collected at a time point other than the observed measurement time points. The former issue is unfortunate, since a meaningful three-mode interaction could be buried in the three-mode data, in the sense of its variance not being large enough to observe compared to the overall variance. The latter issue could be important when the interest is in making predictions given a three-mode data set, rather than interpreting the C-mode component structure. To address these issues, a fixed orthonormal polynomial basis can be introduced to a three-mode principal component analysis.

2 Proposed Methods: Three-Mode Principal Component Analysis with a Fixed Orthonormal Polynomial Basis

In this section, we introduce a 3MPCA with a fixed orthonormal polynomial basis. It is obvious that the C-mode component describes the temporal effect, but the overall temporal trend cannot be seen by looking at each of the C-mode components separately. Therefore, we consider a varying core array in which the elements vary with time, as in the varying coefficient model proposed by [2].

Let $t_1, \ldots, t_K$ be observation time points, $\boldsymbol{t}$ be a K-dimensional vector defined by $\boldsymbol{t} = (t_1, \ldots, t_K)'$, and $\boldsymbol{g}_{pq}$ be an R-dimensional vector defined by $\boldsymbol{g}_{pq} = (g_{pq1}, \ldots, g_{pqR})'$, where g_{pqr} is an element of the core array in (1). Moreover, let $\boldsymbol{f}_k'$ be an R-dimensional vector defined by the kth row vector of $\boldsymbol{C}$, i.e., $\boldsymbol{C} = (\boldsymbol{f}_1, \ldots, \boldsymbol{f}_K)'$, where $\boldsymbol{f}_k$ is some function of t_k. Then, the varying core array representation of (1) is as follows:

$$x_{ij}(t_k) = \sum_{p=1}^{P}\sum_{q=1}^{Q} g_{pq}(t_k) a_{ip} b_{jq} + e_{ij}(t_k), \quad (k = 1, \ldots, K), \tag{3}$$

where $g_{pq}(t_k) = \boldsymbol{f}_k' \boldsymbol{g}_{pq}$. Unfortunately, the above representation of the three-mode data cannot describe the core array's elements outside of the observational time points. Therefore, we consider a continuous representation of g_{pq} at a time t by using a fixed $\boldsymbol{C}$ with a known function of t.

Let $\boldsymbol{\phi}(t) = (\phi_1(t), \ldots, \phi_R(t))'$ be values of an R-dimensional vector of a known basis function, e.g., the polynomial basis $\phi_r(t) = t^{r-1}$ $(r = 1, \ldots, R)$, and $\boldsymbol{T}_\phi$ be a matrix of which the kth row consists of the transpose of $\boldsymbol{\phi}(t_k)$, i.e., $\boldsymbol{T}_\phi = (\boldsymbol{\phi}(t_1), \ldots, \boldsymbol{\phi}(t_K))'$. In addition, let $\boldsymbol{Q}_\phi$ be an $R \times R$ matrix that transforms $\boldsymbol{T}_\phi$ into an orthonormal basis. Such a $\boldsymbol{Q}_\phi$ can be obtained from a singular value decomposition of $\boldsymbol{T}_\phi$, for example. Using $\boldsymbol{Q}_\phi$, we write an orthonormal basis based on $\boldsymbol{T}_\phi$ as $\boldsymbol{C}_\phi = (\boldsymbol{c}_1, \ldots, \boldsymbol{c}_R)$, i.e., $\boldsymbol{C}_\phi = \boldsymbol{T}_\phi \boldsymbol{Q}_\phi$. If we use a polynomial basis for $\boldsymbol{\phi}(t)$, i.e.,

$$\boldsymbol{\phi}(t) = (1, h(t), \ldots, h(t)^{R-1})', \tag{4}$$

then $\boldsymbol{C}_\phi$ is given by

$$\boldsymbol{c}_r = \begin{cases} \boldsymbol{1}_K/\sqrt{K} & (r = 1) \\ \left(\boldsymbol{I}_K - \sum_{j=1}^{r-1} \boldsymbol{c}_j \boldsymbol{c}_j'\right) \boldsymbol{h}_r \Big/ \left\| \left(\boldsymbol{I}_K - \sum_{j=1}^{r-1} \boldsymbol{c}_j \boldsymbol{c}_j'\right) \boldsymbol{h}_r \right\| & (r = 2, \ldots, R) \end{cases}, \tag{5}$$

where $\boldsymbol{h}_r = (h(t_1)^{r-1}, \ldots, h(t_K)^{r-1})'$, $\boldsymbol{1}_K$ is a K-dimensional vector of ones, and $\|\boldsymbol{v}\| = \sqrt{\boldsymbol{v}'\boldsymbol{v}}$ for some vector $\boldsymbol{v}$. Using $\boldsymbol{\phi}(t)$ and $\boldsymbol{Q}_\phi$, the continuous varying core array representation of (1) can be given by

$$x_{ij}(t) = \sum_{p=1}^{P}\sum_{q=1}^{Q} g_{pq}(t)a_{ip}b_{jq} + e_{ij}(t), \quad (t \in \mathcal{T} = \{t_1, \ldots, t_K\}), \tag{6}$$

where $g_{pq}(t)$ is a continuous varying element of the core array given by

$$g_{pq}(t) = \boldsymbol{\phi}(t)'\boldsymbol{Q}_\phi \boldsymbol{g}_{pq}. \tag{7}$$

Alternatively, it may be easier to use $g_{pq}(t) = \boldsymbol{\phi}(t)'(\boldsymbol{T}_\phi'\boldsymbol{T}_\phi)^{-1}\boldsymbol{T}_\phi'\boldsymbol{C}_\phi\boldsymbol{g}_{pq}$ as an expression for $g_{pq}(t)$ instead of (7), since it is sometimes possible to find $\boldsymbol{C}_\phi$ without explicitly defining $\boldsymbol{Q}_\phi$, as in (5). Needless to say, since $g_{pq}(t)$ is continuous with respect to t, it is also possible to compute $g_{pq}(t)$ for points other than the observation time points $\mathcal{T}$.

Proposal 1: Use a usual polynomial basis, i.e., (4) with $h(t) = t$.
Proposal 2: Use a log-transformed polynomial basis, i.e., (4) with $h(t) = \log t$.

Using a log-transformed polynomial basis often yields more realistic results than using a usual polynomial basis since the change with increasing time in the log-transformed polynomial basis is smaller than that in the usual polynomial basis. Note that the interpretation of the core array with our proposed methods differs from that of the classical 3MPCA model in [4]. As described in [4], the elements of the core array in the classical 3MPCA model are interpreted as indicators of the strength of the corresponding three-mode interactions. However, with our proposed methods, the core array can be interpreted as constituting a coefficient of change.

Introducing an orthonormal polynomial basis for the 3MPCA can be expected to have at least three advantages. First, as described in [1], the generic algorithm to estimate component matrices in (2) requires fixing two of the three component matrices, i.e., $\boldsymbol{A}$, $\boldsymbol{B}$, or $\boldsymbol{C}$, and estimating the other component matrix, one combination at a time, until the estimation converges. However, by fixing the C-mode component matrix, as in our proposed methods, the estimation requires fewer steps to converge. The second advantage of the proposed models is that by describing measurement time points using a polynomial basis, it is possible to predict data of other time points besides the observed time points. This allows us to use the resulting component structure of 3MPCA for prediction, rather than merely as a description of the three-mode data. Finally, the third advantage is that the proposed methods could be useful for identifying variables with three-mode interactions, even though the variances of those interactions may not be the largest components of the overall variance. This is because, by fixing the C-mode component structure, we remove interpretation of the C-mode component as a subject of interest. As a result, even when a three-mode data set contains a minor proportion of the overall variance with respect to a three-mode interaction, our proposed methods can be used to examine whether any three-way interaction is buried in a given three-mode data set. This can be useful, for instance, when analyzing psychological longitudinal data, since there could be a meaningful three-mode interaction buried in the three-mode data, even if its variance is not very large. In the following section, we will demonstrate the use of our proposed model.

3 Applications of the Proposed Method

The 3MPCA model with our proposed methods (Proposals 1 and 2) was applied to a clinical data set to demonstrate the use of the methods. In all the methods, the same algorithm, as described in [1] was used. The details of the analyzed data can be found elsewhere [8] and are briefly summarized below. The data are of 219 subjects who were diagnosed with major depressive disorder at the baseline. The Beck Depression Inventory (BDI) was administered every 3 months for 2 years to assess the severity of depression. The BDI is a 21-item multiple-choice inventory in which symptom severity is scored 0 to 3 for each item, based on the severity of the symptom during the previous week.

In short, the analyzed three-mode data set has a size of 219 (subjects) $\times$ 21 (symptoms) $\times$ 9 (time points). Prior to the analysis, the data were preprocessed by centering across the A-mode (subjects) and normalizing within the B-mode (symptoms). The complexity of the model was chosen as (3, 2, 2) for the numbers of components for A-, B-, and C-modes, respectively, as suggested by the generalized scree test [9]. The same complexity was set for our proposed model. Furthermore, after the components and the core array were obtained, joint orthomax rotation [3] was applied to the estimated components to facilitate their interpretation. For the 3MPCA and our proposed method, i.e., Proposals 1 and 2, fit percentages were 32.09%, 31.88%, and 32.09%, respectively, indicating minimal differences between the three models. These fit percentages were calculated from the sum of squares of the estimated values by a model, divided by the total sum of squares of the data.

The resulting B-mode and C-mode matrices and core arrays are presented in Tables 1, 2, and 3, respectively. The A-mode matrix was also estimated. However, since the A-mode matrix itself is not interpretable, we chose to display this matrix as shown in Fig. 3. All the structures of the resulting components were comparable between the three methods. Furthermore, their interpretation was stable between the three tested models. From Table 1, the first and second components for B-mode were interpreted as "Cognitive" and "Somatic-affective" components, respectively, considering the symptoms with high loadings on each component. Likewise, Table 2 indicated that the first and second components can be interpreted as "Beginning" and "Later" for the C-mode components.

Figures 2, 3, and 4 illustrate the estimated elements of varying core arrays obtained from 3MPCA, and Proposals 1 and 2, respectively. The figures are separately presented according to the A-mode component; namely, A-Comp 1, A-Comp 2, and A-Comp 3 correspond to Figs. 2, 3, and 4, respectively. In each figure, the upper plots represent the estimated $g_{pq}(t_k)$ $(k = 1, \ldots, K)$ obtained from 3MPCA (3), denoted as $\bar{g}_{pq}(t_k)$, while the middle and lower plots are the estimated $g_{pq}(t)$ $(t \in \mathcal{T})$ in (7) obtained from Proposals 1 and 2, respectively, denoted as $\hat{g}^{*}_{pq}(t)$ and $\hat{g}^{**}_{pq}(t)$. Moreover, the columns are arranged based on the B-mode components, that is, with the left column indicating B-Comp 1 and the right column indicating B-Comp 2. For instance, in Fig. 2, the associations between A-Comp 1 and B-Comp 1 $((p, q) = (1, 1))$ are presented in the left column, while the right column shows the associations between A-Comp 1 and

Table 1. B-mode components

Item	Symptoms	3MPCA		Proposed 1		Proposed 2	
		Comp 1	Comp 2	Comp 1	Comp 2	Comp 1	Comp 2
5	Guilty feelings	**0.418**	−0.027	**0.421**	−0.034	**0.419**	−0.027
3	Past failure	**0.399**	0.010	**0.398**	0.007	**0.398**	0.011
8	Self-criticism	**0.399**	−0.036	**0.400**	−0.042	**0.399**	−0.035
14	Body image	**0.374**	−0.047	**0.375**	−0.052	**0.374**	−0.046
7	Self-dislike	**0.369**	0.025	**0.371**	0.020	**0.368**	0.025
6	Feeling punished	**0.246**	0.056	**0.245**	0.056	**0.246**	0.056
9	Suicidal thoughts	**0.235**	0.112	**0.232**	0.112	**0.235**	0.113
1	Sadness	**0.219**	0.133	**0.220**	0.129	**0.219**	0.132
15	Work difficulties	−0.076	**0.374**	−0.070	**0.374**	−0.077	**0.374**
17	Tiredness	−0.051	**0.361**	−0.047	**0.361**	−0.053	**0.362**
4	Loss of pleasure	−0.022	**0.342**	−0.016	**0.343**	−0.022	**0.342**
13	Indecisiveness	−0.036	**0.349**	−0.032	**0.351**	−0.037	**0.349**
21	Loss of interest in sex	−0.079	**0.308**	−0.075	**0.311**	−0.080	**0.308**
12	Loss of interest	−0.001	**0.289**	0.004	**0.289**	0.000	**0.289**
11	Agitation	−0.045	**0.281**	−0.040	**0.279**	−0.044	**0.280**
16	Changes in sleeping	0.012	**0.246**	0.014	**0.246**	0.011	**0.246**
10	Crying	0.010	**0.225**	0.013	**0.222**	0.011	**0.223**
2	Pessimism	0.160	0.187	0.159	0.186	0.159	0.188
18	Changes in appetite	0.103	0.147	0.100	0.144	0.102	0.148
20	Somatic preoccupation	0.064	0.160	0.068	0.159	0.065	0.159
19	Changes in weight	0.091	0.048	0.092	0.047	0.093	0.049

Table 2. C-mode components

Time points	3MPCA		Proposed 1		Proposed 2	
	Comp 1	Comp 2	Comp 1	Comp 2	Comp 1	Comp 2
Baseline	**0.725**	−0.028	**0.611**	−0.069	**0.775**	−0.005
3M	**0.487**	0.142	**0.511**	0.013	**0.468**	0.144
6M	**0.366**	0.212	**0.411**	0.094	**0.289**	0.231
9M	0.158	**0.274**	**0.311**	0.176	0.161	0.292
12M	0.111	**0.324**	0.211	**0.258**	0.063	**0.340**
15M	−0.071	**0.407**	0.111	**0.340**	−0.018	**0.379**
18M	−0.045	**0.395**	0.011	**0.421**	−0.086	**0.412**
21M	−0.187	**0.461**	−0.088	**0.503**	−0.145	**0.441**
24M	−0.154	**0.469**	−0.188	**0.585**	−0.198	**0.466**

Table 3. Core arrays

Method	A-mode (person)	C-Comp 1		C-Comp 2	
		B-Comp 1	B-Comp 2	B-Comp 1	B-Comp 2
3PCA	A-Comp 1	23.91	32.27	1.33	3.00
	A-Comp 2	15.29	25.94	39.97	79.34
	A-Comp 3	25.61	2.49	46.49	5.93
Proposed 1	A-Comp 1	23.70	35.12	−2.99	3.39
	A-Comp 2	20.02	38.68	36.01	70.97
	A-Comp 3	36.03	6.47	41.91	7.39
Proposed 2	A-Comp 1	23.49	31.65	1.59	2.74
	A-Comp 2	14.31	23.90	40.44	80.24
	A-Comp 3	23.65	2.35	47.26	5.88

B-Comp 2 ($(p, q) = (1, 2)$). Needless to say, $\bar{g}_{pq}(t_k)$ is represented by a line graph for comparison, since the values can only be computed at each observation time point t_k, while $\hat{g}_{pq}(t)^*$ and $\hat{g}_{pq}(t)^{**}$ can be computed continuously in the region $\mathcal{T}$.

As reflected in Figs. 2, 3, and 4, the overall trend for each of the A-mode components was the same, regardless of the choice of method. Overall, A-Comp 1 has an acute decreasing trend with respect to the severity of depression over time in both of the B-mode components (Fig. 2), while A-Comp 2 indicates a gradual worsening over time in both of the B-mode components (Fig. 3). Interestingly, A-Comp 3 hardly showed any trend for either of the B-mode components (Fig. 4). However, it should be noted that the interaction between A-comp 3 and B-Comp 1 had consistently high scores above 0, while the interaction between A-Comp 3 and B-Comp 1 had scores consistently around 0. This implies that with respect to A-Comp 3, B-Comp 2 hardly has any influence on the severity of depression, while B-Comp 1 has a constant worsening trend. To further facilitate the interpretation of the A-mode components, the Pearson correlations were calculated between each of the A-mode components and auxiliary variables. The resulting correlations were highly comparable across the three methods. Briefly, A-Comp 1 was associated positively with psychopathology (e.g., anxiety ($r = 0.4$) or depression ($r = 0.6$)) and negatively with quality of life (e.g., mental health ($r = -0.5$)), A-Comp 2 was moderately correlated with somatic complaints ($r = 0.44$) and physical functions ($r = -0.45$), and A-Comp 3 was moderately associated positively with depression ($r = 0.31$) and neuroticism personality ($r = 0.38$) and negatively with low self-esteem ($r = -0.47$). These characteristics of the A-mode components and their trajectories depicted in Figs. 2, 3, and

4 agree with the clinical interpretation. In general, the characteristics of psychopathology are known to improve over time for most patients with depressive disorder, as depicted in Fig. 2. Given that B-Comp 1 is a "Cognitive" component, as represented by the symptoms, such as "guilty feelings", and B-Comp 2 is a "Somatic-affective" component, represented by the symptom "work difficulties", the change in B-Comp 1 is expected to be less acute than that of B-Comp 2. This tendency can be also seen in Figs. 2, 3, and 4.

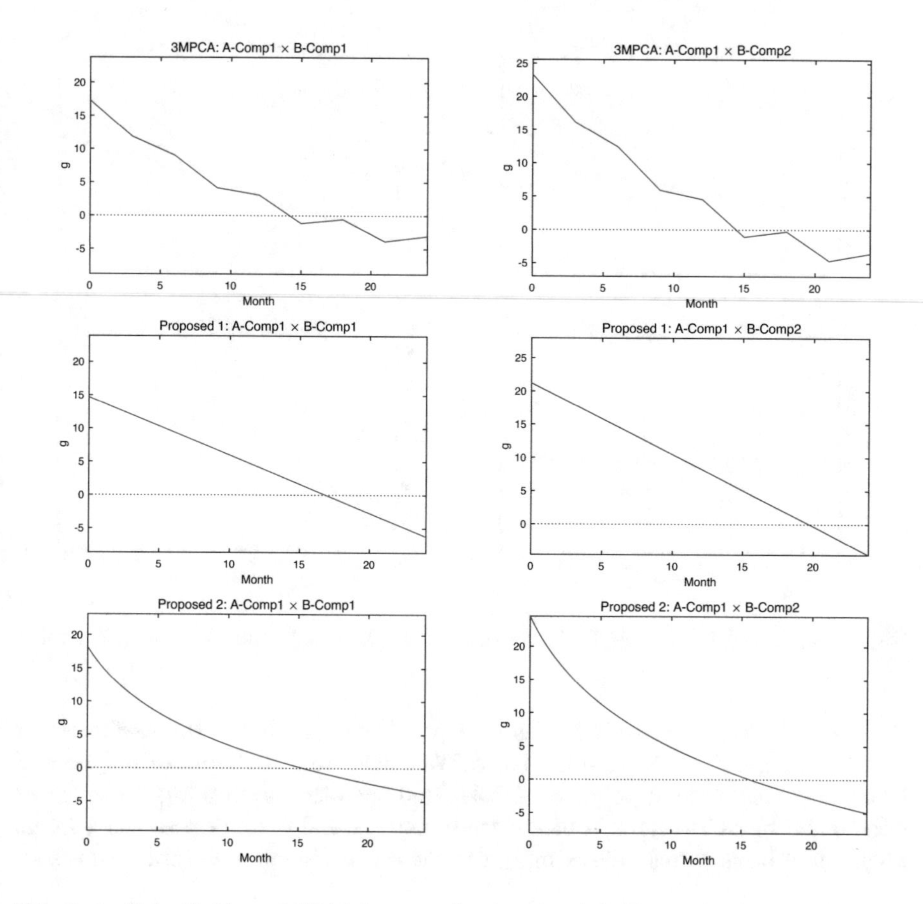

Fig. 2. $\bar{g}_{pq}(t_k)$, $\hat{g}^*_{pq}(t)$, and $\hat{g}^{**}_{pq}(t)$ for cases $(p, q) = (1, 1)$ (left) and $(p, q) = (1, 2)$ (right)

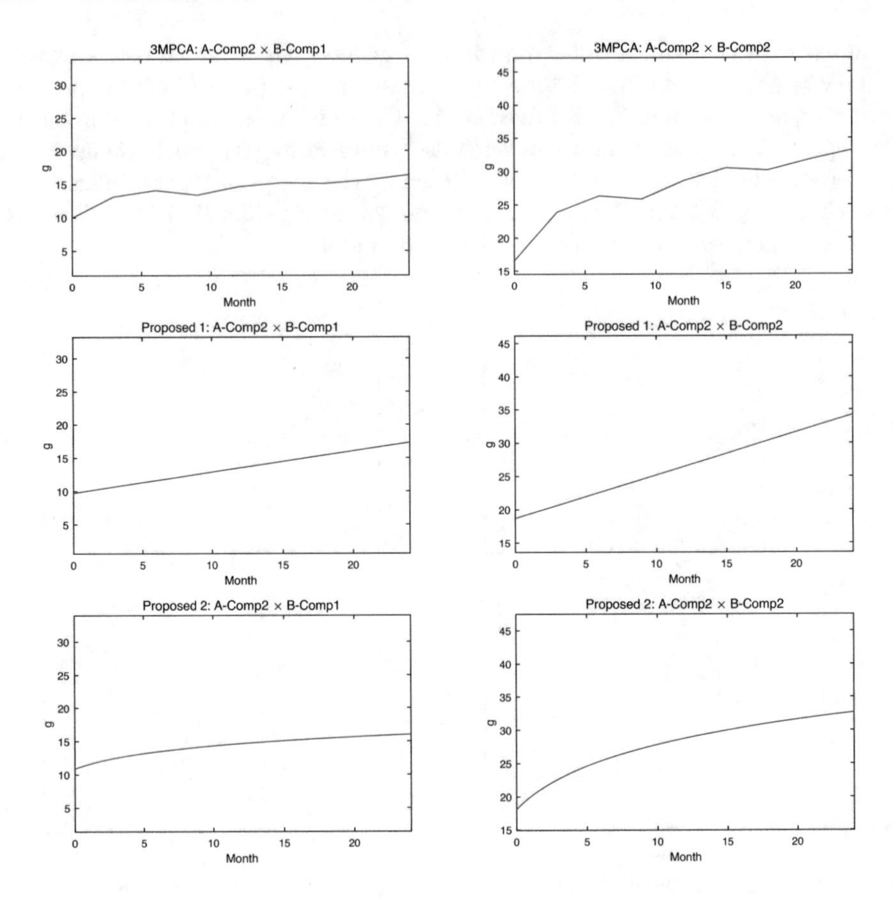

Fig. 3. $\bar{g}_{pq}(t_k)$, $\hat{g}^*_{pq}(t)$, and $\hat{g}^{**}_{pq}(t)$ for cases $(p, q) = (2, 1)$ (left) and $(p, q) = (2, 2)$ (right)

In conclusion, this study serves as a proof-of-principle for the usefulness of 3MPCA with a fixed polynomial basis. We considered two types of polynomial basis, one with a usual polynomial basis and the other with a log-transformed polynomial basis. However, a future study may extend the model to use a spline polynomial basis, which allows more flexible estimation of a longitudinal trend.

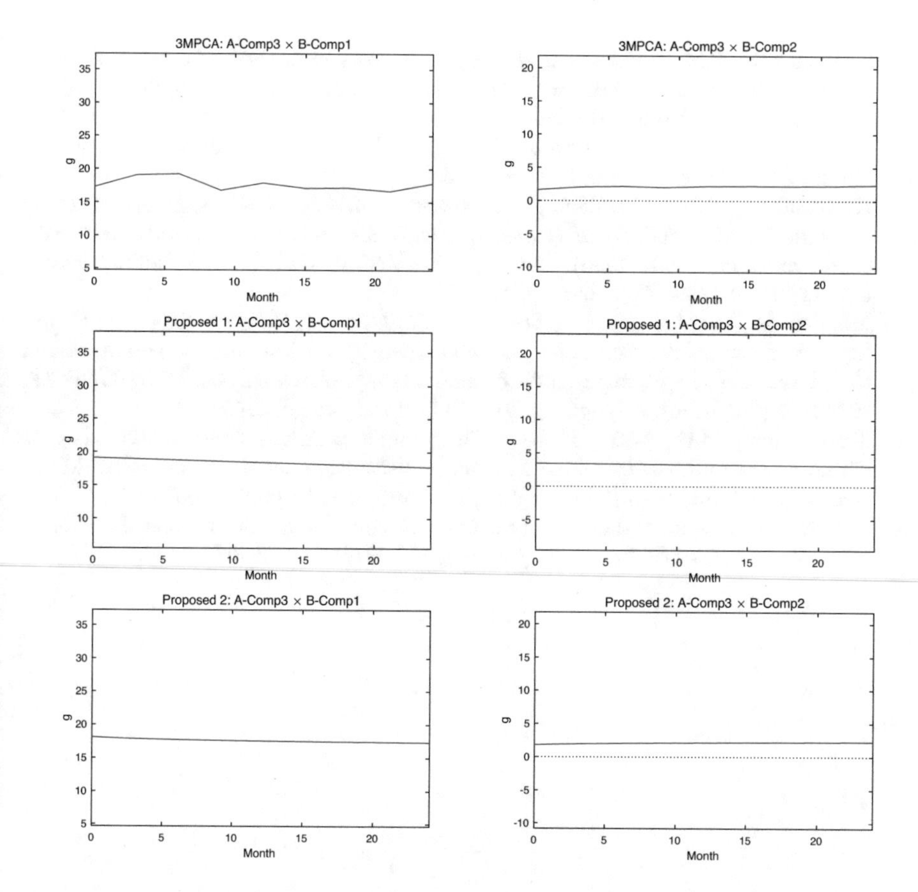

Fig. 4. $\bar{g}_{pq}(t_k)$, $\hat{g}^*_{pq}(t)$, and $\hat{g}^{**}_{pq}(t)$ for cases $(p, q) = (3, 1)$ (left) and $(p, q) = (3, 2)$ (right)

Acknowledgments. The second and last authors' research was partially supported by JSPS Bilateral Program Grant Number JPJSBP 120219927, and the last author's research was partially supported by JSPS KAKENHI Grant Number 20H04151.

References

1. Andersson, C.A., Bro, R: Improving the speed of multi-way algorithms: Part I. Tucker 3. Chemom. Intell. Lab. Syst. **42**, 93–103 (1998). https://doi.org/10.1016/S0169-7439(98)00010-0
2. Hastie, T., Tibshirani, R.: Varying-coefficient models. J. Roy. Stat. Soc. Ser. B **55**, 757–796 (1993). https://doi.org/10.1111/j.2517-6161.1993.tb01939.x
3. Kiers, H.A.: Joint orthomax rotation of the core and component matrices resulting from three-mode principal components analysis. J. Classif. **15**(2), 245–63 (1998). https://doi.org/10.1007/s003579900033
4. Kiers, H.A., van Mechelen, I.: Three-way component analysis: principles and illustrative application. Psychol. Methods **6**, 84–110 (2001). https://doi.org/10.1037/1082-989x.6.1.84

5. Kroonenberg, P.M., de Leeuw, J.: Principal component analysis of three-mode data by means of alternating least squares algorithms. Psychometrika **45**, 69–97 (1980). https://doi.org/10.1007/BF02293599
6. Kroonenberg, P.M.: Three-Mode Principal Component Analysis: Theory and Applications. DSWO Press, Leiden (1983)
7. Kroonenberg, P.M., Murakami, T., Coebergh, J.W.: Added value of three-way methods for the analysis of mortality trends illustrated with worldwide female cancer mortality (1968–1985). Stat. Methods Med. Res. **11**, 275–92 (2002). https://doi.org/10.1191/0962280202sm287ra
8. Monden, R., Wardenaar, K.J., Stegeman, A., Conradi, H.J., de Jonge, P.: Simultaneous decomposition of depression heterogeneity on the person-, symptom-and time-level: the use of three-mode principal component analysis. PLoS ONE **10**, e0132765 (2015). https://doi.org/10.1371/journal.pone.0132765
9. Timmerman, M.E., Kiers, H.A.L.: Three-mode principal components analysis: Choosing the numbers of components and sensitivity to local optima. Br. J. Math. Stat. Psychol. **53**, 1–16 (2000). https://doi.org/10.1348/000711000159132
10. Tucker, L.R.: Some mathematical notes on three-mode factor analysis. Psychometrika **31**, 279–311 (1966). https://doi.org/10.1007/BF02289464

Spatio-Temporal Analysis of Rates Derived from Count Data Using Generalized Fused Lasso Poisson Model

Mariko Yamamura[1], Mineaki Ohishi[2], and Hirokazu Yanagihara[3](✉)

[1] Department of Statistics, Radiation Effects Research Foundation, Hiroshima 732-0815, Japan
[2] Center for Data-driven Science and Artificial Intelligence, Tohoku University, Sendai 980-8576, Japan
[3] Mathematics Program, Graduate School of Advanced Science and Engineering, Hiroshima University, Higashi-Hiroshima, Hiroshima 739-8526, Japan
yanagi-hiro@hiroshima-u.ac.jp

Abstract. We propose a method to statistically analyze rates obtained from count data in spatio-temporal terms, allowing for regional and temporal comparisons. Generalized fused Lasso Poisson model is used to estimate the spatio-temporal effects of the rates; the coordinate descent algorithm is used for estimation. The results of an analysis using data on crime rates in Japan's Osaka Prefecture from 1995 to 2008 confirm the validity of the approach.

Keywords: Spatio-temporal proportion data · Adaptive fused lasso · Coordinate descent algorithm

1 Introduction

Statistical data representing regional characteristics are often of interest for purposes of comparison across regions. In such cases, rather than the observed values themselves, which are commonly expressed as count data, the values of interest are counts that have been converted to rates by dividing the counts by the corresponding population size or total geographic area. For example, aging or crime rates are likely to be more important than the number of elderly or the number of crimes when comparing conditions in two or more regions. The reason for this is straightforward: If the number of elderly or the number of crimes were to be used for the comparison, we would expect the count values to be larger in the more populated areas, rendering use of the raw numbers alone unhelpful in establishing which areas have a larger proportion of elderly citizens or higher crime rates. Moreover, using raw counts to assess changes over time without taking into account increases or decreases in population could be highly misleading. Again, using rates would seem to provide a much more effective alternative. This study proposes a method for the spatio-temporal statistical analysis of rates derived

I. Czarnowski et al. (Eds.): KESIDT 2023, SIST 352, pp. 225–234, 2023.
https://doi.org/10.1007/978-981-99-2969-6_20

from count data to allow for regional and temporal comparisons. Since Poisson model of generalized linear models is used to analyze rates obtained from count data, the focus is on spatio-temporal Poisson model.

Geographically weighted regression analysis (GWR), as defined by [3], is well known in spatial statistics and can be performed using ESRI's ArcGIS Pro spatial statistics toolbox or the R spgwr package. GWR is characterized by determining bandwidths or weights and their optimal size and location, and analyzing them using a local regression model. The estimated regional effects are therefore represented by a smoothed heat or cold map. The weakness of GWR is that its estimated effect is smoothed, making it difficult to detect its hot and cold spots on such maps. GWR is expected to be accurate when the space is large, the samples are evenly distributed in space, and the number of samples is above a certain level. However, if the space is intricate and the samples are sparsely distributed, GWR may not provide accurate estimates. The reason for this is that it is difficult to set bandwidths and weights for intricate terrain, and the effect of small sample areas is not well reflected in the smoothing estimation results.

To address this problem, [8] proposed spatial effect estimation by generalized fused Lasso (GFL), a generalization of the fused Lasso approach proposed in [11]. From the estimation results, areas having the same effect are unified, and the effect is color-coded and shown on a choropleth map. Since no smoothing is performed, hot and cold spots can be easily detected, and the effects of large and small sample size areas are not overestimated or underestimated. Thus, GFL can be expected to provide more accurate estimates than GWR, even in intricate terrain.

Furthermore, [8] reported that estimation with the coordinate descent algorithm is computationally faster and the estimation results are more accurate than using the R genlasso package [2] proposed in [1], which solves the dual problem suggested by [10]. In this study, we offer a method for analyzing spatio-temporal effects using GFL Poisson model, with reference to [8].

The remainder of the paper is organized as follows. Section 2 describes the GFL Poisson model for spatio-temporal analysis. Section 3 details the estimation method for the GFL Poisson model. Section 4 gives examples of actual data analysis and concludes the discussion.

2 Model

Suppose that the observations of interest, such as the number of crimes or the number of elderly residents, are available as count data in space-time. Further, assume there are p periods of time and q spaces, and that the number of samples is $n = p \times q$. Let y_i be the count data in the form of a non-negative integer, including zero, i.e., $y_i \in \mathbb{Z}_0^+$, $i = 1, \ldots, n$. Suppose that the rate y_i/m_i is of primary interest, where $m_i \in \mathbb{R}^+$ is a positive real number for the population, the area, etc., that adjusts y_i to a rate per person, per unit area, etc. Then we can use the following Poisson model:

$$f(Y_i = y_i \mid \theta_i) = \frac{\exp(-\theta_i)\theta_i^{y_i}}{y_i!}, \quad i = 1, \ldots, n. \tag{1}$$

Let parameter μ_i, $i = 1, \ldots, n$ be the spatio-temporal effect of the ith space-time for the rate y_i/m_i. The link function in the generalized linear model of the Poisson model focused on the rate y_i/m_i is introduced in [5,6] as

$$E(Y_i) = \theta_i = m_i \exp(\mu_i). \tag{2}$$

Here, m_i is called the "offset". In this study, we focus only on the spatio-temporal effects, not dealing with other explanatory variables; that is, the link function consists only of μ_i. Let $\varphi(x|a, b)$ be the following function with respect to x depending on positive a and b, and C_n be the following constant, independent of the unknown parameters:

$$\varphi(x|a, b) = b\exp(x) - ax, \quad C_n = -\sum_{i=1}^{n} (y_i \log m_i - \log y_i!). \tag{3}$$

The negative log-likelihood function of (1) substituting (2) for θ_i is as follows:

$$\ell(\boldsymbol{\mu}) = \sum_{i=1}^{n} \varphi(\mu_i \mid y_i, m_i) + C_n,$$

where $\boldsymbol{\mu} = (\mu_1, \ldots, \mu_n)'$. Let $w_{jk} = |\log(y_j/y_k) - \log(m_j/m_k)|^{-1}$, which is the inverse of the absolute value of the difference between the maximum likelihood estimators of μ_j and μ_k. The penalized log-likelihood function for GFL is as follows:

$$\ell_{\mathrm{PL}}(\boldsymbol{\mu}) = \ell(\boldsymbol{\mu}) + \lambda \sum_{j=1}^{n} \sum_{k \in D_j} w_{jk} |\mu_j - \mu_k|, \tag{4}$$

where $\lambda \geq 0$ is a regularization parameter. The weight w_{jk} is required for the adaptive Lasso [13]. The reason for adding the weights for the adaptive Lasso is that it improves the accuracy of the estimation by satisfying the oracle property, which is said to be a desirable property of sparse estimation (for details, see [13]). $D_j \subseteq \{1, \ldots, n\} \backslash \{j\}$ is the index set of the spatio-temporal μ_k adjacent to μ_j in (4). The index set is explained with figures in [7]. We also show it in detail in Sect. 4. If we can estimate that the jth effect is equal to the kth effect in space-time for (4), then the two are unified and have the same effect.

3 Optimization Procedure

We use the coordinate descent algorithm to obtain estimates of $\boldsymbol{\mu}$ that minimize the penalized negative log-likelihood functions (4). [8] used the coordinate descent algorithm for minimizing the penalized residual sum of squares with the adaptive fused Lasso penalty. [12] applied the same estimation method proposed in [8] to the logistic model by linearly approximating the penalized negative log-likelihood function with a Taylor expansion up to the second order term. Unfortunately, Poisson model does not allow us to apply the method in [8] using the linear approximation as in [12]. This is because it is impossible to derive such a quadratic approximation for the negative log-likelihood function in Poisson model, which suppresses the negative log-likelihood function from above for any parameter. In our approach, we use an algorithm to minimize the objective function without approximation by deriving the update equation of the coordinate descent algorithm in closed form, as proposed in [7] for logistic model.

3.1 Methods for Minimizing $\ell_{\mathrm{PL}}(\boldsymbol{\mu})$

The $\boldsymbol{\mu}$ that minimizes $\ell_{\mathrm{PL}}(\boldsymbol{\mu})$ is obtained by the coordinate descent algorithm. The estimation process is briefly described below. (See [4,8] for details of the estimation process.)

The coordinate descent algorithm is an optimization algorithm that performs sequential minimization along the coordinate axes to find the minimum value of a function. In (4), sequential minimization is performed along n-coordinate axes from μ_1 to μ_n; when minimizing μ_j $(j = 1, \ldots, n)$, μ_k $(k \in \{1, \ldots, n\}\backslash\{j\})$ is taken as a given. To avoid the problem of stagnation without reaching a minimum, the calculation process consists of two stages: a descent cycle and a fusion cycle. In the descent cycle, the n-dimensional vector $\boldsymbol{\mu}$ is estimated using (4) as the objective function to be minimized. Let $\hat{\boldsymbol{\mu}} = (\hat{\mu}_1 \ldots, \hat{\mu}_n)'$ be the estimated value of $\boldsymbol{\mu}$. If $\hat{\mu}_j = \hat{\mu}_k$ $(j, k = 1, \ldots, n,\ j \neq k)$ is obtained, then we treat the two as a single parameter by unifying them. As a result, the space of $\boldsymbol{\mu}$ is reduced from n dimensions to b dimensions. Let $\boldsymbol{\xi} = (\xi_1, \ldots, \xi_b)'$, $b \leq n$ be the parameter vector reduced in dimension. The fusion cycle then performs the same estimation as the descent cycle for $\boldsymbol{\xi}$ reduced to b dimensions. If there is a unified estimate, the dimension of the parameter space is further reduced from b. The descent and fusion cycles are repeated until the dimension of the parameters does not reduce any further. The calculation algorithm is summarized in Table 1. Theorem 1 in the Table 1 will be described later.

Objective Functions and Update Equations. An expression for sequential minimization in the coordinate axes of the objective function in (4) is shown below. By focusing on μ_j, (4) can be rewritten as

$$\ell_{\mathrm{PL}}(\boldsymbol{\mu}) = \varphi(\mu_j \mid y_j, m_j) + 2\lambda \sum_{k \in D_j} w_{jk}|\mu_j - \mu_k| + \sum_{i \neq j} \varphi(\mu_i \mid y_i, m_i) + C_n + \lambda \sum_{\ell \neq j}^{n} \sum_{k \in D_\ell \backslash \{j\}} w_{\ell k}|\mu_\ell - \mu_k|, \quad (5)$$

where the function φ and the constant C_n are given by (3). By omitting the terms that do not include μ_j from (5), the objective function for the μ_j-coordinate axis can be expressed as

Table 1. Coordinate descent algorithm for GFL Poisson model (4)

Output : $\boldsymbol{\xi} = (\xi_1, \ldots, \xi_b)'$
Input : initial vector $\boldsymbol{\mu} = (\mu_1, \ldots, \mu_n)'$, $\lambda > 0$

1: **repeat**
2: (Descent cycle)
3: **for** $j = 1, \ldots, n$ **do**
4: update μ_j by Theorem 1
5: **end for**
6: define the number of distinct values of current solutions b for $\boldsymbol{\xi}$
7: **if** $b < n$ **then**
8: (Fusion cycle)
9: **for** $j = 1, \ldots, b$ **do**
10: update ξ_j by Theorem 1
11: **end for**
12: **until** the $\boldsymbol{\xi}$ converges

$$\ell_j(\mu) = \varphi(\mu \,|\, y_j, m_j) + 2\lambda \sum_{k \in D_j} w_{jk} |\mu - \hat{\mu}_k|, \tag{6}$$

where $\hat{\mu}_k$ is a given value.

A subdifferential of $\ell_j(\mu)$ at $\tilde{\mu} \in \mathbb{R}$ is defined by $\partial \ell_j(\tilde{\mu}) = [g_-(\tilde{\mu}), g_+(\tilde{\mu})]$, where $g_-(\tilde{\mu})$ and $g_+(\tilde{\mu})$ are left and right derivatives. The left and right derivatives at $\hat{\mu}_k$ are given by

$$g_-(\hat{\mu}_k) = \dot{\varphi}(\hat{\mu}_k \,|\, y_j, m_j) - 2\lambda \left\{ w_{jk} - \sum_{\ell \in D_j \setminus \{k\}} w_{j\ell} \operatorname{sign}(\hat{\mu}_k - \hat{\mu}_\ell) \right\},$$

$$g_+(\hat{\mu}_k) = \dot{\varphi}(\hat{\mu}_k \,|\, y_j, m_j) + 2\lambda \left\{ w_{jk} + \sum_{\ell \in D_j \setminus \{k\}} w_{j\ell} \operatorname{sign}(\hat{\mu}_k - \hat{\mu}_\ell) \right\},$$

where $\dot{\varphi}(x|a, b)$ is the derivative of $\varphi(x|a, b)$ with respect to x as

$$\dot{\varphi}(x|a, b) = \frac{d}{dx}\varphi(x|a, b) = b \exp(x) - a.$$

If there exists $k_\star \in D_j$ such that $0 \in \partial \ell_j(\hat{\mu}_{k_\star})$, then (6) is minimum at $\mu = \hat{\mu}_{k_\star}$. Since (6) is a convex function, $k_\star$ is unique if it exists.

Next, consider the case where $k_\star$ does not exist, i.e., there is no minimizer of (6) at a non-differentiable point. Let t_h $(h = 1, \ldots, r)$ be the hth order statistic of $\hat{\mu}_k$ $(k \in D_j)$. (S1), (S2), and (S3) show that the interval $\mu \in \mathcal{I}$ contains the optimal solution of (6), i.e., the optimal solution is contained in one of them.

$$\mathcal{I} = \left(t_{(\mathrm{L})}, t_{(\mathrm{R})}\right) = \begin{cases} (-\infty, t_1) & \text{(S1)} \\ (t_r, \infty) & \text{(S2)} \\ (t_{h_*}, t_{h_*+1}) & \text{(S3)} \end{cases},$$

where $t_{(\mathrm{L})}$ and $t_{(\mathrm{R})}$ are left and right endpoints, respectively, and $h_* \in \{1, \ldots, r-1\}$.

In $\mu \in \mathcal{I}$, we can rewrite the $|\mu - \hat{\mu}_k|$ in (6) without using absolute values. Let K_+ and K_- be sets of k such that $\mu - \hat{\mu}_k$ is positive or negative as

$$K_+ = \left\{k \in D_j \mid \hat{\mu}_k \leq t_{(\mathrm{L})}\right\}, \quad K_- = \left\{k \in D_j \mid t_{(\mathrm{R})} \leq \hat{\mu}_k\right\}.$$

By using the sets K_+ and K_-, we have

$$\begin{aligned}\sum_{k \in D_j} w_{jk}|\mu - \hat{\mu}_k| &= \sum_{k \in K_+} w_{jk}(\mu - \hat{\mu}_k) + \sum_{k \in K_-} w_{jk}(\mu - \hat{\mu}_k) \\ &= \tilde{w}_j \mu - u_j,\end{aligned} \tag{7}$$

where $\tilde{w}_j$ and u_j are given by

$$\tilde{w}_j = \sum_{k \in K_+} w_{jk} - \sum_{k \in K_-} w_{jk}, \quad u_j = \sum_{k \in K_+} w_{jk}\hat{\mu}_k - \sum_{k \in K_-} w_{jk}\hat{\mu}_k.$$

Then, (6) can be rewritten with (7) as follows:

$$\ell_j(\mu) = \varphi(\mu \mid c_j, m_j) - 2\lambda u_j, \quad c_j = y_j - 2\lambda \tilde{w}_j. \tag{8}$$

It follows from a simple calculation that c_j is positive when $k_\star$ does not exist. The first-order differential equation for (8) is $d\ell_j(\mu)/d\mu = \dot{\varphi}(\mu_j \mid c_j, m_j)$, $(\mu \in \mathcal{I})$. This means that the optimal solution is $\mu = \log(c_j/m_j)$. The derivation of the optimal solution of (6) is summarized below as a theorem.

Theorem 1. *The optimal solution of* (6), $\hat{\mu}$, *is obtained in a closed form as*

$$\hat{\mu} = \begin{cases} \hat{\mu}_{k_\star} & (k_\star \text{ exists}) \\ \log(c_j/m_j) & (k_\star \text{ does not exist}) \end{cases}.$$

Optimization of $\boldsymbol{\lambda}$. The λ in the objective function (6) is an unknown value, and (6) is not a convex function with respect to the λ. Since λ is not uniquely solvable, we list candidates for λ, obtain $\boldsymbol{\mu}$ estimates under each candidate, and select the best estimates among them using a consistent information criterion, i.e., the Bayesian Information Criterion (BIC) of [9]. For a candidate λ, if its minimum and maximum values are known, we can focus only on the values in between. The minimum value of λ, $\lambda_{\min}$, is considered to be λ when there is no unification at all among the elements of $\boldsymbol{\mu}$. On the other hand, the maximum λ, $\lambda_{\max}$, is the λ when the $\boldsymbol{\mu}$ are unified into one scalar, $\mu_{\max}$. As shown in standard textbooks on sparse estimation, e.g., [5], we know that as λ increases from 0, the unification of $\boldsymbol{\mu}$ increases.

The estimated value $\hat{\mu}_{\max}$ of $\mu_{\max}$ is obtained by minimizing $\boldsymbol{\mu} = \mu$ from (4) and is $\hat{\mu}_{\max} = \log(\sum_{i=1}^n y_i / \sum_{i=1}^n m_i)$. In the situation where $\hat{\mu}_{\max}$ is obtained in the objective function (6), the following subdifferential is satisfied for all j:

$$\begin{aligned}&0 \in \partial \ell_j(\hat{\mu}_{\max}) \\ \Leftrightarrow\ &\dot{\varphi}(\hat{\mu}_{\max} \mid y_i, m_i) - 2\lambda \sum_{k \in D_j} w_{jk} \leq 0 \leq \dot{\varphi}(\hat{\mu}_{\max} \mid y_i, m_i) + 2\lambda \sum_{k \in D_j} w_{jk}.\end{aligned}$$

Therefore, λ must satisfy the following:

$$\lambda \geq \lambda_{j,\max} = \frac{|\dot{\varphi}(\hat{\mu}_{\max} \mid y_i, m_i)|}{2\sum_{k \in D_j} w_{jk}}.$$

Let $\lambda_{\max}$ be the largest value among $\lambda_{j,\max}$. Then,

$$\lambda_{\max} = \max_{j \in \{1,\ldots,n\}} \lambda_{j,\max}.$$

The interval $[0, \lambda_{\max}]$ can be subdivided to list candidate λs. The BIC is obtained from the candidate λ and the $\boldsymbol{\mu}$ estimate obtained under that λ.

$$\mathrm{BIC} = 2\ell(\boldsymbol{\xi}) + df \log n. \tag{9}$$

Notice that, since $\ell(\boldsymbol{\xi})$ is a negative log-likelihood function, there is no need to put a minus before 2 in (9). The $\boldsymbol{\xi}$ is the realization of "Output" $\boldsymbol{\xi}$ in Table 1, and df is the degree of freedom, $df = b$. The λ with the smallest BIC and the $\boldsymbol{\xi}$ obtained under that λ are the estimation results.

4 Application and Conclusion

To illustrate the proposed method, crime rate data for 43 municipalities in Osaka Prefecture, observed over a 14-year period from 1995 to 2008, were used. Here, crime rate is the number of crimes divided by the population; in (1) and (2) the number of crimes is represented by y_i and the population by m_i. Space-time μ_i, $i = 1, \ldots, n$ represents 14 years in time and 43 municipalities in space, $n = 14 \times 43 = 602$. The estimation results are not displayed as μ_i itself, but as an exponential function $\exp(\mu_i)$ according to the link functions (2). The adjacencies of a given space-time are the space itself at times before and after the given time, and the space bordering the given space at the given time. From Fig. 1, taking the adjacencies of the "city 301 in 2000" as an example, city 301 around the year 2000, i.e., "city 301 in 1999" and "city 301 in 2001", and the municipalities adjacent to city 301 in 2000, i.e., "city 207 in 2000" and "city 210 in 2000" are considered to be adjacent. In this case, the μ_j is the effect of the "city 301 in 2000", and the identification numbers of "city 301 in 1999", "city 301 in 2001", "city 207 in 2000", and "city 210 in 2000" are included in D_j as elements. Space-time and their adjacencies are limited to the space-time in Osaka Prefecture between 1995 and 2008 within the years of observation. For example, the year 2009 is not added as an adjacency to the endpoint 2008. In addition, adjacencies to municipalities in prefectures other than Osaka are not considered.

In the estimation results, the optimal model selected for the BIC is $\lambda = 0.618$, unifying the space-time of 602 to 408. The most unified space-times include 10 municipalities: Kishiwada City (202) and Izumi City (219) from 2005 to 2008, Izumiotsu City (206) in 2008, and Tadaoka Town (341) in 2007. The numbers in parentheses are the municipal codes. Specific areas can be located by referring to Fig. 1, which shows the 43 municipalities of Osaka Prefecture. As the map

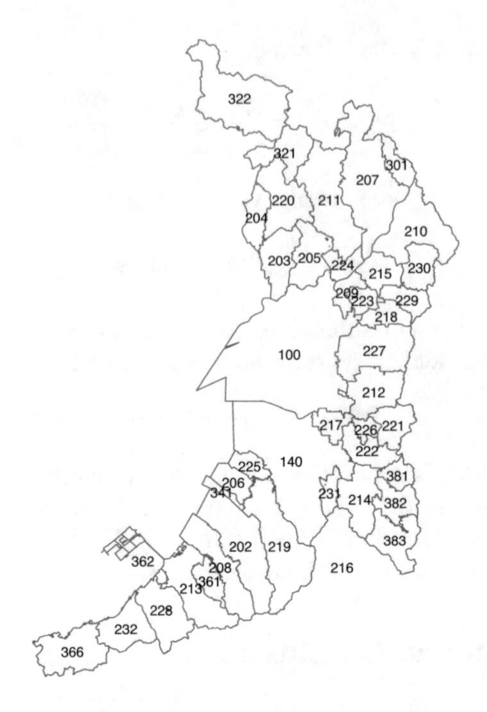

Fig. 1. White map of Osaka Prefecture and municipality codes

indicates, the 10 unified municipalities noted above border each other in the southern part of the prefecture. As a summary of the estimated results $\exp(\mu_i)$, $i = 1, \ldots, 602$, the minimum and maximum values were respectively 0.004 for Toyono-cho (321) in 2007 and 0.052 for Osaka City (100) in 2001. The respective median and mean values were 0.019 and 0.020. The results closest to the median were 0.019 for Sennan City (228) in 1999 and Izumisano City (213) in 2000. Those closest to the mean were 0.020 for Izumiotsu City (206) in 1996, Hirakata City (210) in 2000, and Takatsuki City (207) in 2002 and 2003. The estimated spatio-temporal effects were displayed in 14 maps of Osaka Prefecture for the years 1995 to 2008. To illustrate, Fig. 2 shows the maps for the first and last years, 1995 and 2008, as well as the year 2001, which had the highest crime rate in the entire space-time. In addition, five municipalities were selected as representative of the 43 municipalities in Osaka Prefecture. The chart in the lower right corner of Fig. 2 shows the crime rates for the five municipalities over the 14-year period. The vertical axis identifies the municipal codes for the five municipalities: Nose Town (322), Minoh City (220), Izumisano City (213), Sakai City (140), and Osaka City (100). In the map for 2001 (top right corner of the figure), the deeper red color in the area in and around Osaka City (100), the second largest city in Japan after Tokyo and Yokohama, indicates a high crime rate for the area during that year. In fact, as shown in the chart in the lower right corner of the figure, the crime rate in Osaka City was consistently higher than

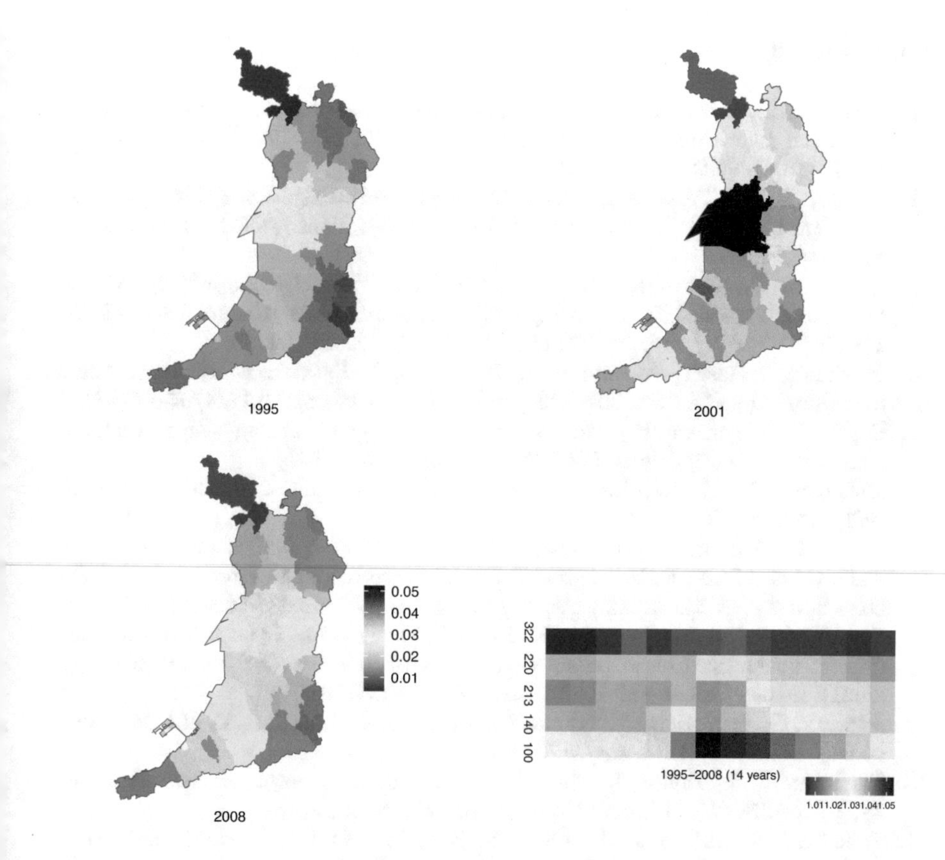

Fig. 2. Estimates of spatio-temporal effects

in the other areas throughout the 14-year period. Careful examination of the three maps in Fig. 2 reveals that some of the municipalities are indeed unified. Particularly in 2008, a large light-blue-to-near-white unification can be seen in the south. This corresponds to the 2008 unification status of the aforementioned 10 municipalities.

Analysis of the data showed that crime rates are higher in urban areas such as Osaka City and lower as one moves away from the urban areas and into the countryside. Thus, the results obtained from the GFL Poisson model are similar to those typically observed in daily life in Japan. Based on this study, we suspect that coarsening the estimates to, for example, five decimal places would make the evaluation more gradual, unify more municipalities, and may be more suitable for actual data analysis. This is an issue left to future study.

Acknowledgments. This research was supported by JSPS Bilateral Program Grant Number JPJSBP 120219927 and JSPS KAKENHI Grant Number 20H04151.

References

1. Arnold, T.B., Tibshirani, R.J.: Efficient implementations of the generalized Lasso dual path algorithm. J. Comput. Graph. Stat. **25**, 1–27 (2016). https://doi.org/10.1080/10618600.2015.1008638
2. Arnold, T.B., Tibshirani, R.J.: Genlasso: path algorithm for generalized Lasso problems. R Package version 1.6.1 (2022). https://CRAN.R-project.org/package=genlasso
3. Brunsdon, C., Fotheringham, A.S., Charlton, M.E.: Geographically weighted regression: a method for exploring spatial nonstationarity. Geogr. Anal. **28**, 281–298 (1996). https://doi.org/10.1111/j.1538-4632.1996.tb00936.x
4. Friedman, J., Hastie, T., Höfling, H., Tibshirani, R.: Pathwise coordinate optimization. Ann. Appl. Stat. **1**, 302–332 (2007). https://doi.org/10.1214/07-AOAS131
5. Hastie, T., Tibshirani, R., Wainwright, M.: Statistical Learning with Sparsity. The Lasso and Generalizations. CRC Press, Boca Raton (2015)
6. McCullagh, P., Nelder, J.A.: Generalized Linear Models, 2nd edn. Chapman & Hall, London (1989)
7. Ohishi, M., Yamamura, M., Yanagihara, H.: Coordinate descent algorithm of generalized fused Lasso logistic regression for multivariate trend filtering. Jpn. J. Stat. Data Sci. **5**, 535–551 (2022). https://doi.org/10.1007/s42081-022-00162-2
8. Ohishi, M., Fukui, K., Okamura, K., Itoh, Y., Yanagihara, H.: Coordinate optimization for generalized fused Lasso. Comm. Statist. Theory Methods **50**, 5955–5973 (2021). https://doi.org/10.1080/03610926.2021.1931888
9. Schwarz, G.: Estimating the dimension of a model. Ann. Stat. **6**, 461–464 (1978). https://doi.org/10.1214/aos/1176344136
10. Tibshirani, R.J., Taylor, J.: The solution path of the generalized lasso. Ann. Stat. **39**, 1335–1371 (2011). https://doi.org/10.1214/11-AOS878
11. Tibshirani, R., Saunders, M., Rosset, S., Zhu, J., Knight, K.: Sparsity and smoothness via the fused Lasso. J. Roy. Stat. Soc. Ser. B **67**, 91–108 (2005). https://doi.org/10.1111/j.1467-9868.2005.00490.x
12. Yamamura, M., Ohishi, M., Yanagihara, H.: Spatio-temporal adaptive fused Lasso for proportion data. Smart Innov. Syst. Tec. 238, Intelligent Decision Technologies 2021: Proceedings of the 13th KES International Conference on Intelligent Decision Technologies KES-IDT-2021 (eds. I. Czarnowski, R. J. Howlett & L. C. Jain), 479–489 (2021). https://doi.org/10.1007/978-981-16-2765-1_40
13. Zou, H.: The adaptive Lasso and its oracle properties. J. Amer. Statist. Assoc. **101**, 1418–1429 (2006). https://doi.org/10.1198/016214506000000735

Large-Scale Systems for Intelligent Decision Making and Knowledge Engineering

An Ontology Model to Facilitate Sharing Risk Information and Analysis in Construction Projects

Heba Aldbs[1(✉)], Fayez Jrad[1], Lama Saoud[1], and Hadi Saleh[2]

[1] Tishreen University, Lattakia, Syria
heba.aldbs@gmail.com, lama.saoud@tishreen.edu.sy
[2] HSE University, Moscow, Russia
hsalekh@hse.ru

Abstract. Construction projects face a high level of dynamic and various risks. Risks may result in deviation from pre-determined construction project' objectives. Systematic risk analysis is critical for sharing risk information related to decision making thus effective risk management. In this study, a Risk Analysis Ontology RA-Onto is proposed that may facilitates development of databases and information sharing for risk analysis. A detailed review of the literature on construction risks has been carried out to development of the RA-Onto that organizes risk knowledge into unified classes together with corresponding properties and relations. Ontology is evaluated theoretically and practically by using five case studies. RA-Onto could be used to support decision-making during the risk management. It enables companies to corporate memories, create databases, and develop a model to support the systematic risk analysis for better decision making.

Keywords: Risk Analysis · ontology · taxonomy · construction project

1 Introduction

Construction projects consist of different phases and involve many cooperative participants, and have to be performed under conditions 'variations. As consequence, projects become even more complex and the risks may be increased. Recently, the methods of the risk management process have made great progress in dealing with the risks. During the risk management process, the causal and consequences of risks should be analyzed in detail during all stages of the project and the allocation of the risks to each party should be determined properly [1]. The simple methods of risk analysis; such as, "checklist" or "risk breakdown structure"; no longer satisfy the analysis of risk requirements. Because, they are not generic and based on case-dependent models; they suffer from the weak representation of knowledge; and don't take into account the interrelationships among risks [2, 3] It is necessary to understand the causal relations between risk events, and their consequences. Within this context, authors have proposed approaches based on cause-effect diagrams, risk paths and maps, such as ontology. Several types of research

I. Czarnowski et al. (Eds.): KESIDT 2023, SIST 352, pp. 237–249, 2023.
https://doi.org/10.1007/978-981-99-2969-6_21

proposed ontologies in the construction domain for the representation, and sharing of knowledge. An ontology is "a formal specification of a shared conceptualization of a domain." [4]. An ontology can facilitate semantic interoperability [5]. Ontology can be used to represent human knowledge, and can be understood by the computer [5]. Thus, using ontology may increase the chance of sharing risk knowledge. Especially the information is recorded in text form-based documents in the project. An ontology may provide a systematic method to structure construction risks.

2 Ontology and Risk Management in AEC Industry

Ontology technology has played a vital role in formalizing knowledge and used for various research purposes: [6] developed an ontology in multiple detailed levels with represent the common vocabulary in construction but relations between concepts aren't represented in the ontology. [7] developed an ontology-based risk management framework for contractors, he has adopted Information Retrieval algorithm to develop the dynamic ontology extraction tool and update it. [2] proposed an ontology as a database system of risk concepts in projects and developed common concepts to explain the interrelationships between risk paths and consequences on cost overruns. [8] developed an ontology for delay analysis to support decision-making during claim management. [9] developed a method to predict the propagation of change within a building information model environment in order to identify interdependencies between design parameter relations. [3] developed an ontology to support safety risk knowledge. The studies presented above focused on ontology applications with different research purposes and established their benefits in describing and retrieving the knowledge. But most of those studies' scope focuses on cost overruns, and there is no generic ontology represent concepts in construction in multiple detailed levels with related relations between concepts. There is a need for a systematic approach to capture realistic risk scenarios. It can overcome these gaps, by constructing risk analysis ontology RA-Onto. This systematic knowledge may guide risk causes and consequences and enable its users to deal with the correct response. In practical terms, the development of a database on risk analysis for use by companies may make the process easier to understand while also serving as a knowledge reference for other projects. Accordingly, constructing a decision support system may increase the chance of prevention of risk occurrence. The motivation for this study is hence the possible application of the ontology in such company-specific attempts at risk analysis. The main objective is to introduce a RA-Onto that aims to offer a semantic knowledge base. RA-Onto captures and shares the risk information that are available in the literature efficiently. So, forming a basis for the construction of databases; decision support systems for qualitative or quantitative risk.

3 Research Methodology and Stages

There is no one specific way to construct ontologies, as well as the possibility of creating several ontologies for the domain [10]. However, the main rule for ontology building is the consideration of nouns and verbs in the domain representing objects and relations respectively. To develop and maintain consistency, Methonology is the most mature approach which is adopted in this study [11].

3.1 Steps of Ontology Development

The five main steps of METHONOLOGY as detailed by [11] are displayed with corresponding deliverables at each step in Fig. 1 as follow: *Specification* (Identify the purpose of an ontology); *Conceptualization* (representation of knowledge by the conceptual model and review existing ontologies); *Formalization* (represent the knowledge in a formal language by enumerate classes and its hierarchy); *Implementation* (construct ontology in editor tool); *Maintenance* (evaluate the ontology).

Specification. The competency questions are used in the Methontology to identify the ontology requirements, purpose, and end-users. They are used to develop the model and at the evaluation stage [12]. The purpose of RA-Onto is deemed to treat one of the important problems in construction sector. As mentioned before about the importance of risk analysis, so sharing the information of risk causes through the ontology is aimed to enhance risk analysis by integrating the knowledge of risk that is available in the literature. This improved knowledge can serve as a checklist for the prevention of risks in the preconstruction phase of project. Experts and risk assessors are main users of the ontology. Further questions are asked to identify the requirements of building ontology. While these questions are used to materialize the right set of concepts and relations [12]. In the questions, terms adjectives, verbs, and names are represented as properties, relations, and individuals consecutively [13]. In the study, some examples of questions are developed: "What is risk?", "What are causes of risk?", "How can risks caused by several sources happen?", "How can changes in the risks lead to opportunities or threats?" , "What affects project implementation outputs (cost, quality or time)?".

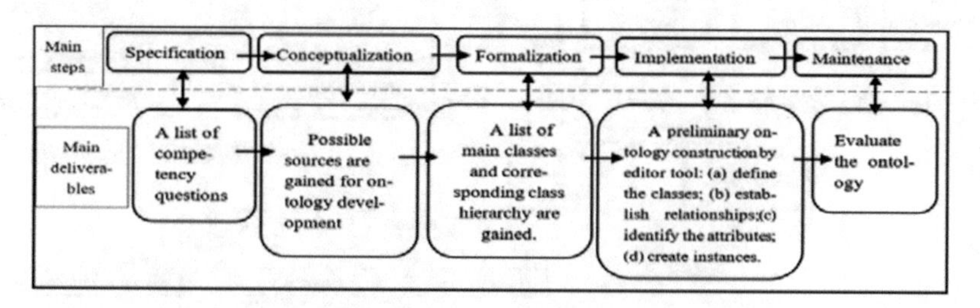

Fig. 1. Methodology of building ontology [11].

Conceptualization. It focuses on reviewing existing models as a shorthand for building and transformed into a conceptual model. This step was carried out in two stages:

- *First Stage: A conceptual model for the taxonomy.* Summing up the definition of risk, "*an uncertain event affects one of the project objectives*" [1]. Thus, each term "*sources, events,* and *consequence of risk*" are basic concepts in the conceptual model, Fig. 2. With taking into account the integrated processes of project and integration of stages based on communication flow and interactive between the processes; participants and outputs of each process under the internal and external environment surrounding of the project. RA-Onto is proposed to represent the formalized domain of risk sources.

A set of risk sources is collected from the related research work papers of different classifications of risk. These sources may directly affect the project objectives or interact with each other as a risk path.

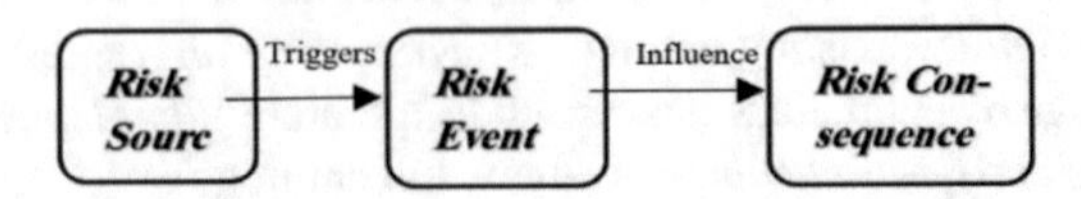

Fig. 2. Illustration of the Conceptual Model for the Taxonomy.

- *Second Stage: Review previous models.* Several influential conceptual models and taxonomies previously within the construction have been reviewed to identify commonly used root concepts to initiate the conceptualization process, they are (Fig. 3.):

  1. *Information Reference Model for Architecture, Engineering, and Construction IRMA.* [14]. *IRMA* identifies some main concepts and relations of projects Fig. 3-a.
  2. *Industry Foundation Classes* [15]. as the data model for describing construction industry data. In sub-levels are divided into 6 concepts Fig. 3-b.
  3. *Domain* taxonomy for construction concepts (*e-COGNOS (*[6]. it uses seven significant domains to classify construction concepts Fig. 3-c.

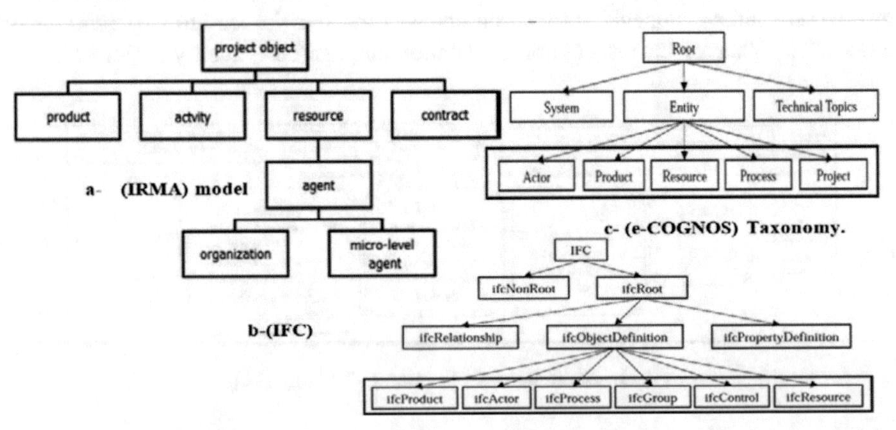

Fig. 3. Excerpt of the models and corresponding hierarchy: a- (IRMA) model [14]; b- (IFC) model [15]; c- (e-COGNOS) Taxonomy [6].

All of these models can be used as a guideline for establishing root classes of the taxonomy of the construction domain. The justifications for deciding on the required concepts is explained as follows:

- Choose the most commonly used classes among the classifications.
- Concepts must be applied in the construction domain and classes that are irrelevant to it should be discarded. E.g., "*ifc group*" represents collections of objects that are

compiled for a particular purpose such as for its role, and it should not be picked. As well as "IfcControl" which are the rules controlling time, cost, quality.
- If a class is equivalent to a possible concept, then that class should be chosen. E.g., "*Agent*" is equivalent to *"Actor"*.

Based on the rationales for capturing concepts from existing classifications, and the review in the two stages, six root classes are selected for the taxonomy development. They are: *Process, Participant, Product, Resource, Contract and Environment.*

Formalization. Frame-based models or description logic are used to formalize the conceptual model and formalizing the ontology [10]. Where concepts are represented as frames that help easily to define classes; properties; relations and instances of the ontology [13]. Concepts represent a set of entities that describe a domain. Taxonomy (or class hierarchy) is a concept scheme for organizing terms and concepts in domain knowledge, and it needs to utilize some relationships among the concepts to organize them at many detailed levels such as "Subclass-Of" & "Is-A" relations. A middle-out strategy [13] is used for building a class hierarchy, and adopt a taxonomy that organizes the root concepts into three levels L(1), (2), (3).

Implementation. Several ontology editors have been reviewed to choose a suitable one for coding. Protégé (v.5) is free, open-source ontology editor. It can provide a visual environment to edit, save an OWL ontology [16].

a. *Define the Classes.* Main classes of the ontology are:

PARTY (participant): includes the participants involved in the processes of project. They perform activities by roles and are linked to each other under contracts [2, 7].

PROCESS: includes procedures that should be done or followed by participants for certain activities. It consists of three subclasses:Phase; Activity &Task [6].

RESOURCE: an entity is needed to perform the work and used by a participant to achieve an outcome such a product [2, 6].

PRODUCT: refers to all elements that a specific participant of the project is required by the contract to provide. It can include tangible construction products such as a column or a beam. or knowledge elements such as documents, reports [6].

CONTRACT: is physical document that specifies the contractual relationships between two parties, e.g., the contract between an owner and a contractor [7].

EXTERNAL ENVIRONMENT: causes that emphasize things out of the project Parties' control (such as nature, Economic conditions, Legal environment…) [3, 6].

b. *Relations Among Concepts:* Relations are ways in which concepts and individuals can be related to one another. They provide the basis of the ontology and background information. Three categories of relations are used in RA-Onto:
   (1) "*Subsumption*": "**is-a**" relationships. It creates sub-class trees.
   (2) "*Meronomy*": "**is-a part** "describe concepts that are parts of other concepts.
   (3) *"Associative"*: uses to link a concept with another in the same level. E.g., "**influence**" & "**cause**" are causality relationships among concept.

The relations between concepts are available in Table 1. And Fig. 4. The relations "is a" and "is a part of" used to link root concepts to their sub-concepts. The associative relation is used to link different concepts in the same level, and forms a high-level object model. The associative relations of concepts at sub-levels are inherited from the upper level, as well as it can be restricted to provide a better definition of an ontology [17].

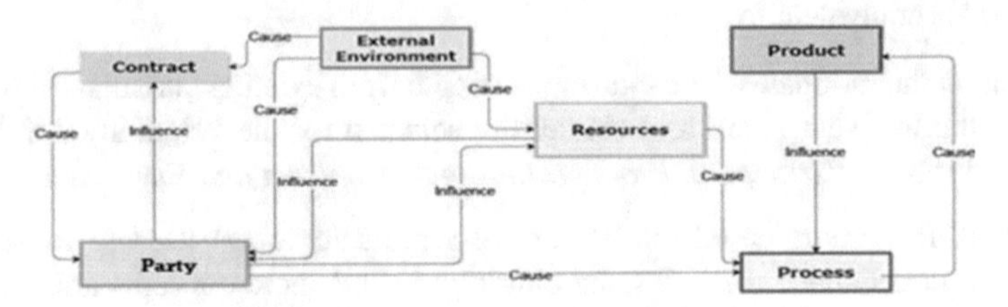

Fig. 4. RA-Onto high-level object model.

For example, the associative relation between concepts parties-resources at levelL1 is inherited to L2 and is restricted, Fig. 5. e.g., *Material and Equipment* influence only suppliers at L2. That is because the suppliers are responsible parties for supplying the construction requirements. While, laborers don't relate to the designer or supplier but to the contractors. Accordingly, relations between concepts developed at the two L 2, 3. Figure 4 & Fig. 6 only display the relationships at L1, 2 due to the complexity of L3.

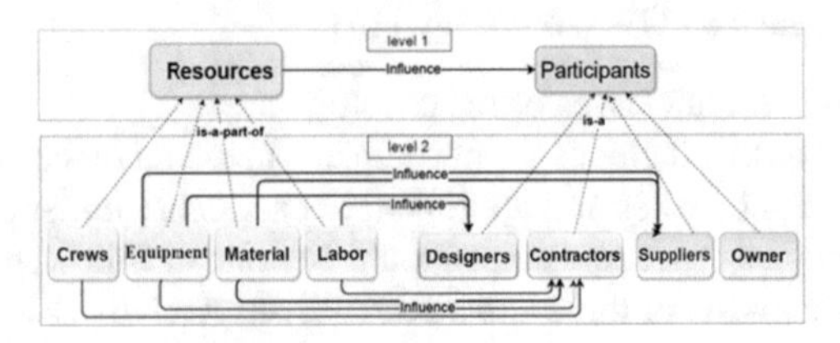

Fig. 5. Party and Resource and their Inherited Relations from Top-Level 1 To Level 2.

c. *Attributes and their relation*:

(1) *Attributes.* Attributes are properties that describe concepts. *R*equired risk information is represented as attributes. For more expression, properties are restricted with either a single value or a set of values. E.g., "*participant*" is characterized with Ability and "*Resource*" with Quantity and "*Process*" with Cost, and the restrictions imposed on the attributes take the values (increase, average, decrease) for Quantity, (long, natural, short) for Duration and (Lack, expected, sufficiency) for Ability. So, the low ability of contractors, like inaccurately calculating resources, influences reducing the number of resources which thus impacts on delay of the process. The attributes of concepts are inherited with sub-concepts. The attribute can be restricted according to the concepts 'relation such as "**is part of**" or "**is-a** "relations [17] Fig. 7:

- In **"is-a"** relations: attributes are inherited with exceptions, where sub-concept inherits properties of its concept with some modifications. E.g., "*Executing Process*" is a "*Process*", hence the Duration of the "*process*" will be inherited, but "*executing* process" has new attributes as follows, Duration to perform work and Duration to adjust work, concerning Duration.

Table 1. Relations at Concepts Level

Relation Name	Source Concept	Target Concept
(Is-a)	Participant	contractor, designer, owner
	Contract	Contractor; supply; Design Contracts
	Product	Information & Construction Products
	Process	Initiating, Planning, Executing …
(Is-a-part- of)	Resources	Crews; Equipment; labors…
	Environment	Economic; Weather Condition
(Cause)	Environment	Contract, Participant, Resources
	Resources	Process
	Contract	Participant
	Participant	Process
	Process	Product
(Influence)	Product	Process
	Resource	Participant
	Participant	Contract, Resources

- In "is a part of" relations: attributes inherited with distinguishing, where it is refined to provide a better definition of concept. E.g., Quantity of "*Resource*" in L1 is refined to: Requirement of unit work and Total work of "*Labor*" in L2. Similar to that; attribute continues to be refined in L3.

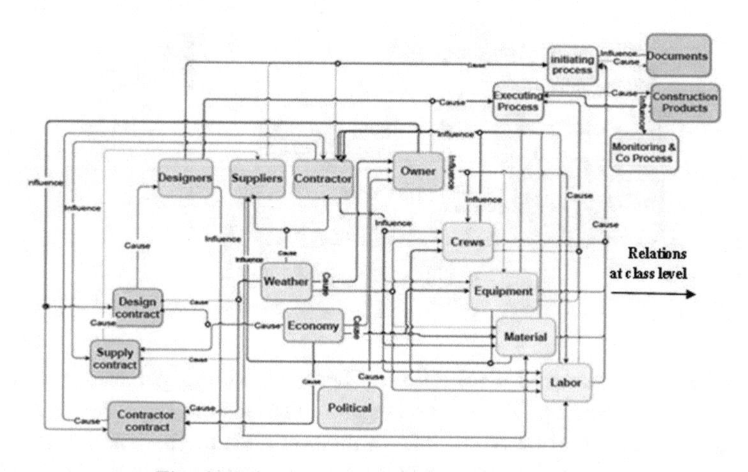

Fig. 6. Relations Between Concepts at Level 2

(2) *Relations at instances level.* Relationships also exist mainly among instances of concept. The associative relations between concepts are inherited to their

attributes between levels with restricted them, as some attributes interact with some others and describe possible risk paths. Relationships at the instances level representing risk cases were determined using face-to-face interviews. Where the basic reliance was made on detailed interviews with implementation engineers. And relations between attributes at levels 1, 2, 3 were developed. Fig. 8 displays the relations between the attributes in L1. Figure 9-(a, b) displays a part of taxonomies and relations between attributes for "*Resource*"; "*Participants*" which stem from Fig. 8, Table 2. Explanation about relations.

(3) *Creating Instances.* In total, 32 different projects performed in Syria are examined to develop a database. The knowledge base regroups more than 65 instances of risk-cases. Each case consists of a risk path. Previewed reports and documents analyzed to extract risk information. Collected risk cases are stored through Protégé 5.0 according to the information expression mechanism of RA-Onto. Risk cases are input according to the different detailed levels of description at levels 1, 2, 3.

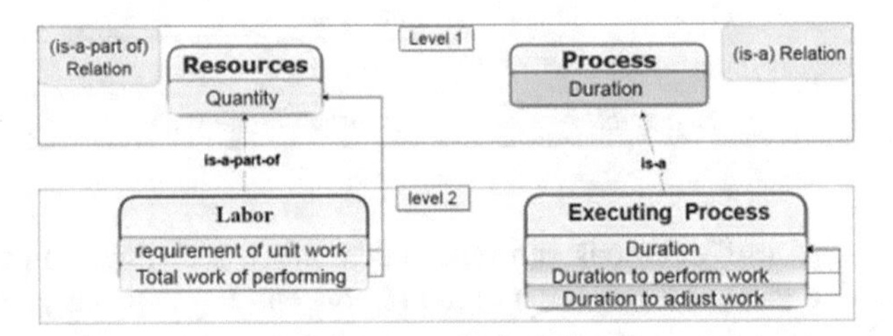

Fig. 7. Attribute inheritance of "*is-a*" & "*is a part of*" Relations.

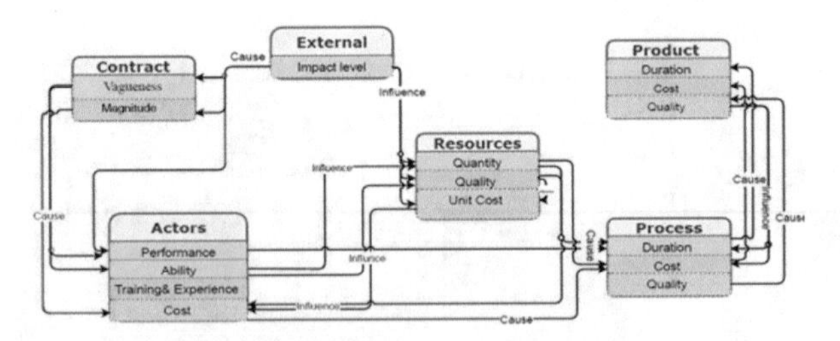

Fig. 8. Relations at Instances Level 1.

Ontology Maintenance and Evaluation. There is no one right way to evaluate an ontology [10]. The ontology evaluation stage emphasizes the importance of verifying the quality and validity of ontology. The evaluation aims to validate that an ontology meets the requirements and purpose of use. The important part is "validation" which means acceptance and suitability of the ontology based on reviewed literature with the real case study, with a point to the need to update the ontology through all phases of its life cycle to keep valid [18]. The evaluation consists verification and validation [11]. Verification is used to check the correctness content of ontology through a set of

Table 2. Examples of Relations between inherited attributes at levels 1 and 2.

"Concept"-Relation- "Concept"	Source Concept 'feature	Target Concept 'feature	EXAMPLE
Level (1)			
"Resource"-Influence- "Party"	*Unit cost*	*Cost*	Party calculates his cost based on *unit cost* of resource
"Party"-Influence-"Resource"	*Ability*	*Quality*	if Party is equipped with all the appropriate resources, then he will be capable of performing the work efficiently
Level (2)			
"Labor"-Influence- "Contractor"	*Work Time Cost*	*Building Cost*	Increasing or decreasing labor' *work time cost* influences contractor' *Building cost*
"Contractor"-Influence-"Labor"	*Technical Ability*	*Requirement of Unit Work*	The contractor's good *technical ability* (such as using computer software) influences the *requirement of unit work*

criteria, such as consistency, Conciseness, and completeness. While validation refers to the accordance of ontology to the real world that is supposed to represent. Validation consists of: - checking by the competency questions proposed in the first step, and validation by using one of the following approaches (expert interview, survey, and case study). In this study, the evaluation criteria predefined by several researchers are deemed [4, 11, 19]. The ontology structure is built through a review of previous studies about analysis of causes and events of risks as well as the interactions between them. A common vocabulary is used to prevent ambiguity and form the uniform representation system with the ability to update it easily. This achieves the clarity and redundancies criteria. Also, the ontology structure was formed using a reliable tool (Protégé). Pellet reasoner was used to verifying the consistency [20]. RA-Onto was confirmed to be coherent and consistent. Finally, the competency questions were checked throughout the phases of the ontology under development. The validation was selected by using a real case between many other options [7, 8].

i. *Testing cases study of ontology validation.*

In this study, the ontology is constructed to form a database for storing risk cases and reusing this knowledge. The ontology is used by a specialist responsible for risk

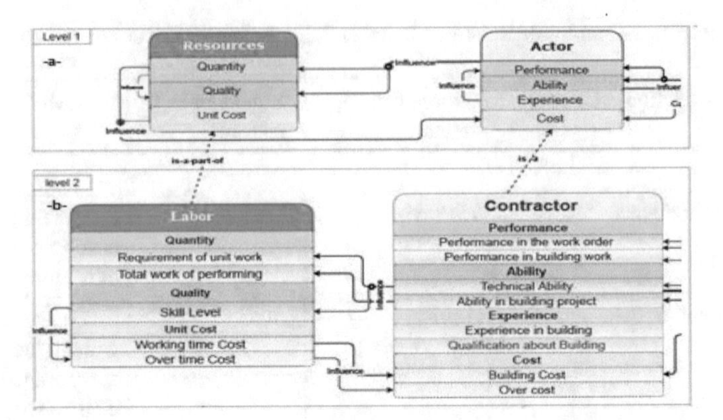

Fig. 9. (a) Relations between (Resource and Party) in level 1, (b) Inherited concepts' attribute relations from L1 to L2.

management in the initiating phases of the project. The utility of ontology checking by testing its applicability to real risk cases for different projects is as follows: Project reports and documents are used as case studies for risks. Five reports to four projects are compared in Table 3. Each report is divided into a group of phrases that are compared with corresponding concepts within the ontology. During the comparison, concepts or attributes that denote the same information are not repeated, but there are some attributes or concepts that are dealt with more than once within the same document. The quotation mark " " is used to place phrases in the document and attributes of concepts. Each concept and its top-level concept within the ontology are determined. The term 'sub' refers to taxonomic relations between attributes or concepts. Associative are expressed by the demonstration "-relation_name-". E.g.,"- relation_influence-". Risk paths form the "phrases in ontology" as: " attribute" – relationship_name - " attribute", which corresponding to the 'cases. The concepts in the document corresponding to the ontology phrases are marked with a dotted line. Ontology phrases are presented according to the order in which the concepts appear in the document. The details of the cases are shown in Table 3.

ii. *Description of the Case Studies.*

- **Case study (A (1, 2)) (shopping mall).** (2013), cost overruns was 97%, the most important risks occurred due to the economic conditions. With review the reports, contractor requested to increase the cost of excavation and backfilling works. Due to the increasing cost of fuel used for drilling and transportation mechanisms. From Table 3. The underlying causes can be referred occurrence of the risk "increase cost fuel" is "national economic inflation". By ontology, it can infer other cases that were caused by inflation, case A (2). Where the inflation caused an increased salary of the workers who used in the executing works. Then increase building cost of contractor.
- **Case Study (B) (Military Police Building).** (2012), the cost overrun and delay was 80%, 31%. Con. Owner ordered to change the design several times, which caused changes in design parameters by the designer, and construction works was delay.

Table 3. Examples from Case Studies A (1, 2), B, C, D

PHRASE IN THE REPORT	PHRASE IN THE ONTOLOGY
Case (A1): According to the monthly reports provided to us, the contractor informs the owner about the variations in cost, where $_1$contractor demanded an increase of estimated cost of $_2$excavation and backfilling works by 25% of bill quantities, $_3$due to increase the cost of fuel necessary for the required transportation machines that caused by $_4$bad economic conditions.	"$_1$cost to perform construction work" -sub attribute of -"cost to perform work"-sub-attr. of-"cost". "$_2$ Engineering and construction tasks"-subclass of "Executing process"- subclass of-"Process". "$_{3-a}$Cost of fuel-energy of transport process"-sub attr. of -"Transportation cost"-sub-attr. of-"cost". "$_{3-b}$Costruction Equipment"-subclass of-"Equipment "-subclass of-" resources". "$_{4-a}$Cost push inflation"-sub attr. of -"Inflation"-sub-attr. of-"Impact level". "$_{4-b}$ National economy"-subclass of- "Economy"- sub-class of-" External Environment"
RISK PATH: "$_{3-a}$Cost push inflation(value=Increasing)"-causes-"$_{2-a}$Cost of fuel-energy of transport process(V=High)"-influence-"$_{1-a}$ cost to perform construction work (V=High)".	
Case (A2): Based on the bill of quantities document and the price schedule, $_1$the building contractor responsible for executing the structure demanded the consulting engineer to $_2$increase the bill of quantities due to $_3$the increase in workers' wages.	"$_1$ Building cont." - subclass of -"Cont."- subclass of -"Party". "$_2$ Main cost building" -sub attr. of -"Building cost"-sub-attr. of-"cost". "$_{3-a}$ worker" - subclass of -"Labor"- subclass of -"Resource". "$_{3-b}$ Salary of each unit of time performing" -sub attr. of -"Working time cost"-sub-attr. of-"Cost".
RISK PATH: "Cost push inflation (V=Increase)"-causes-"$_{3-b}$ Salary of each unit of time performing (V=High)"-influence-"$_2$ Main cost building (V=High)".	
case (B): From analyzing the project report, it is obvious that the contractor waits for the permission of his request in reply to design changes ordered $_1$by the owner. The contractor requests a $_2$extension of time (50 days as a minimum) until the $_3$new design report is issued.	"$_{1-a}$ Public "- subclass of -"Owner"- subclass of -"Party". "$_{1-b}$ Time management ability" -sub attr. of -"managerial ability"-sub-attr. of-"Ability". "$_{2-a}$ Duration of re-calculating" -sub attr. of -"Duration to adjust work"-sub-attr. of-"Duration". "$_{2-b}$ Preliminary design activity" - subclass of -"Initiating process "- subclass of -"Process". "$_{3-a}$ Report of Preliminary design "- subclass of -"information product"- subclass of -"Product". "$_{3-b}$ Duration to have drawing of re-designing " -sub attribute of -"duration of adjust works"-sub-attribute of-"Duration".
RISK PATH: "$_{1-b}$Time management ability (V=Inefficient)"-causes-"$_{2-a}$ Duration of re-calculating (V=Long))"-influence-"$_{3-b}$ Duration to have drawing of re-designing (V=Long)".	
Case C:; $_1$the rejection of the geological report and $_2$unclear contract clauses decreased delays	$_{1,2}$(-The term will be added-)
Case (D): There are $_1$construction delays due to $_2$heavy precipitation. The contractor claims a time extension of a week due to the $_3$pump being stopped.	"$_{1-a}$ Engineering and construction task"- subclass of -"Executing process"- subclass of -"Process". "$_{1-b}$ Duration to perform construction work " -sub attr. of Duration to perform work"-sub-attr. of-"Duration". "$_{2-a}$ Precipitation"- subclass of -"Weather conditions"-subclass of -"External environment". "$_{2-b}$ Level of precipitation " -sub attr. of -"Amount"-sub-attr. of-"Impact level". "$_{3-a}$ Productivity in performing construction work" -sub-attr.of -"Productivity"-sub-attr. of-"Quality".
	"$_{3-b}$ Construction heavy equipment" - subclass of -"Equipment "- subclass of -"Resource".
RISK PATH: "$_{2-b}$ Level of precipitation (V= too much)"-causes- "$_{3-a}$ Productivity in performing construction work (V=slow)"-influence- "$_{1-b}$ Duration to perform construction work (V=Long)".	

- **Case Study (C) (General building of Endowments),** (2016). According to documents, there may be disputes and concurrent delays. Due to the privacy area surrounding the building, the geological study was rejected due to inaccuracy and lack of identifying the required contract clauses about this.
- **Case Study (D) (Additional building in the industrial zone),** (2018). One reason for the delay that was (62%) is the bad weather that occurred. When "frequency" & "amount" of precipitation are known, it can infer possible risk caused by a climate.

iii. *Result of evaluation.*

In total, 69 concepts & 642 attributes specified concepts were entered into the developed ontology. 215 attributes that corresponding to the risk cases are matched with the corresponding ontology phrases. Since only 7 attributes are added, 96.7% match among the phrases in reports and the developed ontology. This process ensures comprehensiveness of AR-Onto for the cases that have been investigated [21]. As noted, it can infer possible risk paths semantically. This process ensures usability and the generality of the ontology for the ability to represent a variety of cases [6]. On the other hand, that achieves completeness criteria because it covers all the necessary attributes and concepts to relate risk factors with a cost; time overrun &poor quality.

4 Conclusions and Recommendations

In research, the ontology aims to extend previous studies in risk management by presenting relating risk-related concepts to cost overrun, time delay, and poor quality. The core contributions of the RA-Onto are: providing formalized risk knowledge for the domain of construction projects that can promote Knowledge sharing among different parties and applications without semantic ambiguity. On the other hand, it can be used to develop databases for risk cases that facilitate risk management or act as a decision support system. The knowledge was obtained from reference studies and interviews with experts during ontology development. While the ontology evaluation was done, but the validation had been done using a limited number of case studies that may be identical to the ontology, so it is better to do more case studies, and the validation using case studies may be insufficient. Reliable feedback can be obtained from the active usage of the ontology and may make actual validation of the ontology. In the ontology, concepts, and attributes are general and represent the domain of risk knowledge, and it can be applied to all types of projects. Risk information may be collected by feedback obtained from direct usage of the ontology during project stages. For future work, the proposed ontology can be used to develop IT-support tools that can facilitate learning from previous projects. And that is done by retrieving similar risk cases or risk assessment by developing mathematical forms and supporting decision-making.

References

1. Project Management Institute PMI.: A Guide to the Project Management Body of Knowledge (PMBOK-Guide)-Sixth V, Pennsylvania, USA (2017)

2. Fidan, G., Dikmen, I., Tanyer, A.M., Birgonul, M.T.: Ontology for relating risk and vulnerability to cost overrun in international projects. J. Comput. Civ. Eng. **25**(4), 302–315 (2011)
3. Xing, X., Zhong, B., Luo, H., Li, H., Wu, H.: Ontology for safety risk identification in metro construction. Comput. Ind. **109**, 14–30 (2019)
4. Gruber, T.R.: Toward principles for the design of ontologies used for knowledge sharing. Int. J. Hum. Comput. Stud. **43**(5–6), 907–928 (1995)
5. Keet, M.: An introduction to ontology engineering, vol. 1. Cape Town (2018)
6. El-Diraby, T.A., Lima, C., Feis, B.: Domain taxonomy for construction concepts: toward a formal ontology for construction knowledge. J. Comput. Civ. Eng. **19**, 394–406 (2005)
7. Tserng, H.P., Yin, S.Y., Dzeng, R.J., Wou, B., Tsai, M.D., Chen, W.Y.: A study of ontology-based risk management framework of construction projects through project life cycle. Autom. Constr. **18**(7), 994–1008 (2009)
8. Bilgin, G., Dikmen, I., Birgonul, M.T.: An ontology-based approach for delay analysis in construction. KSCE J. Civ. Eng. **22**(2), 384–398 (2018). https://doi.org/10.1007/s12205-017-0651-5
9. Saoud, L.A., Omran, J., Hassan, B., Vilutienė, T., Kiaulakis, A.: A method to predict change propagation within building information model. J. Civ. Eng. Manag. **23**(6), 836–846 (2017)
10. Breitman, K., Casanova, A., Truszkowski, W.: Knowledge representation in description logic. In: Breitman, K.K., Casanova, M.A., Truszkowski, W. (eds.) Semantic Web: Concepts, Technologies & Applications. NASA Monographs in Systems and Software Engineering, pp. 35–55. Springer, London (2007). https://doi.org/10.1007/978-1-84628-710-7_3
11. Fernández-López, M., Gómez-Pérez, A., Juristo, N.: Methontology: from ontological art towards ontological engineering (1997)
12. Fernandes, P.C., Guizzardi, R.S., Guizzardi, G.: Using goal modeling to capture competency questions in ontology-based systems. Inf. Data Manag. **2**(3), 527–540 (2011)
13. Noy, N.F., McGuinness, D.L.: Ontology development 101: a guide to creating your first ontology (2001)
14. Luiten, G., et al.: An information reference model for architecture, engineering, and construction. In: Mathur, K.S., Betts, M.P., Tham, K.W. (eds.) Proceedings of the First International Conference on the Management of Information Technology for Construction, Singapore, pp. 1–10 (1993)
15. Building SMART. IFC 2x3: Industry Foundation Classes, ⟨http://www.buildingsmarttech.org/ifc/IFC2x3/TC1/html/⟩. Accessed 04 June 2014
16. Protégé. The Protege Project (2020). http://protege.stanford.edu
17. Lassila, O., McGuinness, D.: The role of frame-based representation on the semantic web. Linköping Electron. Art. Comput. Inf. Sci. **6**(5) (2001)
18. Fernandez-Lopez, M., Corcho, O.: Ontological Engineering: with Examples from the Areas of Knowledge Management, e-Commerce and the Semantic Web. Springer, London (2010). https://doi.org/10.1007/b97353
19. Fox, M.S., Grüninger, M.: Ontologies for enterprise modelling. In: Kosanke, K., Nell, J.G. (eds.) Enterprise Engineering and Integration. Research Reports Esprit, pp. 190–200. Springer, Heidelberg (1997). https://doi.org/10.1007/978-3-642-60889-6_22
20. Sirin, E., Parsia, B., Grau, B.C., Kalyanpur, A., Katz, Y.: Pellet: a practical OWL-DL reasoner. J. Web Semant. **5**(2), 51–53 (2007)
21. Park, M., Lee, K.-W., Lee, H.-S., Jiayi, P., Yu, J.: Ontology based construction knowledge retrieval system. KSCE J. Civ. Eng. **17**(7), 1654–1663 (2013)

Designing Sustainable Digitalization: Crisisology-Based Tradeoff Optimization in Sociotechnical Systems

Sergey V. Zykov, Eduard Babkin, Boris Ulitin(✉), and Alexander Demidovskiy

HSE University, Moscow, Russia
{szykov,eababkin,bulitin,ademidosvky}@hse.ru

Abstract. Design and development of modern information systems represent a bright ex-ample of a complex decision problem. In that context our especial interest lays in proposing a new approach to decision support in the tasks of multi-criteria tradeoff optimization of information systems which will facilitate sustainable development of the application domain where the information system will be used. In that work we propose a method and a corresponding decision support software service for linguistic multi-criteria choice among multiple design alternatives. Our method is based on a hierarchy of cross-disciplinary criteria which reflect the concept of crisisology. The decision support service uses an ontology-based mechanism for dynamic customization of user interface. Application of our method and the decision support service are demonstrated for the case of designing a CRM information system for the enterprise.

Keywords: Decision support · multi-criteria choice · Crisisology · software service

1 Introduction

Over the past decades, production in general and software engineering in particular were understood and practiced in different ways. Changeable business constraints, complex technical requirements, and the so-called human factors imposed on the software solutions caused what was articulated as "software crises", which typically result from an imbalance between available resources, business requirements, and technical constraints [1]. These complex sources of trouble require a multifaceted approach (as well as a related software) to address each of their layers.

What is more important, the digital product should meet certain quality level, which consists of a balanced combination of certain sub-qualities, generally known as quality attributes (QAs). Obviously, these QA are often contradictory (e.g., promoting security inhibits performance, and vice versa). Therefore, to guarantee that a certain QA combination is really well balanced, tradeoff analysis and management is required [1]. Multi-criteria choice of alternatives in IT system architecture facilitates sustainable development of the application domain. The general idea of the research scheme

I. Czarnowski et al. (Eds.): KESIDT 2023, SIST 352, pp. 250–260, 2023.
https://doi.org/10.1007/978-981-99-2969-6_22

within the framework of crisisology is presented as follows: 1) identification of key BTH factors (business, technology, human); 2) categorization of factors and establishment of dependencies (category theory); 3) description of factors and dependencies in the object language; 4) primary "rough" optimization (AHP, ACDM/ATAM, etc.) with semi-automatic evaluation and ranking of alternatives; 5) secondary "fine" optimization (if required); 6) immersion in an applied object environment based on a virtual machine (VM based on category/combinator theory); 7) final search for the optimal solution (DSS).

However, the most important limitation for the application of this scheme is the fact that the proposed assessments of various factors by experts are often not formalized, but in the form of linguistic information that requires further processing.

In this paper, we will focus on the description of a method for a linguistic-based multi-criteria choice of alternatives in IT system architecture. We also propose a software service that allows the users processing such assessments in a linguistic form to make a multi-criteria decision on the use of DPD in the enterprise's processes.

This article presents our results as follows. In Sect. 2 we observe core principles of crisisology which lead to the design of a hierarchically structured system of criteria for assessment of IT-architecture alternatives. Section 3 is devoted to the description of our method for linguistic multi-criteria choice and proposed hierarchy of criteria. Section 4 and 5 contain information on software design and usage scenarios of the decision support service, as well as the results of evaluating its quality and efficiency. We conclude the article with an analysis of results presented and further research steps.

The research is supported by grant of the Russian Science Foundation (project № 23–21-00112 "Models and methods to support sustainable development of socio-technical systems in digital transformation under crisis conditions").

2 Crisisology as a Conceptual Framework for Multi-criteria Decision Support in Information Systems Design

As we mentioned before, in their processes, enterprises use various types of information systems. Once a system is designed, organizations spring up around its structures. In a very real sense technical designs beget organizations for better or worse. Eliminating or adding organizations that support systems or elements of systems can have very real impact to the structure of a system that is built or in the process of being built [1]. Given the large number of alternative information systems, the most difficult for the enterprise is the stage of choosing the target system, based on BTH factors.

At the same time, the description of the system and reviews about it are often not enough to assess the human factors. If the assessment of the technical capabilities of the system and the financial costs of its implementation can be automated to some extent, then taking into account user requirements is a non-trivial process.

User requirements are formulated in the form of linguistic assessments, while each user can use his own system of criteria for assessment (compare: bad-good-excellent and slow-fast). The main task is to take into account all the necessary assessments, expressed in a heterogeneous form, when making the final decision on a system.

There are certain attempts to solve this problem in the literature and the practical environment. For example, in [16] Martínez et al. offer a generalized decision-making model based on linguistic information with the help of an operator that takes into account the number of matching and non-matching expert assessments. [17] provides a system prototype that allows users to enter linguistic scores without modification. However, this system involves preliminary work in the form of creating a single 'terminological' base for experts and ranking all available assessments using unified principles.

As a result, these solutions are highly specialized and still require a certain unification of the rating systems of various users. What is more crucial, both solutions contradict the fact that experts can evaluate different criteria using different scales, selected individually for each of them.

To solve this problem, in the following sections we describe a multi-criteria selection system based on dynamic interfaces. The proposed system allows for each user (or groups of users) to define their own system of evaluation criteria (in linguistic form) with their own gradation scales for these evaluations. This approach makes it possible to unify the procedure for making a decision on the choice of an information system, while not requiring the introduction of a unified system (and scale) of assessment.

3 A Proposed Method for Hierarchical Multi-criteria Choice

There are numerous attempts to elaborate new decision-making approaches or adopt existing ones to real-life cases, like healthcare [2], performance evaluation of partnerships, fiber composites optimization, reverse logistics evaluation [3], project resources scheduling [4], supplier selection, aircraft incident analysis [5]. Usually traditional approaches like TOPSIS [2], ELECTRE, VIKOR [3] are used.

The considerable drawback is that these methods rely mostly on quantitative evaluations, even given in a form of fuzzy sets [6]. On the other hand, estimations that are given by experts during problem discussion can be both quantitative and qualitative. In comparison with quantitative evaluations, qualitative evaluations become more and more preferable in complex situations because of their ability to express fuzzy information (e.g. hesitation). However, according to our rigorous analysis of the field, there is an emerging trend of combining traditional decision-making approaches with methods of processing qualitative evaluations. The combination of TOPSIS methodology and 2-tuple model represents a bright example of such a combination [7].

Reliable and flexible means for analysis of qualitative evaluations are provided within the scientific area of "linguistic decision making" and "linguistic multi-attribute decision making" [8]. These and other methods of processing qualitative evaluations now are generally called "computing with words, among which the most popular are [8] linguistic computational model based on: 1) membership functions; 2) ordinal scales or 3) convex combinations and max-min operators.

In many cases, information that comes from the experts is heterogeneous due to its multigranularity and there are approaches (and methods) to work with such information: the fusion approach for managing multigranular linguistic information [9], the linguistic hierarchy approach and the method of extended linguistic hierarchies [10].

It is important to emphasize that existing approaches concentrate either on analysis of only quantitative/qualitative assessments. Very few approaches focus on both types of estimations. At the same time, modern methodologies are likely to assume that there are a number of experts without capturing the area of their expertise as well as the fact that criteria also belong to different abstraction levels (BTH in our case). More importantly, existing methods for decision making are demonstrated on artificial cases with very few experts and alternative solutions. This brings us to the point to propose a new methodology which could incorporate most of the gaps described above.

We call our approach multilevel multi-attribute linguistic decision making (ML–MA–LDM). The proposed approach consists of several consecutive steps starting from defining the estimation rules and finishing with the communication stage. It is important to note that these steps can be found individually in various papers describing the decision-making process, for example in [9], but never were fused in a consistent way. The proposed approach includes:

1. Setting up rules for providing estimations and distribution of criteria weights.
2. Defining available linguistic sets, a context-free grammar and transformation function;
3. Multi-level definition of the desired state, criteria and alternatives: a) analyzing the desired state on each level of abstraction; b) formulating criteria for each level of abstraction; c) formulating alternatives.
4. Giving multi-level and multi-criteria evaluations: a) aggregating information; b) searching for the best alternative; c) communicating the solution found.

After criteria and alternatives were defined, all experts start giving evaluations of each alternative for each available criterion. Let $x = \{x_1, x_2, \ldots, x_N\}$ is the list of alternatives, $c = \{c_1, c_2, \ldots, c_M\}$ is the list of criteria, $e = \{e_1, e_2, \ldots, e_T\}$ is the list of experts. Each expert e_k can evaluate alternatives using different linguistic scales S_{g_k} with granularity g_k. In the case of comparative evaluations, we also have the grammar G_H which can be also used for creation of linguistic evaluations. Moreover, the criteria are given for each level of abstraction in the meta-decision framework, i.e. let $l = \{l_1, l_2, \ldots, l_Z\}$ be the list of the levels of abstraction. Therefore, one evaluation for each given alternative is obtained and the best alternative can be found by sorting these evaluations according to rules of comparing hesitant 2-tuple fuzzy sets. As a result, for each expert we get a matrix of evaluations $R_k = \left(T^{ij}_{S_{g_k}}\right)_{N \times M}$, where $T^{ij}_{S_{g_k}}$ – an evaluation of the expert e_k for the i-th alternative on the j-th criterion in the format of HFLTS on the scale S_g.

Carrying out successively several aggregation of evaluations for each level of abstraction and transformations of these estimates, described in detail in [18], finally we obtain the total evaluation for each i-th alternative and for each level of abstraction as $T_i = MHTWA^q_{S_{g_k}}\left(T^{i_1}_{S_{g_k}}, T^{i_2}_{S_{g_k}}, \ldots, T^{i_Z}_{S_{g_k}}\right)$, where i – the index of alternative; q – the vector of weights of levels of abstraction, $q = (q_1, q_2, \ldots, q_Z)^T$, $q_j \geq 0$, $\sum_{j=1}^{Z} q_j = 1$. So, we get the following vector of evaluations $r = \left(T^i_{S_{g_k}}\right)_N$, where $T^i_{S_{g_k}}$ is the aggregated evaluation for i-th alternative in a form of HFLTS on the scale S_{g_k}.

As a result, we get assessments that draw insights on how each alternative is measured on each abstraction level, that can be used by a decision maker to better understand the scope of alternatives and their influence on each aspect of the problem situation.

4 A Proposed Hierarchical Structure of the Criteria for Selection of IT System Architecture

In this paper, we do not consider the decision-making process to automate the company's business processes, considering that it has already been successfully completed, and it remains to decide on the IS architecture itself (which generally consists of clients (user devices) and a more powerful server that clients access).

Depending on the location of the components of the 3 layers of IS on the client and server, the following types of distributed architectures are distinguished [11]: a ***file-server*** architecture, a ***client-server*** architecture, a ***monolithic*** architecture, a ***microservice*** architecture and a ***cloud*** architecture. Each architecture has its own characteristics, advantages and disadvantages, so it is extremely important to take into account all of them when choosing and make the best choice for the company, which will positively affect its KPIs. When deciding on an architecture, we take into account both quantitative and qualitative factors, reflecting the opinion not only about the architecture as a whole, but also about the experience of using it by other companies and the possible risks of implementing an architecture implementation project [1]:

- ***Cost.*** This criterion is quantitative, and the evaluation scale is inversely proportional to its values.
- ***Learning curve.*** The easier it is to learn how to use, configure, install, and maintain a product, the more attractive it is. Therefore, low or shallow learning curves are given higher ratings, and steeper learning curves are given lower ratings.
- ***Support.*** Better support is more attractive and is therefore given higher ratings, and weaker support is given lower ratings.
- ***Provider reputation.*** The more solid the provider reputation, the higher the ratings.
- ***Volatility.*** A highly stable product with a long track record in the marketplace is given higher ratings than an emerging product.
- ***Schedule.*** Describes the possible total delay in the project schedule (in days). The greater the possible delay, the less preferred the alternative.
- ***Quality.*** Serves to assess the impact of a potential decline in quality on the vital functions of the enterprise. The greater the assessment of a given risk, the less preferred the alternative.

Arbitrarily fractional (counting) scales can be used to evaluate the indicated criteria. Within the framework of this study, we will adhere to the following scales (Table 1, letter in Criterion parentheses - type of BTH criterion). In this case, the possibility of replacing the rating scale is important, as well as the use of not a single scale, but its adaptation for each individual expert in accordance with the principles described in the Sect. 3. This is also fully consistent with object models (including in terms of category theory), which make it possible to define an integrated assessment as a categorical (multi-argument) valuation in various correlations.

Table 1. IT system architecture evaluation criteria evaluation scales

Criterion	Type	Scale (from least preferred to most preferred)
Cost (*B*)	Quantitative	
Learning curve (*H*)	Qualitative	none, complex, medium, simple, intuitive
Support (*T*)		none, under the contract, short-term, long-term
Provider reputation (*B*)		negative, indefinite, positive
Volatility (*B*)		unstable, volatile, stable
Schedule (*B, T*)		10 segments: from 1 month to 0 days
Quality (*T*)		worse/comparable/higher/is the standard for analogues

5 Description of the Decision Support Service

In order to work with the criteria listed above and evaluate the architecture of IT systems according to them, we use a software prototype that includes a backend responsible for the ranking of alternatives and a frontend (GUI) necessary for setting all the components required for evaluation (hierarchy of criteria, alternatives, assessments of alternatives by experts, etc.). As mentioned earlier, the first step in deciding on the choice of a particular information system is to determine the criteria for its selection. At the same time, it is important to remember that the criteria, by their nature, can be presented in various forms: numerical, textual (linguistic), etc. At the same time, even numerical criteria may differ in assessment scales in terms of quality (from lower to higher and vice versa). Therefore, in the created service, the first stage is the creation of a system of criteria and the setting of scales for their evaluation.

Using the appropriate GUI (Fig. 1a), the user is able to first determine and save the name of the criteria system, then the name of each individual criterion, and select the most appropriate data type to represent it. In this case, the supported data types are selected based on the nature of the possible criteria for selecting information systems: *integer* (int) and *fractional* (float) *numbers* – to quantify information systems (e.g. allowable number of users), *date* – in the case of distinguishing newer information systems from older ones, *textual* (varchar) – to represent linguistic assessments. At the same time, a unique name within the criteria system is assigned for each criterion. What is more important, we can use not only the data types, but full-fledged domains for each individual evaluation criterion, that is fully consistent with the general theory of categories, the main provisions of which are used in the approach proposed.

A subsequent input of assessments on alternatives by experts must initially be stored in some structured form for subsequent transfer to the decision-making service (alternatives ranking). For these purposes, it is optimal to use an object-relation database (ORDB) structure, which allows both to prevent the input of contradictory are (in terms of type to the corresponding criteria) values, and to save all assessments in the form of a single object (table). In addition, in comparison with NoSQL solutions, ORDB has a higher performance and structurally corresponds to the previously identified structure of the assessments hierarchy according to various criteria (multidimensional objects).

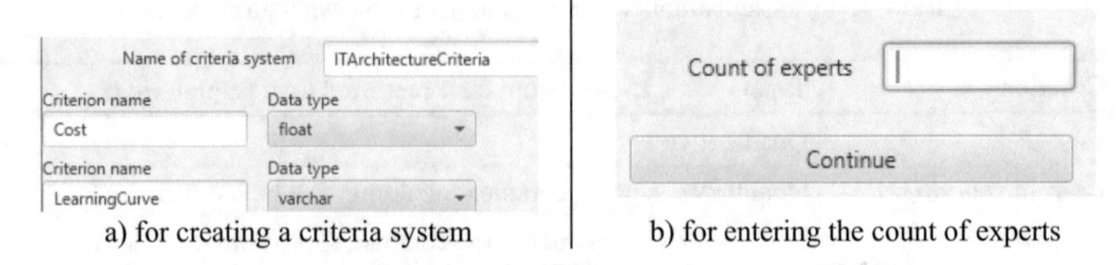

a) for creating a criteria system b) for entering the count of experts

Fig. 1. GUI fragments

In the next step, we need to refine the rating scale for each created criterion using an appropriate interface is used. After the system of criteria and the scale of its evaluation are set, the system switches to the mode of introducing alternatives by experts (count of which is entered Fig. 1b). The count of experts must be natural. Otherwise, the system generates an error and ask the user to enter the value again. Next, the GUI for entering alternative estimates for each expert opens (Fig. 2 - top). In this case, the expert sees only his own estimates for all alternatives, the estimates of other experts are not available to him. To check this limitation, the table shows the identifier of the expert along with the scores for each of the alternatives.

id_expert	cost	learningCurve	support	provider
1	15	intuitive	short-term	positive
1	6	simple	under the contract	positive
1	22	medium	short-term	indefinite
1	65	simple	long-term	positive

id_expert 1
cost 65
learningCurve medium
support long-term

Alternative name	OveralResult	OveralRatingPosition
Brizo	9.8	1
Microsoft Dynamics CRM	9.2	2
WireCRM	8.6	3
KOMMO CRM	8.4	4

Cost - 3, Learning curve - 2.5, Support - 2...

Fig. 2. GUI for entering (top) and displaying ranked (bottom) alternatives by the experts

The most significant advantage of the implemented interface is its adaptability. The GUI is generated completely automatically, adjusting to a previously created set of criteria and their data type. If the system of criteria is changed (new ones are added, some of the previously introduced ones are deleted/modified), the GUI will be updated automatically to allow the expert to make assessments in accordance with the updated state of the criteria system. After one expert enters estimates for all alternatives, using the button of the same name, he can transfer this ability to the next expert.

After all experts have entered their own ratings of alternatives, a button will appear to transfer the entered ratings to the service for comparing and ranking alternatives (see Sect. 6). Data transfer in this case is carried out using a JSON package containing all the information necessary for evaluation: a set of criteria, evaluation scales for individual criteria, a set of alternatives, and the results of evaluation by experts of all alternatives. As a result of processing the package by the alternative evaluation service, a response JSON packet arrives containing ranked alternatives in descending order of their preference. This ranked set of alternatives is displayed on the screen using the appropriate interface component (Fig. 2 - bottom). In this case, when selecting each individual alternative, the results of the assessment for each individual criterion are displayed.

6 Case Study of the Proposed Prototype

In order to demonstrate the application of the described approach and a software prototype, we will compare CRM systems for an enterprise using as an aspect of the comparison the architecture of these systems in the context of the criteria described earlier.

Table 2. Brief description of the systems under consideration

Criterion	Brizo	WireCRM	KOMMO CRM	Microsoft Dynamics CRM
Architecture	client-server/*cloud*	client-server/*monolithic*	client-server/*monolithic*	*client-server/cloud*
Cost (per user)	15$/m	6$/m	22$/m	65$/m
Learning curve	Online Help Docs	Online Help Docs	Support Online Help Docs	Ccourses Support Online Help Docs
Support	Online chat	Online chat	24/7	24/7
Provider Reputation (the last year)	several failures (up to 24 h)	several failures (up to 24 h)	no open reports	1 failure (up to 2 h)
Volatility	several updates with bugs, fix within a day	several updates with bugs, fix within a day	No updates, only fixes	several updates with bugs, fix within a week
Schedule	no open reports	no open reports	the update release plan is published; the timing depends on the specific enterprise	the update release plan is published; the timing is fixed
Quality	10% negative reviews	10.6% negative reviews	1% negative reviews	6% negative reviews

To simplify, we restrict ourselves to 4 systems of this class: Brizo [12] (a CRM system with advanced functionality for management accounting), WireCRM [13] (a multi-functional 'single window' platform for comprehensive control of business processes), KOMMO CRM [14], Microsoft Dynamics CRM [15]. A summary of the systems features in the context of the previously identified criteria is given in Table 2. Tis information can be interpreted differently by experts and lead to the fact that their assessments will differ according to identical criteria.

In our case, we work with two experts who have expertise in the field of CRM systems, and obtain the following estimates of alternatives represented in Table 3 (when experts' ratings match, one value is given). For entering these values, the experts use GUI, described in Fig. 2 (the order of the rows in the table corresponds to the order of the columns with systems). After all the ratings are entered, we send them for processing to the rating ranking service via a JSON file containing all the necessary information both on the alternatives and their evaluation criteria, and on the specific values of the experts' ratings. In the response JSON file, we receive a ranked list of alternatives (Fig. 2 - bottom). In this case, the system, in the course of comparing alternatives according to all criteria, chose Brizo as the optimal system, and KOMMO CRM as the least optimal. At the same time, the discrepancies in expert estimates on such criteria as Learning curve, Volatility and Schedule had the greatest impact on the decision. In this case, when ranking, we use the entire structure of the BTH criteria described earlier (Table 1), obtaining a socio-technical (not purely technical) assessment.

Table 3. Results of evaluation of alternatives by experts (1st expert/2nd expert)

Criterion	Brizo	WireCRM	KOMMO CRM	Microsoft Dynamics CRM
Cost	15	6	22	65
Learning curve	intuitive/simple	simple	medium/simple	simple/medium
Support	short-term	under the contract	short-term	long-term
Provider reputation	positive	positive	indefinite	positive
Volatility	stable/volatile	stable/volatile	volatile/unstable	volatile
Schedule	14/28	14/28	3/3	7/7
Quality	higher	comparable	comparable/worse	standart

7 Conclusion

In this work, an approach and tools were considered that allow for a multi-criteria assessment of various alternatives based on linguistic data. The proposed approach is strikingly different from the existing approaches described in [10, 16], which are highly specialized and require significant prior preparation to evaluate alternatives. In contrast, the proposed approach consists in setting individual rating scales for each criterion and does not require a transition from qualitative to quantitative criteria. The approach is universal and can be applied to various subject areas, which was demonstrated on the example of choosing the optimal IT-architecture of the CRM-system.

It is important to note that all manipulations with the criteria and assessments of experts are carried out through a GUI that is adaptive and automatically adjusts to the system of criteria that is set at the initial stage of evaluating alternatives and can be changed in real time without the need to recreate the interface and without the need re-entry of assessments by experts. This favorably distinguishes the proposed approach and software prototype from existing systems [16, 17], which require to unify the used scales for evaluating various criteria in advance and do not allow changing them in the process of working with the system. This makes the system applicable for organizing digital transformation in crisis conditions, since it allows you to set a complete system of BTH criteria in the original linguistic form, without pre-processing. In the future, we plan to introduce into the prototype support for more complex rules for accounting for linguistic assessments, as well as the ability to vary scales based on the mechanisms of category theory. This can help not only changing the set of evaluation criteria through the interface, but also the rules for taking them into account in the final evaluation of alternatives, making the system more universal and flexible.

References

1. Zykov, S.V.: IT Crisisology: Smart Crisis Management in Software Engineering: Models, Methods, Patterns, Practices, Case Studies. Springer, Singapore (2021).https://doi.org/10.1007/978-981-33-4435-8
2. Dehe, B., Bamford, D.: Development, test and comparison of two multiple criteria decision analysis (MCDA) models: a case of healthcare infrastructure location. Expert Syst. Appl. **42**(19), 6717–6727 (2015)
3. Senthil, S., Srirangacharyulu, B., Ramesh, A.: A robust hybrid multi-criteria decision-making methodology for contractor evaluation and selection in third-party reverse logistics. Expert Syst. Appl. **41**(1), 50–58 (2014)
4. Markou, C., Koulinas, G.K., Vavatsikos, A.P.: Project resources scheduling and leveling using multi-attribute decision models: models implementation and case study. Expert Syst. Appl. **77**, 160–169 (2017)
5. Skorupski, J.: Multi-criteria group decision making under uncertainty with application to air traffic safety. Expert Syst. Appl. **41**(16), 7406–7414 (2014)
6. Igoulalene, I., Benyoucef, L., Tiwari, M.K.: Novel fuzzy hybrid multi-criteria group decision making approaches for the strategic supplier selection problem. Expert Syst. Appl. **42**(7), 3342–3356 (2015)
7. Cid-Lopez, A., Hornos, M.J., Carrasco, R.A., Herrera-Viedma, E., Chiclana, F.: Linguistic multi-criteria decision-making model with output variable expressive richness. Expert Syst. Appl. **83**, 350–362 (2017)
8. Xu, Z.: A method based on linguistic aggregation operators for group decision making with linguistic preference relations. Inf. Sci. **166**(1–4), 19–30 (2004)
9. Herrera, F., Herrera-Viedma, E., Martinez, L.: A fusion approach for managing multi-granularity linguistic term sets in decision making. Fuzzy Sets Syst. **114**(1), 43–58 (2000)
10. Herrera, F., Martinez, L.: A model based on linguistic 2-tuples for dealing with multigranular hierarchical linguistic contexts in multi-expert decision-making. IEEE Trans. Syst. Man Cybern. Part B (Cybern.) **31**(2), 227–234 (2001)
11. Crawley, E., Cameron, B., Selva, D.: System Architecture: Strategy and Product Development for Complex Systems. Pearson, London (2016)
12. BRIZO CRM. https://brizo-crm.com/. Accessed 24 Jan 2023
13. WireCRM. https://wirecrm.com/. Accessed 24 Jan 2023
14. KOMMO CRM. https://www.kommo.com/. Accessed 24 Jan 2023
15. Microsoft Dynamics CRM. https://dynamics.microsoft.com/. Accessed 24 Jan 2023
16. Martínez, L., Rodriguez, R.M., Herrera, F.: Linguistic decision making and computing with words. In: Martínez, L., Rodriguez, R.M., Herrera, F. (eds.) The 2-Tuple Linguistic Model, pp. 1–21. Springer, Cham (2015). https://doi.org/10.1007/978-3-319-24714-4_1

17. Wu, J.-T., Wang, J.-Q., Wang, J., Zhang, H.-Y., Chen, X.-H.: Hesitant fuzzy linguistic multi-criteria decision-making method based on generalized prioritized aggregation operator. Sci. World J. **2014** (2014)
18. Demidovskij, A., Babkin, E.: Neural multigranular 2-tuple average operator in neural-symbolic decision support systems. In: Kovalev, S., Tarassov, V., Snasel, V., Sukhanov, A. (eds.) IITI 2021. LNNS, vol. 330, pp. 350–359. Springer, Cham (2022). https://doi.org/10.1007/978-3-030-87178-9_35

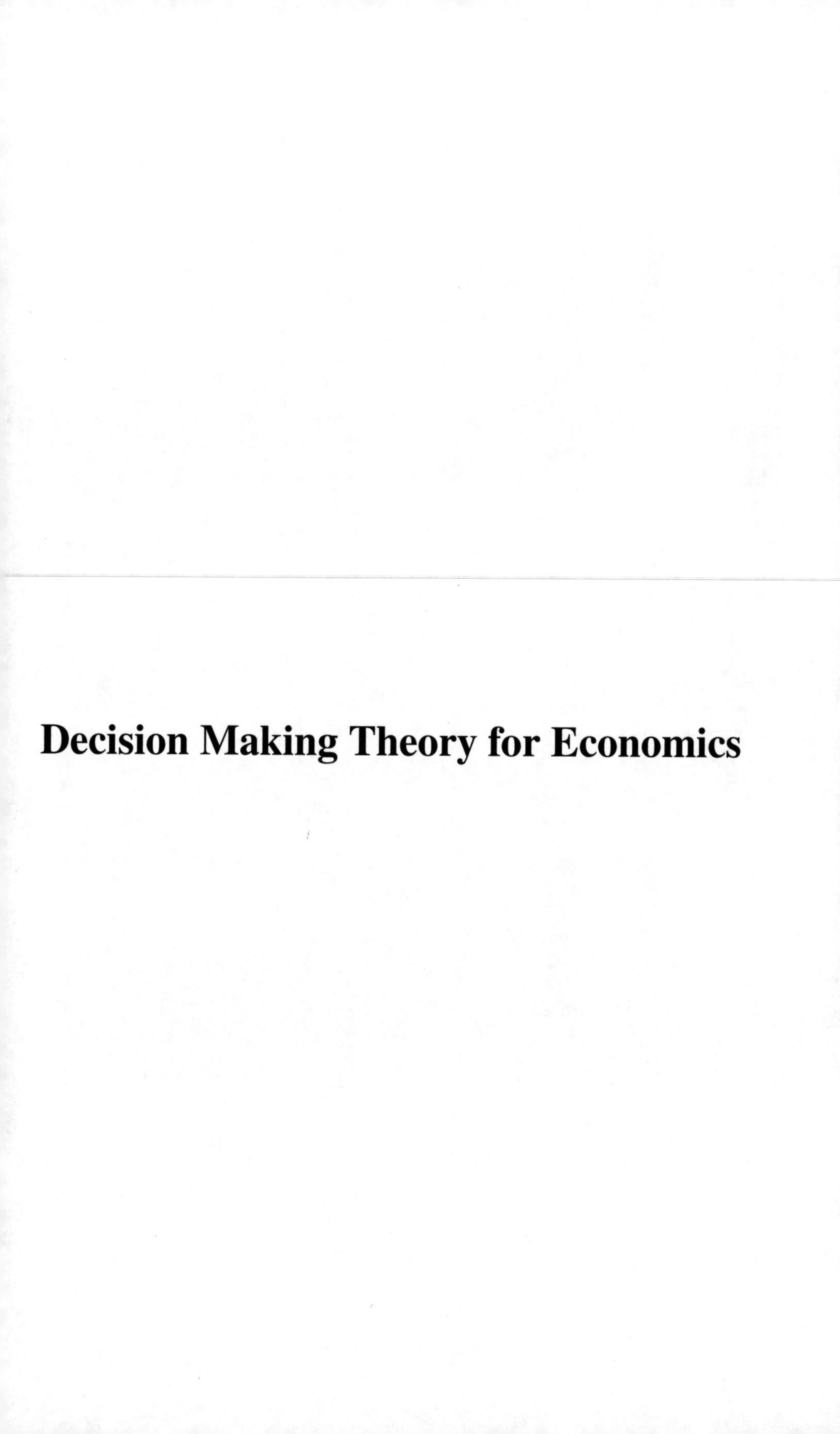

Decision Making Theory for Economics

Calculations by Several Methods for MDAHP Including Hierarchical Criteria

Takao Ohya(✉)

Kokushikan University, 4-28-1, Setagaya, Setagaya-ku, Tokyo 154-8515, Japan
takaohya@kokushikan.ac.jp

Abstract. We have proposed a super pairwise comparison matrix (SPCM) to express all pairwise comparisons in the evaluation process of the dominant analytic hierarchy process (D-AHP) or the multiple dominant AHP (MDAHP) as a single pair wise comparison matrix. This paper shows the calculations for MDAHP including hierarchical criteria by the concurrent convergence method, by the geometric mean MDAHP and from SPCM by the logarithmic least squares method, the Harker method, and the improved two stage method.

Keywords: super pairwise comparison matrix · the multiple dominant analytic hierarchy process · the logarithmic least squares method · the Harker method · the improved two stage method

1 Introduction

In actual decision-making, a decision-maker often has a specific alternative (regulating alternative) in mind and makes an evaluation based on the alternative. This was modeled in D-AHP (the dominant AHP), proposed by Kinoshita and Nakanishi [1].

If there are more than one regulating alternative and the importance of each criterion is inconsistent, the overall evaluation value may differ for each regulating alternative. As a method of integrating the importance in such cases, CCM (the concurrent convergence method) was proposed. Kinoshita and Sekitani [2] showed the convergence of CCM.

Ohya and Kinoshita [3] proposed the geometric mean multiple dominant AHP (GMMDAHP), which integrates weights by using a geometric mean based on an error model to obtain an overall evaluation value.

Ohya and Kinoshita [4] proposed an SPCM (Super Pairwise Comparison Matrix) to express all pairwise comparisons in the evaluation process of the D-AHP or the multiple dominant AHP (MDAHP) as a single pairwise comparison matrix.

Ohya and Kinoshita [5] showed, by means of a numerical counterexample, that in MDAHP an evaluation value resulting from the application of the logarithmic least squares method (LLSM) to an SPCM does not necessarily coincide with that of the evaluation value resulting from the application of the geometric mean multiple dominant AHP (GMMDAHP) to the evaluation value obtained from each pairwise comparison matrix by using the geometric mean method.

I. Czarnowski et al. (Eds.): KESIDT 2023, SIST 352, pp. 263–272, 2023.
https://doi.org/10.1007/978-981-99-2969-6_23

Ohya and Kinoshita [6] showed, using the error models, that in D-AHP an evaluation value resulting from the application of the logarithmic least squares method (LLSM) to an SPCM necessarily coincide with that of the evaluation value resulting obtained by using the geometric mean method to each pairwise comparison matrix.

Ohya and Kinoshita [7] showed the treatment of hierarchical criteria in D-AHP with a super pairwise comparison matrix.

SPCM of D-AHP or MDAHP is an incomplete pairwise comparison matrix. Therefore, the LLSM based on an error model or an eigenvalue method such as the Harker method [8] or two-stage method [9] is applicable to the calculation of evaluation values from an SPCM. Nishizawa proposed Improved Two-Stage Method (ITSM).

Ohya and Kinoshita [10] and Ohya [11, 12] showed calculations of SPCM by each method applicable to an incomplete pairwise comparison matrix for the multiple dominant AHP including Hierarchical Criteria.

Ohya [13] shows the calculations of SPCM by LLSM, the Harker method, ITSM for the multiple dominant AHP including Hierarchical Criteria.

Ohya [14] shows the calculations for D-AHP including hierarchical criteria by the eigenvalue method and the geometric mean method, and with SPCM by LLSM, the Harker method, ITSM including Hierarchical Criteria.

This paper shows the calculations for MDAHP including hierarchical criteria by the CCM, by the geometric mean MDAHP and from SPCM by the logarithmic least squares method, the Harker method and the improved two stage method.

2 SPCM

The true absolute importance of alternative $a(a = 1, \ldots, A)$ at criterion $c(c = 1, \ldots, C)$ is v_{ca}. The final purpose of the AHP is to obtain the relative value between alternatives of the overall evaluation value $v_a = \sum_{c=1}^{C} v_{ca}$ of alternative a.

The relative comparison values $r^{ca}_{c'a'}$ of importance v_{ca} of alternative a at criteria c compared with the importance $v_{c'a'}$ of alternative a' in criterion c', are arranged in a (CA × CA) or (AC × AC) matrix. This is proposed as the SPCM $\mathbf{R} = \left(r^{ca}_{c'a'}\right) or \left(r^{ac}_{a'c'}\right)$.

In a (CA × CA) matrix, index of alternative changes first. In a (CA × CA) matrix, SPCM's $(\mathrm{A}(\mathrm{c}-1)+a,\ \mathrm{A}(\mathrm{c}'-1)+a')$th element is $r^{ca}_{c'a'}$.

In a (AC × AC) matrix, index of criteria changes first. In a (AC × AC) matrix, SPCM's $\left(C(a-1)+c,\ C\left(a'-1\right)+\mathrm{c}'\right)$th element is $r^{ac}_{a'c'}$.

In an SPCM, symmetric components have a reciprocal relationship as in pairwise comparison matrices. Diagonal elements are 1 and the following relationships are true:

If $r^{ca}_{c'a'}$ exists, then $r^{c'a'}_{ca}$ exists and

$$r^{c'a'}_{ca} = 1/r^{ca}_{c'a'} \tag{1}$$

$$r^{ca}_{ca} = 1 \tag{2}$$

SPCM of D-AHP or MDAHP is an incomplete pairwise comparison matrix. Therefore, the LLSM based on an error model or an eigenvalue method such as the Harker method [10] or two-stage method is applicable to the calculation of evaluation values from an SPCM.

3 Numerical Example of Using SPCM for Calculation of MDAHP

Let us take as an example the hierarchy shown in Fig. 1. Three alternatives from 1 to 3 and seven criteria from I to VI, and S are assumed, where Alternative 1 and Alternative 2 are the regulating alternatives. Criteria IV to VI are grouped as Criterion S, where Criterion IV and Criterion V are the regulating criteria.

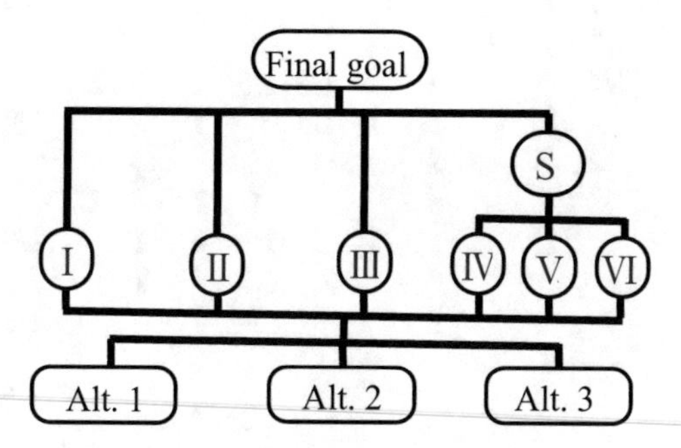

Fig. 1. The hieratical structure.

As the result of pairwise comparisons between alternatives at criterion c(c = I, ..., VI), the following pairwise comparison matrices $\mathbf{R}_c^A$, c = $I, \ldots, VI$ are obtained:

$$\mathbf{R}_{\mathrm{I}}^{A} = \begin{pmatrix} 1 & \frac{1}{3} & 5 \\ 3 & 1 & 3 \\ \frac{1}{5} & \frac{1}{3} & 1 \end{pmatrix}, \mathbf{R}_{\mathrm{II}}^{A} = \begin{pmatrix} 1 & 7 & 3 \\ \frac{1}{7} & 1 & \frac{1}{3} \\ \frac{1}{3} & 3 & 1 \end{pmatrix}, \mathbf{R}_{\mathrm{III}}^{A} = \begin{pmatrix} 1 & \frac{1}{3} & \frac{1}{3} \\ 3 & 1 & \frac{1}{3} \\ 3 & 3 & 1 \end{pmatrix},$$

$$\mathbf{R}_{\mathrm{IV}}^{A} = \begin{pmatrix} 1 & 3 & 5 \\ \frac{1}{3} & 1 & 1 \\ \frac{1}{5} & 1 & 1 \end{pmatrix}, \mathbf{R}_{\mathrm{V}}^{A} = \begin{pmatrix} 1 & \frac{1}{3} & 3 \\ 3 & 1 & 5 \\ \frac{1}{3} & \frac{1}{5} & 1 \end{pmatrix}, \mathbf{R}_{\mathrm{VI}}^{A} = \begin{pmatrix} 1 & \frac{1}{5} & 3 \\ 5 & 1 & 7 \\ \frac{1}{3} & \frac{1}{7} & 1 \end{pmatrix}.$$

With regulating Alternative 1 and Alternative 2 as the representative alternatives, and Criterion IV and Criterion V as the representative criteria, importance between criteria was evaluated by pairwise comparison. As a result, the following pairwise comparison matrices $R_1^C, R_1^S, R_2^C, R_2^S$ are obtained:

$$R_1^C = \begin{bmatrix} 1 & \frac{1}{3} & 3 & \frac{1}{3} & \frac{1}{5} \\ 3 & 1 & 5 & 1 & \frac{1}{2} \\ \frac{1}{3} & \frac{1}{5} & 1 & \frac{1}{5} & \frac{1}{9} \\ 3 & 1 & 5 & 1 & \frac{1}{2} \\ 5 & 2 & 9 & 2 & 1 \end{bmatrix}, R_1^S = \begin{bmatrix} 1 & \frac{1}{2} & 2 \\ 2 & 1 & 5 \\ \frac{1}{2} & \frac{1}{5} & 1 \end{bmatrix},$$

$$R_2^C = \begin{bmatrix} 1 & 5 & 1 & 3 & \frac{1}{9} \\ \frac{1}{5} & 1 & \frac{1}{3} & 1 & \frac{1}{9} \\ 1 & 3 & 1 & 1 & \frac{1}{9} \\ \frac{1}{3} & 1 & 1 & 1 & \frac{1}{9} \\ 9 & 9 & 9 & 9 & 1 \end{bmatrix}, R_2^S = \begin{bmatrix} 1 & \frac{1}{9} & \frac{1}{4} \\ 9 & 1 & 6 \\ 4 & \frac{1}{6} & 1 \end{bmatrix}.$$

The (CA × CA) order SPCM for this example is

$$R_{(CA \times CA)} = \begin{array}{c|cccccccccccccccccc|}
 & I1 & I2 & I3 & II1 & II2 & II3 & III1 & III2 & III3 & IV1 & IV2 & IV3 & V1 & V2 & V3 & VI1 & VI2 & VI3 \\
\hline
I1 & 1 & 1/3 & 5 & 1/3 & & & 3 & & & 1/3 & & & 1/5 & & & & & \\
I2 & 3 & 1 & 3 & & 5 & & & 1 & & & 3 & & & 1/9 & & & & \\
I3 & 1/5 & 1/3 & 1 & & & & & & & & & & & & & & & \\
II1 & 3 & & & 1 & 7 & 3 & 5 & & & 1 & & & 1/2 & & & & & \\
II2 & & 1/5 & & 1/7 & 1 & 1/3 & & 1/3 & & & 1 & & & 1/9 & & & & \\
II3 & & & & 1/3 & 3 & 1 & & & & & & & & & & & & \\
III1 & 1/3 & & & 1/5 & & & 1 & 1/3 & 1/3 & 1/5 & & & 1/9 & & & & & \\
III2 & & 1 & & & 3 & & 3 & 1 & 1/3 & & 1 & & & 1/9 & & & & \\
III3 & & & & & & & 3 & 3 & 1 & & & & & & & & & \\
IV1 & 3 & & & 1 & & & 5 & & & 1 & 3 & 5 & 1/2 & & & 2 & & \\
IV2 & & 1/3 & & & 1 & & & 1 & & 1/3 & 1 & 1 & & 1/9 & & & 1/4 & \\
IV3 & & & & & & & & & & 1/5 & 1 & 1 & & & & & & \\
V1 & 5 & & & 2 & & & 9 & & & 2 & & & 1 & 1/3 & 3 & 5 & & \\
V2 & & 9 & & & 9 & & & 9 & & & 9 & & 3 & 1 & 5 & & 6 & \\
V3 & & & & & & & & & & & & & 1/3 & 1/5 & 1 & & & \\
VI1 & & & & & & & & & & 1/2 & & & 1/5 & & & 1 & 1/5 & 3 \\
VI2 & & & & & & & & & & & 4 & & & 1/6 & & 5 & 1 & 7 \\
VI3 & & & & & & & & & & & & & & & & 1/3 & 1/7 & 1
\end{array}$$

4 Results of Calculation by CCM

This section shows the calculations for MDAHP including hierarchical criteria by the CCM [1] with eigenvalue method or the geometric mean method.

4.1 Results of Calculation by CCM with the Eigenvalue Method

Table 1 shows the evaluation values obtained by CCM with the eigenvalue method.

Table 1. Evaluation values obtained by CCM with the eigenvalue method

Criterion	I	II	III	IV	V	VI	Overall evaluation value
Alternative 1	1	3.5259	0.5291	2.5415	4.4859	1.0792	13.1616
Alternative 2	1.7544	0.4632	1.1006	0.7145	11.0632	2.2448	17.3408
Alternative 3	0.3420	1.2780	2.2894	0.6027	1.8189	0.5188	6.8498

4.2 Results of Calculation by CCM with the Geometric Mean Method

Table 2 shows the evaluation values obtained by CCM with the geometric mean method.

Table 2. Evaluation values obtained by CCM with the geometric mean method

Criterion	I	II	III	IV	V	VI	Overall evaluation value
Alternative 1	1	3.5461	0.5392	2.5896	4.5670	1.0992	13.3411
Alternative 2	1.7544	0.4659	1.1215	0.7280	11.2633	2.2865	17.6196
Alternative 3	0.3420	1.2853	2.3329	0.6141	1.8518	0.5285	6.9545

5 Results of Calculation by GMMDAHP

This section shows the calculations for MDAHP including hierarchical criteria by the GMMDAHP [3] with eigenvalue method or the geometric mean method.

5.1 Results of Calculation by GMMDAHP with the Eigenvalue Method

Table 3 shows the evaluation values obtained by GMMDAHP with the eigenvalue method.

Table 3. Evaluation values obtained by GMMDAHP with the eigenvalue method

Criterion	I	II	III	IV	V	VI	Overall evaluation value
Alternative 1	1	3.3842	0.5252	2.5324	4.4337	1.0773	12.9529
Alternative 2	1.7544	0.4446	1.0925	0.7120	10.9345	2.2408	17.1789
Alternative 3	0.3420	1.2266	2.2725	0.6005	1.7978	0.5179	6.7574

5.2 Results of Calculation by GMMDAHP with the Geometric Mean Method

Table 4 shows the evaluation values obtained by GMMDAHP and the geometric mean method.

Table 4. Evaluation values obtained by GMMDAHP with the geometric mean method

Criterion	I	II	III	IV	V	VI	Overall evaluation value
Alternative 1	1	3.4096	0.5343	2.5789	4.5151	1.0970	13.1349
Alternative 2	1.7544	0.4479	1.1114	0.7250	11.1351	2.2819	17.4559
Alternative 3	0.3420	1.2358	2.3119	0.6115	1.8308	0.5274	6.8594

6 Results of Calculation from SPCM by LLSM

For pairwise comparison values in an SPCM, an error model is assumed as follows:

$$r_{c'a'}^{ca} = \varepsilon_{c'a'}^{ca} \frac{v_{ca}}{v_{c'a'}} \tag{3}$$

Taking the logarithms of both sides gives

$$\log r_{c'a'}^{ca} = \log v_{ca} - \log v_{c'a'} + \log \varepsilon_{c'a'}^{ca} \tag{4}$$

To simplify the equation, logarithms will be represented by overdots as $\dot{r}_{c'a'}^{ca} = \log r_{c'a'}^{ca}$, $\dot{v}_{ca} = \log v_{ca}$, $\dot{\varepsilon}_{c'a'}^{ca} = \log \varepsilon_{c'a'}^{ca}$. Using this notation, Eq. (4) becomes

$$\dot{r}_{c'a'}^{ca} = \dot{v}_{ca} - \dot{v}_{c'a'} + \dot{\varepsilon}_{c'a'}^{ca},\ c, c' = 1, ..., C, a, a' = 1, \ldots, A \tag{5}$$

From Eqs. (1) and (2), we have

$$\dot{r}_{c'a'}^{ca} = -\dot{r}_{ca}^{c'a'} \tag{6}$$

$$\dot{r}_{ca}^{ca} = 0 \tag{7}$$

If $\varepsilon_{c'a'}^{ca}$ is assumed to follow an independent probability distribution of mean 0 and variance σ^2, irrespective of c, a, c', a', the least squares estimate gives the best estimate for the error model of Eq. (5) according to the Gauss Markov theorem.

Equation (5) comes to following Eq. (8) by vector notation.

$$\dot{\mathbf{Y}} = \mathbf{S}\dot{\mathbf{x}} + \dot{\varepsilon} \tag{8}$$

where

$$\dot{\mathbf{x}} = (\dot{x}_{\mathrm{I2}}\ \dot{x}_{\mathrm{I3}}\ \dot{x}_{\mathrm{II1}}\ \dot{x}_{\mathrm{II2}}\ \dot{x}_{\mathrm{II3}}\ \dot{x}_{\mathrm{III1}}\ \dot{x}_{\mathrm{III2}}\ \dot{x}_{\mathrm{III3}}\ \dot{x}_{\mathrm{IV1}}\ \cdots\ \dot{x}_{\mathrm{VI2}}\ \dot{x}_{\mathrm{VI3}})^{\mathrm{T}}$$

$$\dot{\mathbf{Y}} = \begin{bmatrix} \dot{r}_{\mathrm{I2}}^{\mathrm{I1}} \\ \dot{r}_{\mathrm{I3}}^{\mathrm{I1}} \\ \dot{r}_{\mathrm{II1}}^{\mathrm{I1}} \\ \dot{r}_{\mathrm{III1}}^{\mathrm{I1}} \\ \dot{r}_{\mathrm{IV1}}^{\mathrm{I1}} \\ \dot{r}_{\mathrm{V1}}^{\mathrm{I1}} \\ \dot{r}_{\mathrm{I3}}^{\mathrm{I2}} \\ \dot{r}_{\mathrm{II2}}^{\mathrm{I2}} \\ \dot{r}_{\mathrm{III2}}^{\mathrm{I2}} \\ \dot{r}_{\mathrm{IV2}}^{\mathrm{I2}} \\ \dot{r}_{\mathrm{V2}}^{\mathrm{I2}} \\ \dot{r}_{\mathrm{II2}}^{\mathrm{II1}} \\ \vdots \\ \dot{r}_{\mathrm{VI3}}^{\mathrm{VI2}} \end{bmatrix} = \begin{bmatrix} \log(1/3) \\ \log 5 \\ \log(1/3) \\ \log 3 \\ \log(1/3) \\ \log(1/5) \\ \log 3 \\ \log 5 \\ \log 1 \\ \log 3 \\ \log(1/9) \\ \log 7 \\ \vdots \\ \log 7 \end{bmatrix}, \mathbf{S} = \begin{bmatrix} -1 & & & & & & & & & \cdots & & \\ & -1 & & & & & & & & \cdots & & \\ & & -1 & & & & & & & \cdots & & \\ & & & & & -1 & & & & \cdots & & \\ & & & & & & & & -1 & \cdots & & \\ & & & & & & & & & \cdots & & \\ 1 & -1 & & & & & & & & \cdots & & \\ 1 & & & -1 & & & & & & \cdots & & \\ 1 & & & & & & -1 & & & \cdots & & \\ 1 & & & & & & & & & \cdots & & \\ 1 & & & & & & & & & \cdots & & \\ & & 1 & -1 & & & & & & \cdots & & \\ \vdots & \vdots & \vdots & \vdots & \vdots & \vdots & \vdots & \vdots & \vdots & \ddots & \vdots & \vdots \\ & & & & & & & & & & 1 & -1 \end{bmatrix}$$

To simplify calculations, $v_{11} = 1$, that is $\dot{v}_{11} = 0$. The least squares estimates for formula (8) are calculated by $\hat{\mathbf{x}} = \left(\mathbf{S}^{\mathrm{T}}\mathbf{S}\right)^{-1}\mathbf{S}^{\mathrm{T}}\dot{\mathbf{Y}}$.

Table 5 shows the evaluation values obtained from SPCM by LLSM for this example.

Table 5. Evaluation values obtained from SPCM by LLSM

Criterion	I	II	III	IV	V	VI	Overall evaluation value
Alternative 1	1	2.859	0.491	2.536	4.765	1.071	12.723
Alternative 2	1.616	0.492	1.113	0.702	9.403	2.152	15.479
Alternative 3	0.328	1.186	2.219	0.597	1.728	0.506	6.564

7 Results of Calculation from SPCM by the Harker Method

In the Harker method, the value of a diagonal element is set to the number of missing entries in the row plus 1 and then evaluation values are obtained by the usual eigenvalue method.

The SPCM by the Harker method for this example is

$H_{(CA \times CA)} =$

	I1	I2	I3	II1	II2	II3	III1	III2	III3	IV1	IV2	IV3	V1	V2	V3	VI1	VI2	VI3
I1	12	1/3	5	1/3			3			1/3			1/5					
I2	3	12	3		5			1			3			1/9				
I3	1/5	1/3	16															
II1	3			12	7	3	5			1			1/2					
II2		1/5		1/7	12	1/3		1/3			1			1/9				
II3				1/3	3	16												
III1	1/3			1/5			12	1/3	1/3	1/5			1/9					
III2		1			3		3	12	1/3		1			1/9				
III3							3	3	16									
IV1	3			1			5			11	3	5	1/2			2		
IV2		1/3			1			1		1/3	11	1		1/9			1/4	
IV3										1/5	1	16						
V1	5			2			9			2			11	1/3	3	5		
V2		9			9			9			9		3	11	5		6	
V3													1/3	1/5	16			
VI1										1/2			1/5			14	1/5	3
VI2											4			1/6		5	14	7
VI3																1/3	1/7	16

Table 6 shows the evaluation values obtained from the SPCM by the Harker method for this example.

Table 6. Evaluation values obtained from SPCM by the Harker method

Criterion	I	II	III	IV	V	VI	Overall evaluation value
Alternative 1	1	2.673	0.466	2.311	4.350	0.941	11.740
Alternative 2	1.756	0.520	1.152	0.705	9.620	2.011	15.764
Alternative 3	0.348	1.087	2.152	0.518	1.496	0.436	6.038

8 Results of Calculation from SPCM by ITSM

The *ij* element of comparison matrix A is denoted by a_{ij} for $i, j = 1 \ldots n$. Nishizawa [12] proposed the following estimation method ITSM. For unknown a_{ij}:

$$a_{ij} = \left(\prod_{k=1}^{n} a_{ik}a_{kj}\right)^{1/m}, \tag{9}$$

where m is the number of known $a_{ik}a_{kj}$, $i = 1 \ldots n$. If unknown comparisons in factors of $a_{ik}a_{kj}$ are included, then assume $a_{ik}a_{kj} = 1$. If $m \neq 0$, the estimated by (3) is treated as known comparison, and the a_{ij} with $m = 0$ in (3) is treated as unknown in the next level. Repeat above procedure until unknown elements completely estimated.

The complete SPCM T by ITSM for this example is

$T_{(CA\times CA)} =$

	I1	I2	I3	II1	II2	II3	III1	III2	III3	IV1	IV2	IV3	V1	V2	V3	VI1	VI2	VI3
I1	1	0.33	5	0.33	1.97	1	3	0.58	1	0.33	1	1.67	0.20	0.05	0.60	0.82	0.34	1.58
I2	3	1	3	0.85	5	1.67	5.20	1	0.33	1	3	3	0.45	0.11	0.56	1.85	0.71	3.29
I3	0.20	0.33	1	0.07	1.67	0.28	0.60	0.33	0.15	0.07	1	0.53	0.04	0.04	0.20	0.26	0.17	0.59
II1	3	1.18	15	1	7	3	5	1.97	1.67	1	4.58	5	0.50	0.36	1.50	2.24	1.18	5.83
II2	0.51	0.20	0.60	0.14	1	0.33	0.85	0.33	0.11	0.22	1	1	0.15	0.11	0.56	0.63	0.41	1.41
II3	1	0.60	3.61	0.33	3	1	1.67	1	0.53	0.33	3	1.90	0.17	0.33	0.72	1.17	0.58	2.87
III1	0.33	0.19	1.67	0.20	1.18	0.6	1	0.33	0.33	0.20	0.45	1	0.11	0.04	0.33	0.47	0.20	0.91
III2	1.73	1	3	0.51	3	1	3	1	0.33	0.45	1	1	0.33	0.11	0.56	1.07	0.41	1.90
III3	1	3	6.76	0.60	9	1.87	3	3	1	0.6	3	3.56	0.33	0.33	1.34	1.67	1.21	3.95
IV1	3	1	15	1	4.58	3	5	2.24	1.67	1	3	5	0.50	0.24	1.50	2.00	0.71	6
IV2	1	0.33	1	0.22	1	0.33	2.24	1	0.33	0.33	1	1	0.24	0.11	0.56	0.71	0.25	0.75
IV3	0.6	0.33	1.90	0.20	1	0.53	1	1	0.28	0.20	1	1	0.10	0.11	0.35	0.40	0.25	1.16
V1	5	2.24	25	2	6.48	6	9	3	3	2	4.24	10	1	0.33	3	5	1.83	15
V2	20.1	9	27	2.78	9	3	27	9	3	4.24	9	9	3	1	5	16.4	6	18
V3	1.67	1.80	5.04	0.67	1.8	1.40	3	1.8	0.75	0.67	1.80	2.90	0.33	0.20	1	1.67	1.20	3.47
VI1	1.22	0.54	3.86	0.45	1.58	0.85	2.12	0.93	0.60	0.50	1.41	2.50	0.20	0.06	0.60	1	0.33	3
VI2	2.90	1.41	5.87	0.85	2.45	1.72	5.01	2.45	0.83	1.41	4	4	0.55	0.17	0.83	3	1	3
VI3	0.63	0.30	1.70	0.17	0.71	0.35	1.10	0.53	0.25	0.17	1.33	0.86	0.07	0.06	0.29	0.33	0.33	1

Table 7 shows the evaluation values obtained from SPCM by ITSM.

Table 7. Evaluation values obtained from SPCM by ITSM

Criterion	I	II	III	IV	V	VI	Overall evaluation value
Alternative 1	1	2.946	0.495	2.681	5.048	1.075	13.245
Alternative 2	1.800	0.564	1.152	0.731	10.110	2.266	16.624
Alternative 3	0.359	1.283	2.388	0.618	1.762	0.520	6.930

9 Conclusion

SPCM of MDAHP is an incomplete pairwise comparison matrix. Therefore, the LLSM based on an error model or an eigenvalue method such as the Harker method or two-stage method is applicable to the calculation of evaluation values from an SPCM.

This paper shows t the calculations for MDAHP including hierarchical criteria by the CCM, by the geometric mean MDAHP and from SPCM by the logarithmic least squares method, the Harker method and the improved two stage method.

References

1. Kinoshita, E., Nakanishi, M.: Proposal of new AHP model in light of dominative relationship among alternatives. J. Oper. Res. Soc. Jpn. **42**, 180–198 (1999)
2. Kinoshita, E., Sekitani, K., Shi, J.: Mathematical properties of dominant AHP and concurrent convergence method. J. Oper. Res. Soc. Jpn. **45**, 198–213 (2002)
3. Ohya, T., Kinoshita, E.: The geometric mean concurrent convergence method. In: Proceedings of the 10th International Symposium on the Analytic Hierarchy Process (2009)
4. Ohya, T., Kinoshita, E.: Proposal of super pairwise comparison matrix. In: Watada, J., Phillips-Wren, G., Jain, L.C., Howlett, R.J. (eds.) Intelligent Decision Technologies. Smart Innovation, Systems and Technologies, vol. 10, pp. 247–254. Springer, Berlin, Heidelberg (2011). https://doi.org/10.1007/978-3-642-22194-1_25
5. Ohya, T., Kinoshita, E.: Super pairwise comparison matrix in the multiple dominant AHP. In: Watada, J., Watanabe, T., Phillips-Wren, G., Howlett, R., Jain, L. (eds.) Intelligent Decision Technologies. Smart Innovation, Systems and Technologies, vol. 15, pp. 319–327. Springer, Heidelberg (2012). https://doi.org/10.1007/978-3-642-29977-3_32
6. Ohya, T., Kinoshita, E.: Super pairwise comparison matrix with the logarithmic least squares method. In: Neves-Silva, R., et al. (eds.) Intelligent Decision Technologies, vol. 255. Frontiers in Artificial Intelligence and Applications, pp. 390–398. IOS Press (2013)
7. Ohya, T., Kinoshita, E.: The treatment of hierarchical criteria in dominant AHP with super pairwise comparison matrix. In: Neves-Silva, R., et al. (eds.) Smart Digital Futures 2014, pp. 142–148. IOS Press (2014)
8. Harker, P.T.: Incomplete pairwise comparisons in the analytic hierarchy process. Math. Model. **9**, 837–848 (1987)

9. Nishizawa, K.: Estimation of unknown comparisons in incomplete AHP and it's compensation. Report of the Research Institute of Industrial Technology, Nihon University Number 77, 10 pp. (2004)
10. Ohya, T., Kinoshita, E.: Super pairwise comparison matrix in the multiple dominant AHP with hierarchical criteria. In: Czarnowski, I., Howlett, R.J., Jain, L.C., Vlacic, L. (eds.) KES-IDT 2018 2018. SIST, vol. 97, pp. 166–172. Springer, Cham (2019). https://doi.org/10.1007/978-3-319-92028-3_17
11. Ohya, T.: SPCM with Harker method for MDAHP including hierarchical criteria. In: Czarnowski, I., Howlett, R.J., Jain, L.C. (eds.) Intelligent Decision Technologies 2019. SIST, vol. 143, pp. 277–283. Springer, Singapore (2019). https://doi.org/10.1007/978-981-13-8303-8_25
12. Ohya, T.: SPCM with improved two-stage method for MDAHP including hierarchical criteria. In: Czarnowski, Howlett, R.J., Jain, L.C. (eds.) IDT 2020. SIST, vol. 193, pp. 517–523. Springer, Singapore (2020). https://doi.org/10.1007/978-981-15-5925-9_45
13. Ohya, T.: Calculations of SPCM by several methods for MDAHP including hierarchical criteria. In: Czarnowski, Howlett, R.J., Jain, L.C. (eds.) Intelligent Decision Technologies. SIST, vol. 238, pp. 609–616. Springer, Singapore (2021). https://doi.org/10.1007/978-981-16-2765-1_50
14. Ohya, T.: Calculations by several methods for D-AHP including hierarchical criteria. In: Czarnowski, I., Howlett, R.J., Jain, L.C. (eds.) Intelligent Decision Technologies. Smart Innovation, Systems and Technologies, vol. 309, pp. 385–393. Springer, Singapore (2022). https://doi.org/10.1007/978-981-19-3444-5_33

Utilization of Big Data in the Financial Sector, Construction of Data Governance and Data Management

Shunei Norikumo(✉)

Doshisha University, Kyoto 602-8580, Japan
snorikum@mail.doshisha.ac.jp

Abstract. In recent years, efforts have been made in the financial sector to centrally aggregate the large daily amounts of data generated, manage data that contributes to marketing and analysis, and then utilize it for customer services. In the insurance industry, the principle products that have high customer needs are insurance products, for banks, it is loan products, and for securities companies it is investment trust products. These show trends such as converting customer financial transactions into data and encapsulating customer demand into products. The history of data management in the financial sector is long compared to that of other industries, but the current situation is that it is lacking in proper data utilization. Based on the background outlined herein, this research examines what kind of data management and data governance the financial sector should implement in order to utilize Big data, and how information system platforms and data infrastructures should be developed moving forward.

Keywords: Big Data · Data Management · Information Systems · Social Infrastructure · Economic Information Systems · Economic Decision Making

1 Introduction

1.1 History and Current Status of Data Management in the Financial Sector

Corporate information systems are generally constructed within the organizational structure of core operations. Based on the maximization of production efficiency from Taylor's business management ideas and Bernard's theoretical development of an organization, an information system is constructed for each business department in the organization, and data is then managed based on this system.

In the financial sector, the first operation type is where individuals and companies generating large amounts of deposit and withdrawal management transaction data. The second is the "loan" business, in which the money deposited in deposits is loaned to people who need funds. This includes not only corporate loans but also personal loans such as housing loans. The third is a "money exchange" business that sends money to another account at another bank or receives bills and checks. Bank systems operate with these three main operations: deposits, loans, and exchanges. The system that processes

I. Czarnowski et al. (Eds.): KESIDT 2023, SIST 352, pp. 273–281, 2023.
https://doi.org/10.1007/978-981-99-2969-6_24

these operations is the bank's "accounting system," and the systems that process and support other operations are the "information system," "international system," "securities system" and "external system." There is also a "sales office system." In the insurance sector, the "medical calculation/statistics system" and "pension fund system" are to be regarded as the system for processing the operations.

Due to changes in the information systems that have been implemented by each bank over the years, main "accounting systems" exist with deferring business and system requirements. Financial risk management such as revenue management, credit management, asset liability management (ALM), Customer Relationship Management (CRM) and other systems that process customer information are built as information systems, and data processing is performed based on these information systems.

The "first online system" was the starting point of Japan's first online banking system, which was operated by Mitsui Bank (currently Sumitomo Mitsui Banking Corporation as of 2016) in 1965. Since then, more and more city banks have moved ordinary deposits online. In the first online system, both the business programs and the program that controlled them online were bespoke.

In the "second online system," a comprehensive online system was built that could process multiple tasks simultaneously by linking contract information for multiple products based on the collected customer information and then centrally managing the customer information. With this information system, comprehensive account services began, and the system shifted from single task online systems to comprehensive online systems. In addition, CDs (Cash Dispensers) dedicated to paying out cash became popular, and bank online systems expanded rapidly. After that, the CDs were replaced by ATMs (Automatic Teller Machines) that could receive cash deposits.

In addition, the construction of an inter-financial network for the mutual use of CD/ATMs, as well as the establishment of the Zengin Data Communication System (Zengin System), a network between banks, was also established during this period, forming the infrastructure for Japan's economic transactions.

The "third online system" was constructed in the late 1980s to early 1990s. In order to improve system flexibility and reduce development costs, hardware (mainframe) and control programs are left to IT vendors, and packaged products provided by vendors are used, and only business software are developed in-house. The system was developed such that the bank's large-scale information system and data management kept pace with technological innovations in IT simply by upgrading the hardware and control software, with no need to entirely rework the business software.

1.2 The Era of Ultra-large Data and Corporate Expectations

In recent years, with the spread of cloud computing, the use of SNS, the spread of sensor technology and smartphones, the quantity of digital data has massively and rapidly increased, and so the use of Big data is accelerating.

Big data is ultra-large-scale data that cannot be processed by conventional software for data management and analysis. It has the following characteristics: extremely large quantities of data (Volume), data generation speed (Velocity) and diversity (Variety), the 3Vs as it is called. The data collected, stored and analyzed can be heterogeneous or un-structured, or to state it another way: "unlimited data type variety to handle" and "precision is not important, so long as there is a large quantity of data." In addition, a data processing environment for this is being built year on year, and the definition of Big data is also evolving along with AI (artificial intelligence) technology.

From the above technological background, many companies are now showing strong interest and expectations for the utilization of Big data, with the belief that it will bring useful business knowledge that could not otherwise be obtained by existing analysis.

2 Previous Research on Big Data and Frameworks for Data Management

2.1 Utilization of Big Data in an Organization

Big data has been attracting attention since around 2012, which is earlier than AI. In around 2014, surveys at companies have gradually revealed the usefulness of this data in the utilization of online data.

In a study by Hasan Bakhshi et al. by Nesta in the UK, they surveyed 500 companies that are commercially active online in the UK. They collected, analyzed and deployed online customer data on the use of online data from which they were able to quantify the contributions to business productivity. In "Online data activities and firm productivity" of "The analytical firm," the Total Factor Productivity TFP increased by 8%, in step with increased online data use, and the companies belonging to the highest quartile of online data use showed this condition. In the same case, productivity was reported to be 13% higher [2].

Furthermore, as for what complementary data utilization is most related to productivity, it can be seen that, while data analysis and data insights are strongly related to productivity, data accumulation itself has little effect. In "Complementarities between data activities and process innovation," it is reported that the impact of using online data on the utilization of online data and the complementarity of inter-organizational activities is stronger in companies which have "high employee autonomy" and "it is open to the constructive disruption of business processes."

Based on the theory that data utilization in an organization is a source of gaining a competitive advantage, in recent years, as the analysis of the relationship between online data utilization of companies and their business performance deepened, "utilization of online data" has gradually proved to be more productive.

2.2 Data Management and Data Governance

Japan's Ministry of Economy, Trade and Industry and the Ministry of Internal Affairs and Communications are also promoting the comprehensive use of data, and it is hoped that the results of this investment will be seen in the future. Overseas, data is expected to become an asset for companies and the key to acquiring business opportunities in the future. DAMA DACH, an international organization that promotes the use of data on a par with ISO standards, has proposed a framework for data management and data governance. Furthermore, even in Japanese companies that handle large-scale data such as in the financial sector, awareness of data management and data governance has been passed on from consultants and is gradually proliferating. In both cases, in today's corporate management, the active utilization of data in an organization is expected to be a catalyst for improving productivity and organizational reform.

The aforementioned DAMA DACH institution sets out the behavior expected for data professionals who are members of DAMA. It also dispatches guidance for developing architectures, policies, practices, and procedures that successfully manage an enterprise's entire data lifecycle needs for data resource management (Figs. 1 and 2).

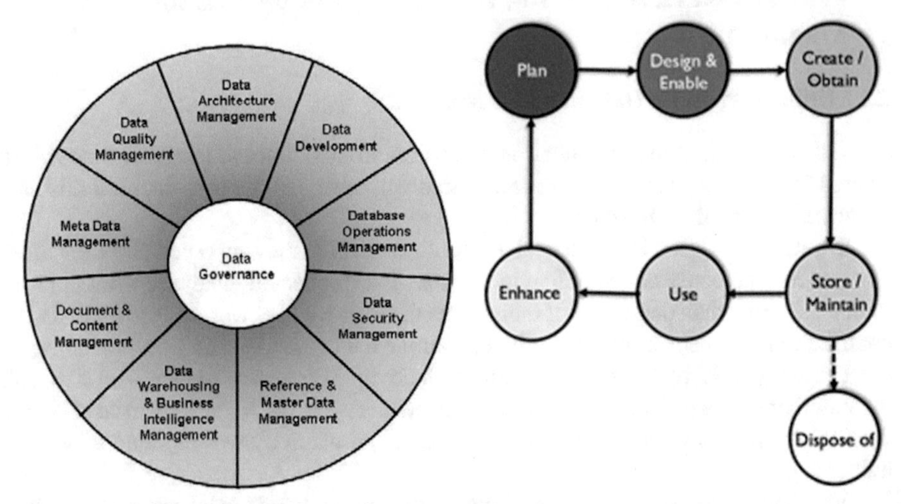

Fig. 1. Data Management Functional Framework Source: DAMA-DMBOK produces

Fig. 2. Data Lifecycle Management Source: DAMA-DMBOK produces

Recommendation from DMBOK2 is that data are a vital enterprise asset. Data and information have been called the 'currency,' the 'life blood,' and even the 'new oil.' Long-standing definitions of data emphasize its role in representing facts about the world. Data represents things other than itself. Traditional Business Intelligence provides 'rear-view mirror' reporting. Data science techniques are used to provide 'windshield' view of the organization. Data governance is about 'Doing the right things' and data management is about 'Doing things right.' Organizations don't change, people change. People don't resist change. They resist being changed.

3 Utilization of Big Data in Japanese Financial Institutions

Sumitomo Mitsui Financial Group (SMFG) established "SMBC Digital Marketing" jointly with the Dentsu Group in July 2021 as a "banking advanced company" recognized under the 2017 revised Banking Act. As a consolidated subsidiary of SMFG, they are developing an advertising and marketing business that utilizes financial Big data. In an era where personal information protection is becoming stricter, they provide advertisements of the services that are demanded by customers and improve their quality of life.

Resona Bank introduced the data warehouse "Teradata Active Enterprise Data Warehouse 6680" in 2013, integrated the CRM database and the MCIF (Marketing Customer

Information File) system, and constructed a new sales support database. As a result, they have built an environment that allows real-time access to the data warehouse from customer channels such as sales sites with approximately 14,000 employees, 2,200 ATMs, a call center with approximately 230 seats, and Internet banking. Resona Bank has reorganized the Omni-Channel Strategy Department and the Payment Business Department to establish the Data Science Department. While receiving support from external partners such as IT companies, it has taken a hybrid strategy of analyzing data on its own and promoting the development of an environment and accumulation of know-how for "in-house production." In addition, in 2020 and 2021, they will be the only bank to be selected as a "digital transformation stock (DX stock)."

The MUFG Group built the "MUFJ Big Data Platform" system with the aim of accumulating a huge amount of data of various types inside and outside the country, and using it freely under appropriate control. Structured and unstructured data can be stored in petabyte-class large volumes, making it easy to search, extract, and link necessary data from the vast amount of stored data. MUFG's unique machine learning model, which realizes more sophisticated and efficient operations, is being used for research and development. In the future, it will support business transformation through digitalization and is positioned as a key to strategy planning and a source of information to strengthen business operations. Japan Digital Design's MUFG AI STUDIO (abbreviation: M-AIS) is an organization established for the purpose of promoting research, development, and implementation of MUFG's unique machine learning models. M-AIS is built to leverage bank transaction data for credit and anti-money laundering.

4 Trend Survey of Global Big Data Utilization

In this study, we reviewed the results of a 2012 IBM Institute for Business Value and a Said Business School at the University of Oxford survey on "Use of Big Data in the Real World in Financial Services." The survey researched 1,144 business and IT professionals in 95 countries, with 124 or 11% responding. In the report, 71% of financial institutions said the use of information (including Big data) and analytics is creating a competitive advantage for their organizations. The financial sector does not manufacture physical products, so financial services are its primary product. While other manufacturing industries use SCM and DCM, the financial sector is expected to utilize Big data to solve various problems.

In Fig. 3, we surveyed the goals of companies working on Big data in all industries. The top is Customer-centric outcomes, and then Operational optimization, Risk/financial management, New business models, and Employee collaboration follows. Comparing all industries with the financial sector, "customer-centric outcomes" is followed by "risk/financial management" which is followed by "new business models." Considering the contents in Sect. 3, it can be inferred that expectations for the use of Big data in the financial sector are customer-oriented services, security measures, and the creation of new businesses. While corporate management is becoming more customer-oriented in all industries, it is difficult for the financial sector to read customer orientation by linking it to products as in manufacturing industries. Due to the current status of information technology utilization, the financial sector is concentrating its efforts on acquiring customer needs through data utilization. These results also relate to the challenge that the financial sector continues to provide traditional organization structures and financial products to its customers. Another difficult problem is that even if the results of Big data analysis suggest a new business model, the decision-making process of traditional organizations may not be able to develop such a new model.

In which case, collaborating with a new company is considered to be a favorable factor in lowering the threshold for promoting new business.

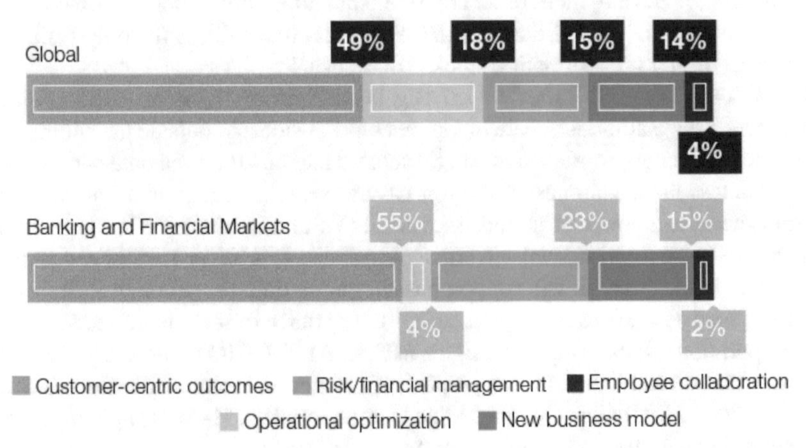

Fig. 3. Big data objectives Source: The real-world use of Big data, a collaborative research study by the IBM Institute for Business Value and the Saïd Business School at the University of Oxford. © IBM 2012.

Figure 4 shows the survey results for Big data infrastructure. 87% of financial institutions say they have Big data infrastructure in place, but information integration is extremely low compared to other industries. Looking at the history of information systems in banking, information sharing has become integrated only recently as departments have been integrated and reorganized repeatedly. However, mergers and consolidations between banks have further complicated this data management. It can be inferred that the situation peculiar to the financial sector has brought the result as above.

Figure 5 shows the survey results for Big data sources. Transactions, logs, events, emails, and geographic information are the sources of data that financial institutions are using more than any other industry. Conversely, sources that financial institutions use less than other industries are SNS, external feeds, free-form text, audio, and still images/video. These results show that financial institutions have an abundance of internal data and tend to use internal data to create new businesses.

The low use of external data is related to the entrenched data analysis habits of financial institutions. In particular, in data analysis using data repositories of financial institutions, external data such as economic indicators as well as economic situations are sometimes reflected in the numerical values of financial products, but weather and regional data are not incorporated into financial products, which is a good contrast with manufacturing and retail industries which actively utilize such data.

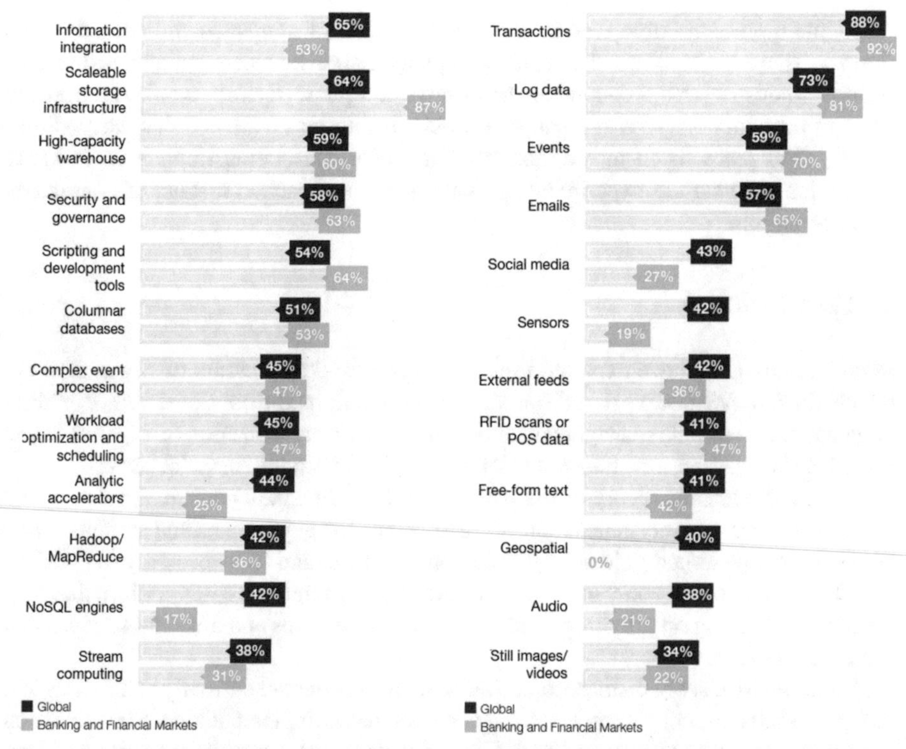

Fig. 4. Big data infrastructure Source: Work survey, a collaborative research survey conducted by the IBM Institute for Business Value and the Saïd Business School at the University of Oxford. © IBM 2012

Fig. 5. Big data sources Source: Work survey, a collaborative research survey conducted by the IBM Institute for Business Value and the Saïd Business School at the University of Oxford. © IBM 2012

Educate	Explore	Engage	Execute
Focused on knowledge gathering and market observations	Developing strategy and roadmap based on business needs and challenges	Piloting big data initiatives to validate value and requirements	Deployed two or more big data initiatives, and continuing to apply advanced analytics
Global 24%	Global 47%	Global 22%	Global 6%
Banking and Finance Management 26%	Banking and Finance Management 47%	Banking and Finance Management 23%	Banking and Finance Management 3%

Fig. 6. Big Data adoption Source: The real-world use of big data, a collaborative research study by the IBM Institute for Business Value and the Saïd Business School at the University of Oxford. © IBM 2012

Figure 6 presents survey results showing the stages of implementation of a Big data strategy. There is almost no difference in implementation between financial institutions and other industries. Most companies are in the review phase. It can be confirmed from the status of introduction cases at Japanese financial institutions in Sect. 3 that the medium- to long-term plans listed in the management strategy are being implemented. Many companies have started experimentations but have not reached a stage of continuous implementation.

5 Conclusion

In this research, we have explored trends in the use and application of Big data in the financial sector. We develop our discussion to introductions and considerations regarding systematization of data management and data governance, utilization of Big data in the Japanese financial sector, and surveys of data utilization in the financial sector.

In recent years, the theory of Big data has been systematizing data management and data governance, and this trend is spreading to Japan. The government and administrative bodies have also issued guidelines for data governance, and it is necessary to consider how it will take root in companies in the future. The introduction cases confirm that they are mainly focusing on providing customer-centered services and financial services that use customer data.

To summarize based on these, most financial institutions are currently in the study and trial phase and have not yet reached a stage of sustainable implementation. Elements such as various data management and governance development, active use of customer data, development of advertising business and the establishment of companies are progressing separately. In order for these individual elements to collaborate, it is necessary to develop a platform. Overseas Chinese Banking Corporation (OCBC) and Alibaba Group's Ant Financial Services Group (AFSG) have successfully used Big data within their group companies or across divisions and sales channels. The benefits of using data within the group include, clear data governing procedures and strong data links. Japanese financial institutions, even if they belong to the same group, face significant barriers in the way data is managed due to their past merging experiences. Regarding the mutual data utilization across institutions, it is thought that the harmful factors are a lack of governance and inability to comprehend customers' personal data.

References

1. IBM Japan Ltd, Takeshi Hoshino. atmarkit.itmedia.co.jp.: "History and basic knowledge of mainframes and banking systems that young people do not know" (2016) https://atmarkit.itmedia.co.jp/ait/articles/1609/07/news007.html.
2. Norikumo, S.: Utilization of big data in the Japanese retail industry and consumer purchasing decision-making. In: Czarnowski, I., Howlett, R.J., Jain, L.C. (eds) Intelligent Decision Technologies. Smart Innovation, Systems and Technologies, vol. 309, pp. 375–383 (2020). https://doi.org/10.1007/978-981-19-3444-5_32
3. Bakhshi, H., Bravo-Biosca, A., Mateos-Garcia, J.: The analytical firm: estimating the effect of data and online analytics on firm performance (2014)

4. DAMA DACH (2023). https://damadach.org/about-us/.
5. Ministry of Internal Affairs and Communications.: White Paper on Information and Communication", Digital transformation promoted by 5G and construction of new daily life. (2020)
6. Sumitomo Mitsui Financial Group.: DX Trend. "A new form of advertising utilizing financial big data" (2022). https://www.smfg.co.jp/dx_link/dxtrend/special/article_22.html.
7. IT media Enterprise.: "Resona Bank optimizes operations by building a new sales support database" (2013). https://www.itmedia.co.jp/enterprise/articles/1307/02/news113.html.
8. NTT DATA Corporation.: "MUFG Big Data Infrastructure The cornerstone of digitalization that supports the development of the entire group" (2020) https://www.nttdata.com/jp/en/case/2020/100100/.
9. Ministry of Internal Affairs and Communications JP.: Survey and research on the current state of measurement and utilization of the economic value of digital data. (2020)
10. IBM Institute for Business Value and the Saïd Business School at the University of Oxford.: "Analytics: The real-world use of big data in financial services" by IBM (2012). https://www.ibm.com/downloads/cas/E4BWZ1PY.

A Block Chart Visualizing Positive Pairwise Comparison Matrices

Takafumi Mizuno(✉)

Meijo University, 4-102-9 Yada-Minami, Higashi-ku, Nagoya-shi, Aichi, Japan
tmizuno@meijo-u.ac.jp

Abstract. This article provides a chart that visualizes pairwise comparison ratios and weights derived from them. The chart consists of piled blocks whose edges represent two weights, and the slope of inclined lines in each block represents the ratio between the two weights. Furthermore how to adjust weights using the chart is described.

Keywords: pairwise comparison matrix · data visualization

1 Introduction

To understand the values of things, we often compare them to each other. In economics, comparing things derives replacement rates that are their values. Especially the rates between money and goods are called the prices. However, we cannot grasp intuitively how comparisons derive values or why the comparisons justify the value. This article treats pairwise comparisons and provides a chart that visualizes the comparisons and values derived from them.

Let us consider that there is a set of items $I = \{1, 2, ..., n\}$, and we want to decide the relative values that help us rank them. To obtain the values, you can try pairwise comparisons. For every pair of items, you show that an item of the pair is how times preferer to another by using a positive real number called ratio in this article. If an item i is three times preferer than an item j, then we denote $a_{ij} = 3$. Notice that $a_{ii} = 1$. The ratios are arranged into a matrix which is reffered as a pairwise comparison matrix:

$$\begin{bmatrix} 1 & a_{12} & \cdots & a_{1n} \\ a_{21} & 1 & \cdots & a_{2n} \\ \vdots & \vdots & \ddots & \vdots \\ a_{n1} & a_{n2} & \cdots & 1 \end{bmatrix} \tag{1}$$

So, our interest is in deciding relative values of the items $w = (w_1, ..., w_n)$ from ratios a_{ij}. In this article, the values derived from the ratios are referred to as weights.

In the field of decision-making, AHP (Analytic Hierarchy Process) [1] uses such pairwise comparisons. The process supports the choice of the best alternative for decision-makers to achieve their goals. For each criterion that decision-makers must care, ratios are shown by pairwise comparisons among the alternatives, and the weights of the alternatives are calculated. A score of an alternative

I. Czarnowski et al. (Eds.): KESIDT 2023, SIST 352, pp. 282–289, 2023.
https://doi.org/10.1007/978-981-99-2969-6_25

is the sum of each product of importance of a criterion and the weight of the alternative for the criterion. The alternative with the highest score is chosen.

In AHP, the eigenvalue method and the geometric mean method are often used to obtain weights. These methods are validated mathematically with an assumption of reciprocal symmetry for ratios, that is, $a_{ij} = 1/a_{ji}$. Notice that ratios may not be reciprocal symmetry in this article.

2 A Block Chart

In this Sect. 1 provide a chart that visualizes ratios and weights. The chart is a pile of rectangular blocks. Each block depicts weights of two items and the ratio between them. The length of each edge of the block represents the corresponding weight, and the incline of a line segment touching the lower left corner represents a ratio. In the block, the inclined line makes a right-angled triangle whose lower acute angle is identified as the tangent of the triangle. The angle is the height if the length of the bottom line of the triangle is 1 (Fig. 1). Under the definition, the angle represents the ratio how times larger the height to the width of the block. So, assigning two weights to the width and the height can visualize the weights of two items and the ratio between them (Fig. 2).

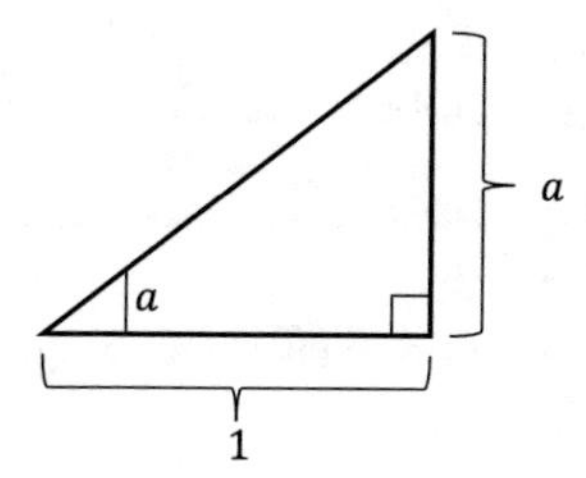

Fig. 1. The lower acute angle is represented as the tangent of the triangle.

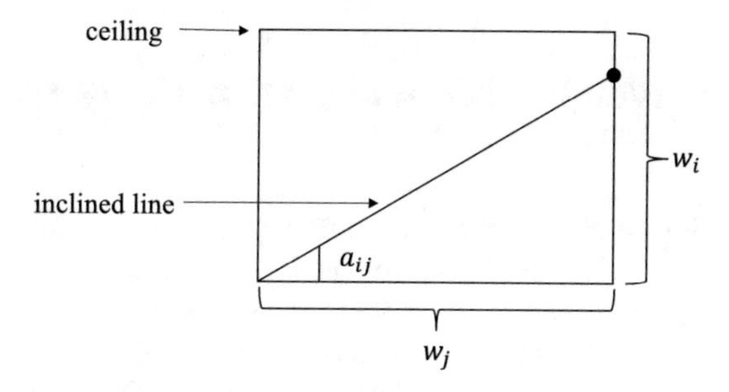

Fig. 2. A block visualizes two weights and the ratio between two items.

For example, let us consider ratios in a pairwise comparison matrix:

$$\begin{bmatrix} 1 & 2 & 3 & 2 \\ \frac{1}{2} & 1 & 7 & 5 \\ \frac{1}{3} & \frac{1}{7} & 1 & \frac{1}{2} \\ \frac{1}{2} & \frac{1}{5} & 2 & 1 \end{bmatrix}. \tag{2}$$

The geometric mean method gives us weights $w = (0.3746, 0.4117, 0.0791, 0.1346)$ from the matrix. The ratios and weights are visualized in the block chart in Fig. 3.

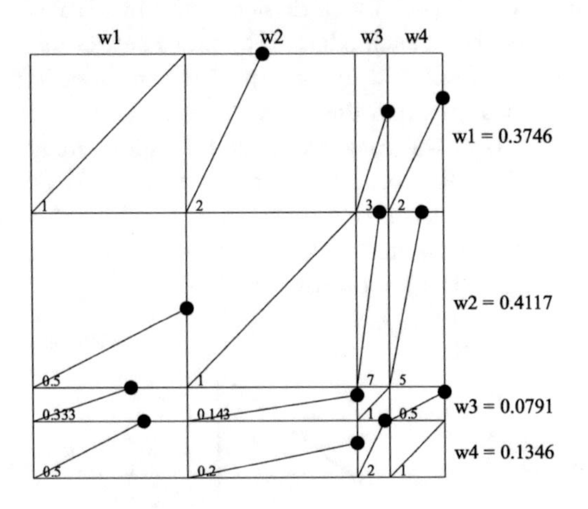

Fig. 3. A block chart visualizes all weights and ratios.

Notice that blocks on the chart's diagonal line are always square because the width and height of blocks represent the same weights.

3 Adjusting Weights Manually with Using the Block Chart

When we draw the block chart on computer graphics, we can seek weights on the chart by using GUI. Enlarging the widths or heights of blocks increases corresponding weights, and shrinking them decreases the weights.

Now, I introduce some terms to describe adjusting weights. While weights are estimated from all ratios, we can derive an item's weight from another item's weight; we can guess i's weight to be $a_{ij}w_j$. We refer to the weight derived from another weight as local weight, and we say that w_i is consistent with w_j when there is no gap between the weight and the local weight; $|w_i - a_{ij}w_j| = 0$. We want to adjust weights to be consistent.

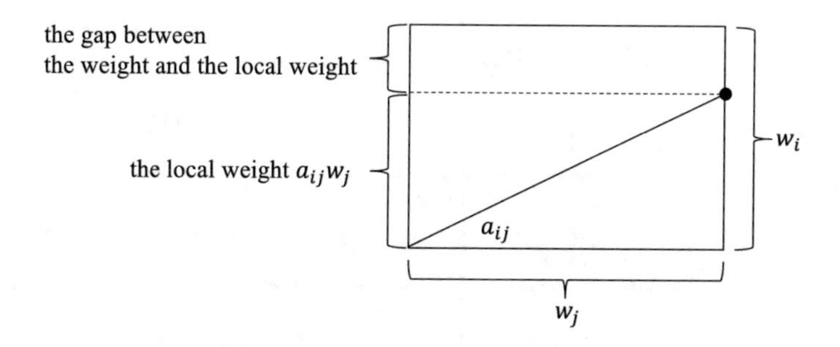

Fig. 4. Weight w_i is larger than its local weight to w_j.

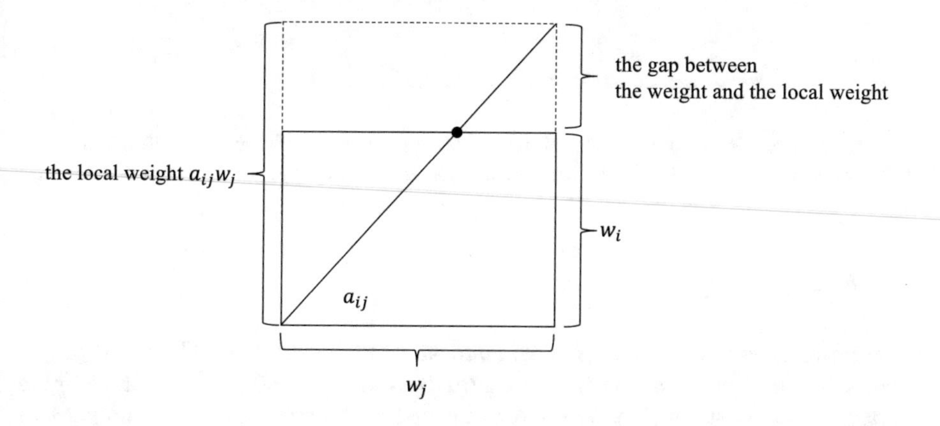

Fig. 5. Weight w_i is smaller than its local weight to w_j.

We say that w_i is overestimated when the weight is larger than all its local weights; $w_i > a_{ij}w_j$, $j \neq i$, and we say that w_i is underestimated when it is smaller than all its local weights; $w_i < a_{ij}w_j$, $j \neq i$. And we say that the weight is acceptable when the weight is not overestimated or underestimated. Shrinking an overestimated weight or enlarging an underestimated weight decreases all the gaps between it and its local weights. But adjusting an acceptable weight must enlarge some gaps. So we can adjust each weight only when the weight is overestimated or underestimated.

If w_i is overestimated, the inclined line does not reach the ceiling in every block in the i-th row in the chart (Fig. 4). And if w_i is underestimated, the inclined line intersects the ceiling in every block in the i-th row (Fig. 5).

A process adjusting weights is shown in Fig. 6 to Fig. 9. In the charts, the thick line on the ceilings tells us the corresponding weight is underestimated, and the thick dashed line on the ceilings tells us the corresponding weight is overestimated.

Figure 6 visualizes ratios in (2) and uniform weights: $w_i = 0.25$. Enlarging the underestimated weight w_1 until it is acceptable causes w_2 be underestimated (Fig. 7). After enlarging w_2 (Fig. 8) and shrinking w_3 until they are acceptable, we can obtain acceptable weights $w = (0.4375, 0.2245, 0.1339, 0.2041)$ (Fig. 9).

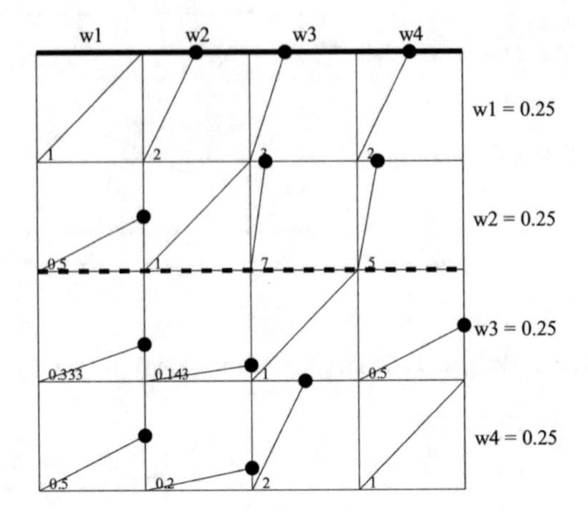

Fig. 6. w_1 is underestimated; all ceilings of blocks whose height is w_1 intersect inclined lines. w_3 is overestimated; all ceilings of blocks whose height is w_1 do not meet inclined lines.

4 Discussions

The adjustment of weights is considered as a n players game. The i-th player wants to put weight w_i with small gaps between w_i and all its local weights $a_{ij}wj$, $j \neq i$. If the weight is underestimated or overestimated, the player can shrink all its gaps, and players do not want to change the acceptable weights. So, we can give preference among weights $w = (w_1, ..., w_n)$ and $w' = (w'_1, ..., w'_n)$ as follows. A player i prefers w to w', denoted as $w \succ_i w'$, when

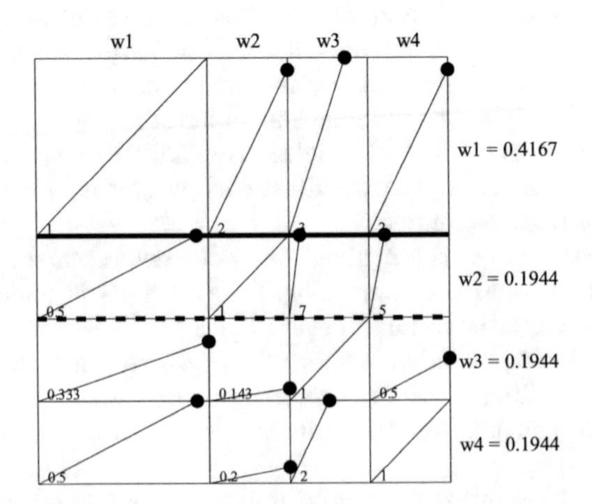

Fig. 7. Increasing w_1 makes w_2 be underestimated.

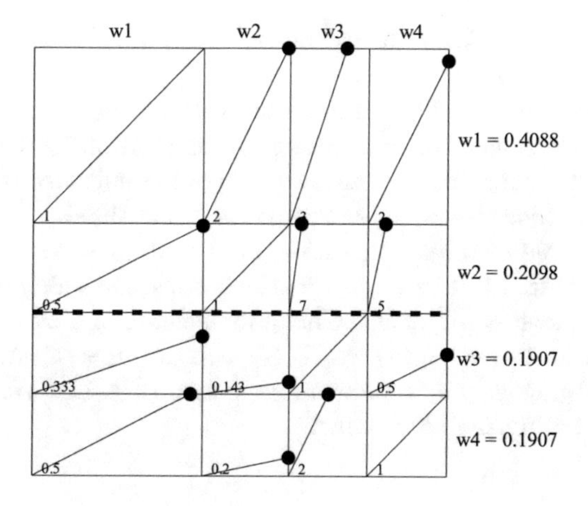

Fig. 8. Increasing w_2 makes w_2 be acceptable.

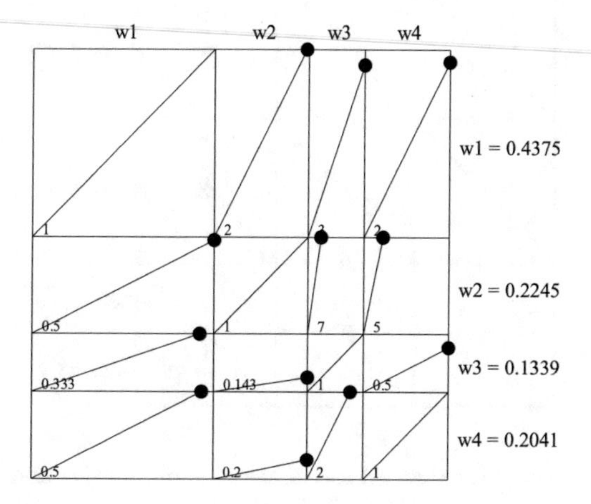

Fig. 9. Decreasing w_3 gives four acceptable weights. They also are a Nash equilibrium.

$$\forall j \neq i, \quad |w_i - a_{ij} w_j| < |w'_i - a_{ij} w'_j|. \tag{3}$$

In the previous section, we adjusted weights one by one until thick lines and thick dashed lines went out. Then the weights w as its result must satisfy a condition: for each player i, there are not weights w' such that

$$(w_1, ..., w_{i-1}, w'_i, w_{i+1}, ..., w_n) \succ_i w. \tag{4}$$

In other words, after the adjustment, each player is not motivated to change the weight if other players do not. Such results are called (pure) Nash equilibria in the game theory [2]. So, the block chart without thick lines represents a Nash equilibrium in the problem of finding weights from ratios.

Also, we can show that Nash equilibria always exist. When every weight has at least one local weight whose gap is zero, the set of weights is a Nash equilibrium. We can make such acceptable weights as $w_1 := 1$ and $w_{i+1} := a_{i+1,i}w_i$, $i = 1, ..., n-1$.

But, unfortunately, the chart cannot represent Pareto optimal weights. When weights are in a Nash equilibrium, there is no motivation to change weights one by one. While changing more than two weights simultaneously may improve weights. It means that the weights are not Pareto optimal.

There is a simple example: all ratios are 1 and $w_1 = w_2$ and $w_3 = w_4$. Its block chart is in Fig. 10. There are not thick lines and thick dashed lines; the weights are in a Nash equilibrium. Changing weights one by one must increase gaps, $|w_1 - a_{12}w_2|$, $|w_2 - a_{21}w_1|$, $|w_3 - a_{34}w_4|$, or $|w_4 - a_{43}w_3|$. But enlarging w_3 and w_4 simultaneously can make all gaps zero (Fig. 11). We cannot find the weights by only seeing the block chart.

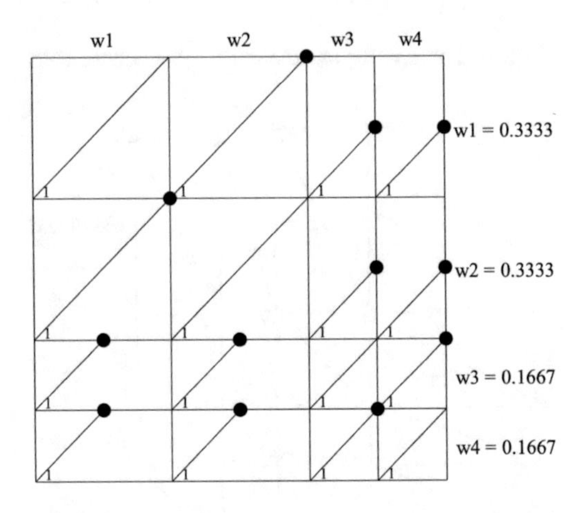

Fig. 10. A chart for that all ratios are 1, and $w_1 = w_2$ and $w_3 = w_4$.

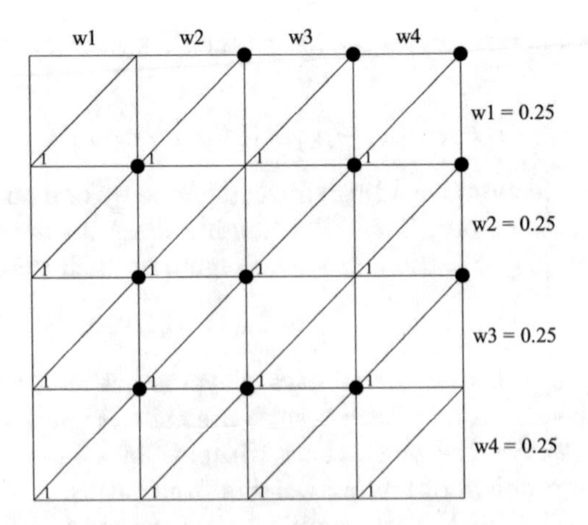

Fig. 11. A chart for that all ratios are 1, and $w_1 = w_2 = w_3 = w_4$. There is no gaps.

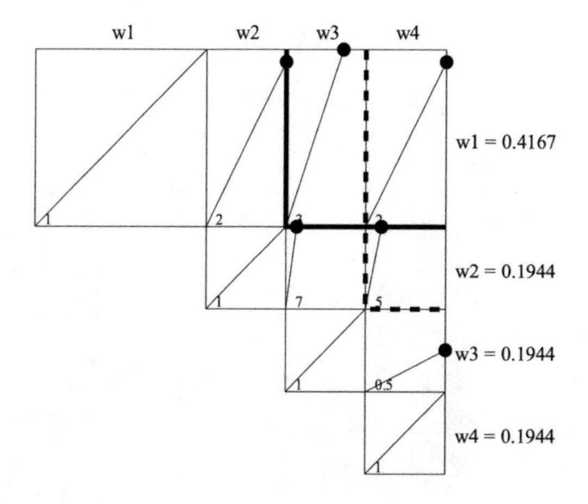

Fig. 12. A block chart for a reciprocal symmetric pairwise comparison matrix (2).

5 Conclusion

This article provides a block chart that visualizes pairwise comparison ratios and weights derived from them. The chart can tell us whether weights are in a Nash equilibrium.

Half of the ratios are unnecessary if we assume pairwise comparisons are reciprocal symmetry. So, the shape of the chart will be an inverted triangle like Fig. 12.

In AHP, the consistency between ratios and weights are represented in C.I. (Consistency Index) calculated with using the priciple eigenvalue of pairwise comparison matrix [3]. The block chart can show consistency as gaps between weights and local weights, which are cleary depicted on blocks.

References

1. Saaty, T.L.: The Analytic Hierarchy Process. McGraw Hill, New York (1980)
2. Nash, J.: Equilibrium points in n-person games. Proc. Natl. Acad. Sci. **36**(1), 48–49 (1950)
3. Brunelli, M.: Introduction to the Analytic Hierarchy Process. Springer, Cham (2015). https://doi.org/10.1007/978-3-319-12502-2

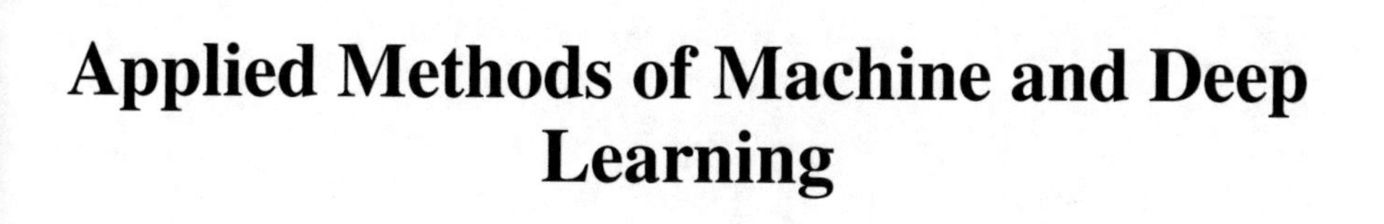

Applied Methods of Machine and Deep Learning

Neural Networks Combinations for Detecting and Highlighting Defects in Steel and Reinforced Concrete Products

Nikita Andriyanov[1](✉), Vitaly Dementiev[2], and Marat Suetin[2]

[1] Financial University Under the Government of the Russian Federation, Leningradsky av. 49, 125167 Moscow, Russia
nikita-and-nov@mail.ru

[2] Ulyanovsk State Technical University, Severny Venets st. 32, 432027 Ulyanovsk, Russia

Abstract. This paper considers the problems of detecting and evaluating defects on real images of the steel surface and reinforced concrete products. To solve these problems, an approach is described based on the preliminary application of the YOLO neural network to localize defect areas and further segmentation of cracks in localized image areas. It is shown that the detection of defect areas in the original images allows one to obtain an acceptable quality of segmentation and significantly reduces the number of false positive errors. The paper investigates different segmentation models such as TernausNet and Deeplabv3+. The augmentation techniques provides more general models after training which has good result on test data. The DICE metrics was used for comparison and evaluation different neural network models. The best results are on the 76% level. It is possible to reach using special encryptor training in TernausNet. In the future it will be possible to use more efficient preliminary models.

Keywords: Image Segmentation · Steel Defects · Deep Learning

1 Introduction

Maintenance of steel and reinforced concrete structures during their operation is based on the timely detection of surface defects for the purpose of subsequent assessment of their parameters and decision-making on localization. Recently, most popular solutions became the approaches to detecting defects based on the processing of images obtained during the flight of buildings by unmanned aerial vehicles. Such approaches based on image processing with devices make it possible to inspect a structure in hard-to-reach places and minimize the risks to the life of experts performing the inspection. However, the quality of crack detection using classical segmentation algorithms is significantly affected by a complex of various interfering factors, such as fine cracks, shadows, concrete surface roughness, etc. A number of works shown [1–3] that deep learning-based approaches can successfully cope with interfering factors and obtain results that are superior in quality to the work of a trained expert.

I. Czarnowski et al. (Eds.): KESIDT 2023, SIST 352, pp. 293–301, 2023.
https://doi.org/10.1007/978-981-99-2969-6_26

One of the problems associated with the use of artificial neural networks is the limitations on the size of the processed image. This limitation forces one to resort to some tricks [4] related to the preliminary division of the original images into separate areas (tiles) with sizes suitable for processing by the neural network, their segmentation and subsequent assembly of the segmentation results into a mask with the dimensions of the original image. The use of this approach as an alternative to resizing (with compressing) the original image makes it possible to prevent loss of segmentation quality. At the same time, the presence of a significant imbalance between the pixels of defects and the background in the original images inevitably leads to the appearance of a significant number of "false" cracks in the results. The good idea is preliminary detection of regions of interest [5]. This paper describes an approach based on the preliminary application of the YOLOv3 neural network for the localization of defect areas and further segmentation of cracks in localized image areas.

2 Detection and Segmentation Artificial Neural Networks

An analysis of works devoted to image segmentation suggests that neural network architectures based on auto-encoders are the most suitable for selecting objects with a complex shape (medical anomalies, surface defects). The paper compares the quality of image segmentation of cracks on the surface of concrete structures using TernausNet and Deeplabv3 + neural networks. These architectures have incorporated the best techniques used in other image segmentation neural networks [6].

TernausNet (baseline) is an improved U-Net that uses the VGG16 neural network as a cipher, pre-trained in classification on ImageNet [7]. The decoder learns segmentation on a specialized set of images of cracks on the surface of structural elements of a railway bridge. In order to eliminate the problems associated with the fading/exploding gradient, the convolutional layers of the decryptor are initialized by the Kaiming algorithm before training. Figure 1 shows the architecture of TernausNet [7].

As a second approach to solving the problem of crack segmentation on steel structure elements, the last convolutional layers of the TernausNet encoder are proposed to be trained on a training set of 40 thousand images of defects on the surface of concrete. It is Surface Crack Detection Dataset [8]. The TernausNet cipher, in turn, consists of 5 blocks of convolutional layers. As a basic model, the TernausNet neural network with a cipher was used, the convolutional layers of which are fully initialized with weights obtained by transfer learning from the VGG16 neural network [9]. Further, the TernausNet neural network (decoder) is trained in segmentation on a specialized set of images in the same way as in the previous case.

The YOLOv3 neural network architecture [10] was used to highlight areas of defects in images of steel structure elements. As the backbone of this architecture, the Darknet-53 neural network [10] is used, preliminarily trained in classification on a set of images of real-world objects ImageNet [11]. The YOLOv3 detector is further trained using the transfer learning technology to detect defects on a specialized training set consisting of labeled photographic images of elements of steel structures of railway bridges. The pre-training sample images were labeled using the professional software tool CVAT (Computer Vision Annotation Tool) in the widely used Pascal VOC format. YOLOv3 was trained in the traditional way for this architecture for 100 epochs. To reduce the effect of neural network retraining, augmentation procedures were used: rotation (rotate) and horizontal reflection (horizontal flip) [12].

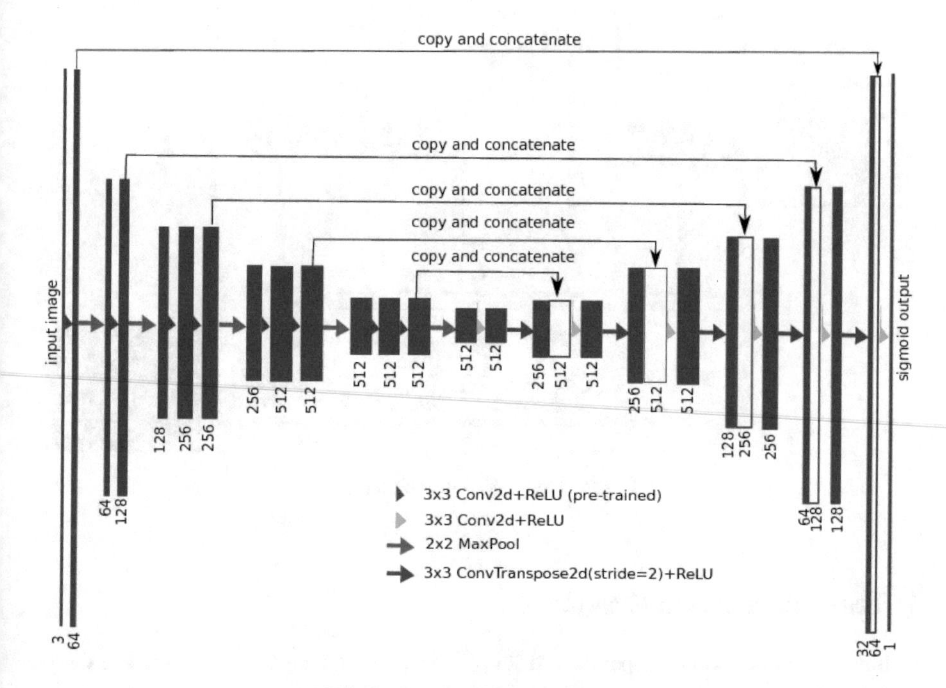

Fig. 1. TernausNet architecture.

It should be noted that convolutional neural networks, which have proven themselves in solving problems of segmentation of surface cracks in images, have an architecture consisting of two parts: an encoder and a decoder. The neural network encoder extracts the characteristic features of a defect (capturing contextual information) from the original image. The decoder performs the selection of fine details (performs a semantic analysis of the selected features) and combines the results into an output mask with the dimensions of the original image. One of the earliest and simplest architectures used for semantic segmentation is the fully convolutional network [13]. Despite its effectiveness, the architecture has a low resolution due to the loss of information in each convolutional layer of the encryptor. The disadvantages of a fully convolutional network associated with data loss can be eliminated by the standard architecture based on the U-Net autoencoder [14] through the use of transposed convolutional layers of the decoder. It is common for the U-Net neural network to achieve high results in various real-world problems when using a small amount of training data. Improving the performance of the U-Net architecture with a small training set is associated with the idea of transferring the weights of a cryptographer previously trained on a large data set. The approach based on the use of pyramidal expanding convolution implemented in DeepLabv3 neural networks [15] allows to improve the quality of segmentation by taking into account different scales of objects. Figure 2 shows Deeplabv3 + [15] neural network, which also provides good result for segmentation tasks and used ResNet50 model for feature extraction.

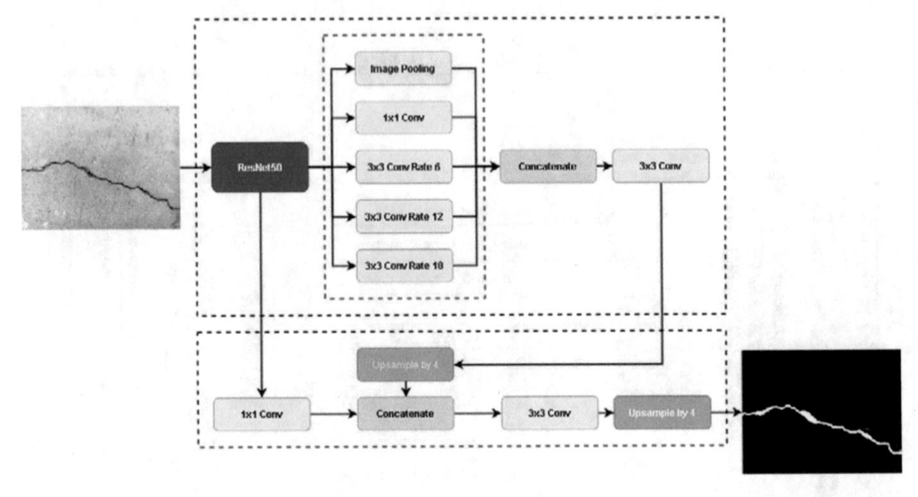

Fig. 2. Deeplabv3 + architecture.

3 Experiments and Results

To train neural networks, a specialized set of images of defects on a steel surface Crack Segmentation Dataset was used. The set consists of 2780 color images of the surfaces of the structural elements of the railway bridge in different resolutions and their black and white masks that separate surface cracks from the background. During training, augmentation procedures (shift, rotation, horizontal reflection) were used to compensate for changes associated with the shooting angle. The investigation compared the quality of image segmentation of cracks on the surface of steel structures for two modern models: TernausNet and Deeplabv3 +. The neural network TernausNet was considered in two versions. In the first case, the standard initialization of the encryptor with the weights of the VGG16 neural network trained on ImageNet is used. In the second case, the last layers of the encoder are trained on the classification on a set of images of cracks on the concrete surface. The models were trained for segmentation on the above-mentioned specialized set of 2780 pixel-level images of cracks on the surface of the structural elements of the railway bridge. The source dataset was splitted in training, validation and test sample in 90% to 5% and 5% proportion. So the number of images was reserved for validation of the results and did not participate in training. Figure 3 shows sample images from the training set.

A tenth of the images from the original set was allocated for validation and testing in equal shares. The segmentation neural networks were trained on the remaining images over 100 epochs.

Only TernausNet and Deeplabv3 + neural network decoders were trained. Before training, images from the training set were divided into fragments with dimensions of

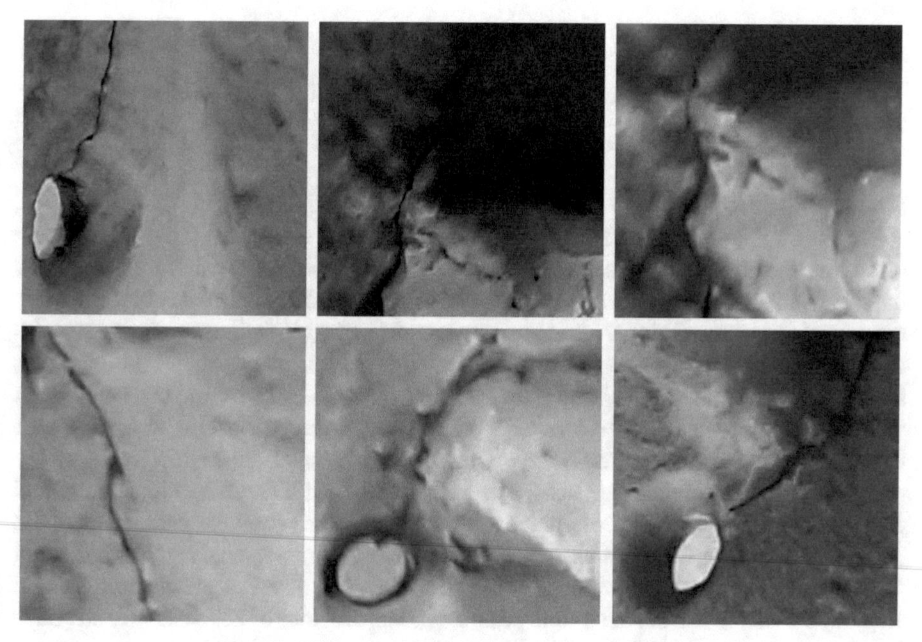

Fig. 3. Examples of processed images with cracks.

256 × 256 pixels to be fed to the input of neural networks. To eliminate the problems associated with the imbalance between the number of background pixels and defects, a technique was used to increase the probability of using images containing a defect at the next training step. When training neural networks, the standard Adam optimization algorithm was used with learning rate lr = 0.0001. As a metric for assessing the quality of fracture segmentation during validation and testing, the DICE coefficient widely used in segmentation problems [16] was chosen:

$$DICE = \frac{2|A \cap B|}{|A| + |B|}, \tag{1}$$

where A are the pixels identified as belonging to the defect during manual marking; B are the results of neural network prediction.

For black and white image segmentation masks, DICE was calculated as follows:

$$DICE = \frac{2\sum_{i=1}^{n} y_i \hat{y}_i}{\sum_{i=1}^{n} (y_i + \hat{y}_i)}, \tag{2}$$

where n is the number of pixels; y_i is the value in the i-th pixel of the black and white mask of the image marked by the expert; $\hat{y}_i$ is the value in the i-th pixel based on the results of neural network predictions.

The binary cross-entropy was used as a loss function:

$$H = -\frac{1}{n}\sum_{i=1}^{n}\left(y_i \log \hat{y}_i + (1 - y_i)\log\left(1 - \hat{y}_i\right)\right). \quad (3)$$

Here all variables have the same sense as in Eq. (2).

At the output of the neural network, a mask image was obtained, each pixel of which represents the probability of a defect. The dimensions of the output mask are the same as the dimensions of the input image. A threshold value of 0.5 is applied to obtain a binary mask from the probabilities. This value is derived from the validation set of images and is universal for a given loss function and different sets of images. For other loss functions, the value is different and must be found independently. All pixels, the probabilities of which are less than the threshold value, were equated to zero, otherwise, to one. Finally, to create a black and white mask, each binary value in a pixel was multiplied by 255.

In the experiment, six TernausNet neural networks with the same architecture and different depths of encryptor pre-training were studied. The convolutional layers of the TernausNet encoder were further trained in classification on a specialized training set of 40 thousand images of cracks on the concrete surface (Surface Crack Detection Dataset Kaggle). Each image from the training sample has dimensions of 227 × 227 pixels. Each image is assigned a label that determines the presence or absence of a defect on it.

Figure 4 shows training curves for TernausNet without augmentation (on the left) and using augmentation (on the right). The line 1 shows losses for validation dataset and the line 2 shows losses for train dataset.

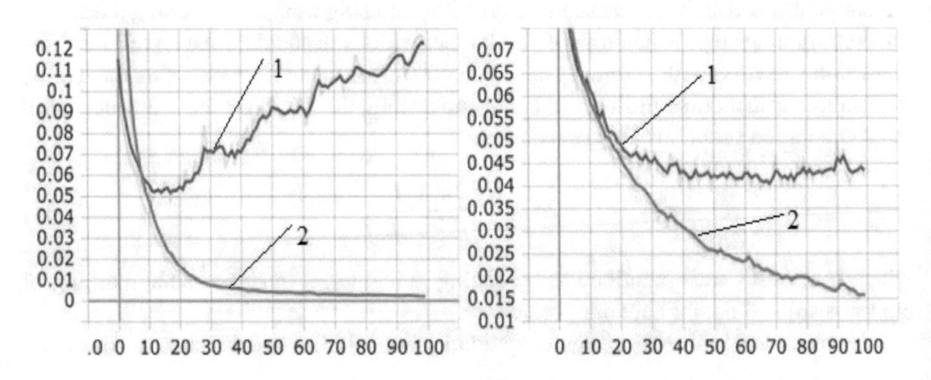

Fig. 4. Training of TernausNet.

Figure 5 shows training curves for Deeplabv3 + without augmentation (on the left) and using augmentation (on the right). The line 1 shows losses for validation dataset and the line 2 shows losses for train dataset.

To ensure the reliability of the training results, a test set of 37 pixel-level images of cracks on the surface of steel structure elements of the railway bridge was prepared, which was not associated with the training set. In the first experiment, images from the test set were divided into square areas of 256 × 256 pixels with a certain step.

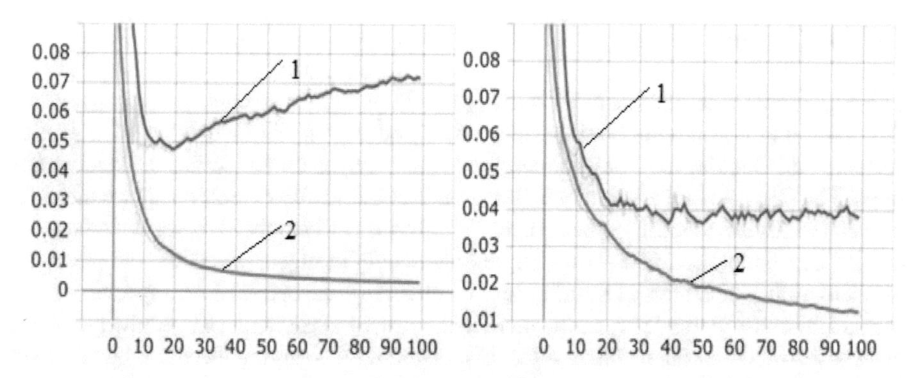

Fig. 5. Training of Deeplabv3 +.

The areas obtained during the separation were processed using pre-trained segmentation neural networks. In the course of the second experiment, using a crack detector based on the YOLOv3 neural network, bounding rectangles with a certain probability containing defects were selected on test images. Further, the selected areas of defects were reduced to the size of 256 × 256 pixels and fed to the input of segmentation neural networks for processing. The results of neural network predictions were compared with the markup made manually by an expert by calculating the DICE segmentation quality score.

Table 1 shows the average results obtained when processing test images.

Table 1. Comparison of different approaches.

Model	Number of parameters	Tiling of the original image		Preliminary processing by YOLOv3	
		No Augmentation	Augmentation	No Augmentation	Augmentation
TernausNet	13706913	0.67	0.69	0.71	0.74
TernausNet with encryptor training	13706913	0.70	0.72	0.73	0.76
Deeplabv3 +	11819361	0.70	0.72	0.73	0.76

An analysis of the obtained results shows that the use of the YOLOv3 detector for preliminary selection of areas with potential objects of interest (defects) as an alternative to solid tiling makes it possible to improve the subsequent segmentation quality by 5–10%. In addition, due to the preliminary exclusion of areas that obviously do not contain a defect, it is possible to reduce the average frame processing time by at least 40%. It is important that text do not consider different visual attacks [17].

4 Conclusions

The paper compares the results of training neural networks TernausNet (without additional training / with additional training of the coder to detect defects on the concrete surface) and Deeplabv3 + segmentation of crack images on the surface of steel structures. The first model under study is based on the classic U-Net architecture, which has proven itself in medical image segmentation. In turn, Deeplabv3+ takes advantage of expanding convolution and shows impressive results in object detection and semantic segmentation. The results of training neural networks were obtained on a specialized set of pixel-by-pixel marked images of cracks in the structural elements of the railway bridge. To take into account the negative effect associated with changing the shooting angle on images, the training set was strengthened by applying augmentation procedures (shift, rotation, scaling, horizontal reflection). An approach based on the preliminary detection of defect areas in the original images using the YOLOv3 neural network and their subsequent segmentation is proposed. The results of the experiments showed that the use of this approach in the development of systems for automatic monitoring of infrastructure facilities allows maintaining the quality of fracture segmentation, minimizing the number of "false positives" and significantly reducing the computational resources required for processing.

Acknowledgements. The study was supported by the Russian Science Foundation grant No. 23-21-00249.

References

1. Kanaeva, I.A., Ivanova, Y., Spitsyn, V.G.: Deep convolutional generative adversarial network-based synthesis of datasets for road pavement distress segmentation. Comput. Opt. **45**(6), 907–916 (2021)
2. Su, H., Wang, X., Han, T., Wang, Z., Zhao, Z., Zhang, P.: Research on a U-Net bridge crack identification and feature-calculation methods based on a CBAM attention mechanism. Buildings **12**, 1–18 (2022)
3. Pu, R., Ren, G., Li, H., Jiang, W., Zhang, J., Qin, H.: Autonomous concrete crack semantic segmentation using deep fully convolutional encoder-decoder network in concrete structures inspection. Buildings **12**, 1–20 (2022)
4. Huang, B., Reichman, D., Collins, L.M., Bradbury, K., Malof, J.M.: Tiling and stitching segmentation output for remote sensing: basic challenges and recommendations. CoRR arXiv preprint, arXiv: 1805.12219 (2018)
5. Andriyanov, N.A., Dementiev, V.E., Tashlinskii, A.G.: Detection of objects in the images: from likelihood relationships towards scalable and efficient neural networks. Comput. Opt. **46**(1), 139–159 (2022). https://doi.org/10.18287/2412-6179-CO-922
6. Chen, L.-C., Zhu, Y., Papandreou, G., Schroff, F., Adam, H.: Encoder-decoder with atrous separable convolution for semantic image segmentation. In: Ferrari, V., Hebert, M., Sminchisescu, C., Weiss, Y. (eds.) ECCV 2018. LNCS, vol. 11211, pp. 833–851. Springer, Cham (2018). https://doi.org/10.1007/978-3-030-01234-2_49
7. Iglovikov, V., Shvets, A.: TernausNet: U-Net with VGG11 encoder pre-trained on ImageNet for image segmentation. CoRR arXiv preprint, arXiv: 1801.05746 (2018)

8. Concrete Crack Segmentation Dataset.https://data.mendeley.com/datasets/jwsn7tfbrp/1. Accessed 13 Feb 2023
9. Andriyanov, N.A., Dementev, V.E., Vasiliev, K.K., Tashlinsky, A.G.: Investigation of methods for increasing the efficiency of convolutional neural networks in identifying tennis players. Pattern Recognit. Image Anal. **31**, 496–505 (2021). https://doi.org/10.1134/S1054661821030032
10. Redmon, J., Farhadi, A.: YOLOv3: an incremental improvement. CoRR arXiv preprint, arXiv: 1804.02767 (2018)
11. Deng, J., Dong, W., Socher, R., Li, L.-J., Li, K., Fei-Fei, L.: ImageNet: a large-scale hierarchical image database. In: Proceedings of IEEE Computer Society Conference on Computer Vision and Pattern Recognition (CVPR), pp. 248–255 (2009)
12. Dementyiev, V.E., Andriyanov, N.A., Vasilyiev, K.K.: Use of images augmentation and implementation of doubly stochastic models for improving accuracy of recognition algorithms based on convolutional neural networks. In: Proceedings of 2020 Systems of Signal Synchronization, Generating and Processing in Telecommunications (SYNCHROINFO), pp. 1–4 (2020). https://doi.org/10.1109/SYNCHROINFO49631.2020.9166000
13. Long, J., Shelhamer, E., Darrel, T.: Fully convolutional networks for semantic segmentation. In: Proceedings of IEEE Conference on Computer Vision and Pattern Recognition (CVPR), pp. 3431–3440 (2015)
14. Ronneberger, O., Fischer, P., Brox, T.: U-Net: convolutional networks for biomedical image segmentation. In: Navab, N., Hornegger, J., Wells, W.M., Frangi, A.F. (eds.) MICCAI 2015. LNCS, vol. 9351, pp. 234–241. Springer, Cham (2015). https://doi.org/10.1007/978-3-319-24574-4_28
15. Chen, L.-C., Papandreou, G., Schroff, F., Adam, H.: Rethinking atrous convolution for semantic image segmentation. CoRR arXiv preprints, arXiv: 1706.05587 (2017)
16. Taha, A.A., Hanbury, A.: Metrics for evaluting 3D medical image segmentation: analysis, selection, and tool. BMC Med. Imaging **15**(29), 24–36 (2015)
17. Andriyanov, N.A., Dementiev, V.E., Kargashin, Y.: Analysis of the impact of visual attacks on the characteristics of neural networks in image recognition. Procedia Comput. Sci. **186**, 495–502 (2021)

Modern Methods of Traffic Flow Modeling: A Graph Load Calculation Model Based on Real-Time Data

Roman Ekhlakov(✉)

Financial University Under the Government of the Russian Federation, Moscow, Russia
rsekhlakov@fa.ru

Abstract. The problem of regulation and management of traffic flows is considered in connection with the increase in the load on the road transport network. In particular, a mathematical model of traffic jams and the problem of predicting the arrival time of a vehicle are studied. The analysis of predicting methods is carried out. To improve the quality of predictive solutions, it is proposed to use an approach based on numerical probabilistic analysis. A comparison of the effectiveness of the application of mathematical models of analysis for the problems of predicting the characteristics of traffic flows is carried out.

Keywords: traffic flow · data monitoring · arrival predicting methods · traffic jams

1 Modern Methods of Traffic Flow Modeling

1.1 Introduction

Currently, in most developed countries of the world, there is a serious imbalance of supply and demand in the field of transport: traffic volumes, especially in large metropolitan areas, are approaching the capacity limits of existing highways, and even exceed them during periods of peak loads hour. Work to meet transport needs is a set of activities carried out in two directions: extensive and intensive.

For a long time, the actually exclusive way to solve the problem was an extensive path aimed at expanding the existing infrastructure. However, at present this option is not rational. It is being replaced by science-intensive approaches that involve the construction of fundamentally new modes of transport. Evaluation of the efficiency and reliability of such systems is an important task at the design stage. The final decision on construction is made only after the creation and comprehensive study of the mathematical model of the entire transport network. However, despite the fact that the modeling of traffic flows in large cities has been carried out all over the world for more than fifty years, a lot of theoretical and practical material has been accumulated in this direction [1].

The key point of the work is the study of methods for constructing, calibrating and evaluating mathematical models of transport networks, building a mathematical model to determine the average load of the transport network, testing the model for predicting

I. Czarnowski et al. (Eds.): KESIDT 2023, SIST 352, pp. 302–309, 2023.
https://doi.org/10.1007/978-981-99-2969-6_27

traffic flows and displaying the current traffic load. The model being developed should provide information on the average load of the transport network, current and average values of speed and traffic density, and also provide the ability to predict and calculate the average required traffic characteristics when changing in the transport network.

1.2 A New Traffic Flow Model Prediction

Based on the existing methods of mathematical modeling of transport networks and work tasks, we will present the following requirements for the developed mathematical model:

- the model must be predictive;
- the model must be macroscopic formulated in terms of the density and speed of vehicles as functions of time and point of the transport network;
- the model should describe the movement of vehicles along a directed graph (with two-way traffic there will be two opposite edges between the points). At the vertices of the graph there are intersections and changes in the number of lanes of the road.

The model should display a congestion traffic picture in real time using anonymized data. To do this it is necessary to collect data on street congestion from various sources, analyze and display on a map. In the largest cities where traffic jams are a serious problem and not just a nuisance it is necessary to calculate the road score – the average level of congestion taking into account various coefficients. The model needs to take into account information about traffic jams from devices which means that drivers help other drivers avoid traffic jams [2].

At the moment, one of the most demanded problems of traffic flow management is the problem of short-term predicting of the arrival time of transport, for example, public transport to a stop. This problem is quite relevant because in most cities, including large ones even if there is a bus schedule the arrival of a bus is largely determined by the state of the road transport network and the influence of a number of uncertain factors. Many cities are currently developing public transport monitoring systems to inform passengers about the location of buses and when they arrive. Various approaches are used to process monitoring data including modeling methods. These transport monitoring data processing systems use models based on archive data in their predicting modules, the main of which are time series models, regression models; models based on Kalman filtering, artificial neural network models, support vector machines and hybrid models. So the work [3] provides effective transport monitoring.

The historical data model is based on the assumption that road conditions can be described by the model on a daily and weekly basis. It is expected that traffic flow parameters predicted from historical data at a particular time and day of the week will provide a good justification for the corresponding parameters in a similar situation. The results of predicting the traffic flow parameters obtained using these models are reliable only under relatively stable road conditions in the area under consideration; in the event of traffic jams and accidents the accuracy of these models can be severely degraded. Models using archived travel times for road segments use the average travel time directly or in combination with other inputs to predict public transport arrival times (PTOs) [4]. In most studies such models are used for comparison with other methods and in almost all works they are inferior to the proposed basic algorithms. Models using average speed are commonly used to predict the travel time of a network segment from data received from GPS sensors (Fig. 1).

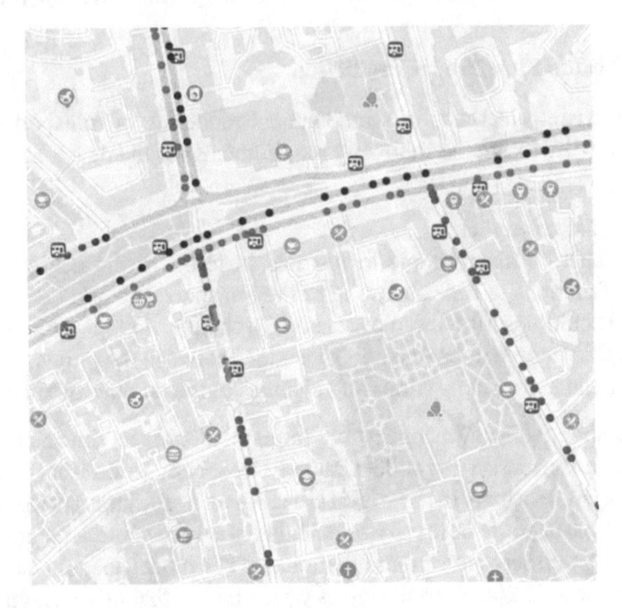

Fig. 1. Device GPS points in real-time

These models use geographic information systems (GIS) to estimate the position of the vehicle with the subsequent determination of the time of arrival at stopping points.

Time series models suggest that the dynamic nature of traffic can be repeated in the future, traffic models can be described by mathematical functions that take into account the cyclicity, seasonality, uncertainty factor and the direction of the general trend in traffic dynamics traffic flow. Archival data can also be used for these purposes. The accuracy of the models depends on the fit function between real-time and historical data, changes in the data or in the relationship between them can lead to a significant error in the prediction.

Regression models are built as dependence functions of input and output parameters, which may include data on the passage of road sections in real time, historical data, traffic conditions, passenger traffic, weather conditions, delays at stops, etc. Unlike models based on historical data regression models show satisfactory results in unstable driving conditions. A set of multilinear regression models is used to estimate the arrival time of a vehicle; as independent variables in the models are the distance to the stop, the number of stops, waiting time, traffic at stops and weather conditions, etc. The necessary condition for the independence of variables limits the applicability of regression models to transport systems where variables can be highly correlated.

Models based on Kalman filtering is widely used to estimate the arrival time. Although the main function of models of this kind is to predict the current state of the system, they can serve as a basis for estimating future values or for correcting previous predictions including noise filtering. There are algorithms for short-term time-of-arrival prediction that combine real-time data with historical data. They use the Kalman filter to determine the location of the vehicle and statistical estimates to predict the time of arrival.

Models of artificial neural networks allow modeling complex non-linear dependencies between the travel time of network segments and variables characterizing the transport situation. Neural network models do not require independence of input variables. It has been noted that the neural network is superior to the regression model.

Support vector machine it is a set of supervised learning algorithms used for classification and regression problems. This method was used to predict the arrival time of public transport and is computationally complex which requires further research on the choice of input variables and the determination of algorithm parameters.

An important direction in predicting traffic flow parameters is the use of hybrid models which are a combination of two or more models for predicting the arrival time [5].

2 Complex Traffic Jams Predicting Method

2.1 Initial Data Processing Algorithm

To predict the change in the average speed of movement on different sections of the road, it is necessary to have up-to-date data in real time. To implement the following mechanism, it is necessary to receive data every few seconds from a device capable of transmitting its geographic coordinates, direction and speed of movement and superimposed on a cartographic substrate to the general system. All data is anonymized and does not contain any information about the user or his car.

Of course, this is not a new scientific problem, and different specialists around the world are engaged in it. All over the world, traffic predictions are used for automatic traffic management in some cities. The first prototypes using predictions appeared in 1998 in the USA. The first pilot use of the system began in 2006 in Singapore. Many countries use fixed flow sensors that measure speed and traffic density on major highways. This is a good source of clean traffic data, but covering a metropolitan area requires a gigantic investment [6]. The most optimal solution would be anonymous GPS tracks that come from users of modern devices. Each track is a chain of signals about the location (latitude and longitude) of the car at a certain time. It is necessary to link GPS tracks to the road graph to determine which streets cars drive or people walk on. Then the speed of all tracks passing along the same edges is averaged. GPS receivers make errors in determining coordinates which makes it difficult to build a track. An error can move an object several meters in any direction such as onto a sidewalk or the roof of a nearby building. The coordinates received from user devices are sent to an electronic map of the city which accurately displays all buildings, parks, streets with road markings and other city objects. For more accurate prediction the Viterbi algorithm is used. Algorithm finding the most suitable list of states which in the context of Markov chains obtains the most probable sequence of events that have occurred (Fig. 2). The algorithm makes several assumptions:

1. Observable and hidden events must be a sequence. The sequence is most often ordered by time;
2. Two sequences must be aligned: each observable event must correspond to exactly one hidden event;
3. The calculation of the most probable hidden sequence up to time t should depend only on the observed event at time t and the most probable sequence up to time $t - 1$.

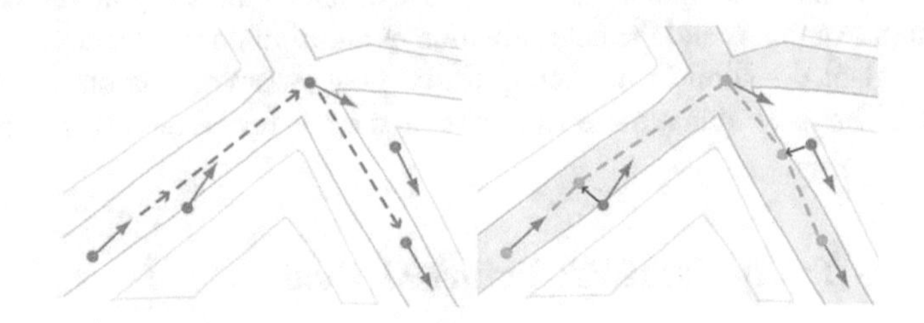

Fig. 2. The object movement trajectory according to the Viterbi algorithm

With the help of such a detail it is possible to determine how the car actually moved. For example, the car could not drive into the oncoming lane and the turn must be made according to the road markings. Then a single route of movement is built with information about the speed of its passage – a track. Technically, a roadmap is a graph. Each intersection is a vertex of this graph and the sections of the road between two intersections are edges. The latter have the attribute length and speed. The length is known in advance and the speed is calculated in real time – depending on it the road section is painted in green, yellow and red (Fig. 3).

2.2 Point Scale

In large cities the algorithm evaluates the situation on a 10-point scale where 0 points means free movement and 10 points means the city stands still. With this estimate drivers can quickly understand how much time they will lose in traffic jams. For example, if the average traffic jam is seven, then the trip will take about twice as long as free traffic.

The score scale is configured differently for each city and has a minimum and maximum coefficient selected empirically. Points are awarded as follows: routes are drawn up in advance along the streets of each city including major highways and avenues. Each route has a control time during which it can be driven on a free road without violating the rules. After estimating the overall load of the city, the algorithm calculates how much the real time differs from the reference one. Based on the difference of all routes the load in points is calculated.

Fig. 3. Average speed on the road

2.3 Choosing a Model Training Method

In the field of predicting future traffic jams a wide variety of methods are used, among them:

Parametric Regression Model. It assumes a certain type of function that depends on $V_{future} = F \times (V_{past}, V_{current})$ and selects parameters in F. The peculiarity of the model is that the choice of function imposes restrictions on the result. On the one hand, this protects against serious errors that are inevitable when using noisy data – parametric function allows you to reduce the spread of predicted values. On the other hand, if the feature is chosen incorrectly, the model may give systematically incorrect answers.

Nonparametric Model ("k Nearest Neighbors"). It just looks for a situation in the past similar to the current one and as a prediction it gives out how events developed further. This model is more flexible and it can be used to say: "If after a calm there is a storm 10 times, then after a new calm there will also be a storm". If, for example, we take a linear function in a parametric model we cannot predict this since we are limited by the hypothesis that the future can only be a linear combination of the past. However, on "noisy" data the linear model provides greater accuracy than the nonparametric one.

Another classification of regression methods is *scalar* and *vector*. In the first one, it is assumed that the situation in the selected area does not depend on what happens in neighboring ones, in the second, on the contrary, dependencies are taken into account. Hence the difference in the calculations amount. If there are 10 thousand sections of roads in the city, then to build a scalar model it is enough to choose 10 thousand parameters. And for a vector model you may need up to 100 million parameters. In our case, it is obvious that situations in different areas influence each other. And the key question is how to limit the circle of neighbors so that the calculation can be made within a reasonable time.

Flow modeling methods take into account the nature of the process, and do not simply operate on abstract numbers. The equation for the flow of a fluid with variable viscosity is used: more density – less speed. The methods are widely used for road planning and are promising for predictions, but are too demanding on the accuracy and completeness of data. They give good quality on streaming sensors, but almost do not work on "noisy" GPS tracks.

After experimenting with different methods and evaluating their accuracy and computational complexity it was decided to stop at linear vector auto regression. It is assumed that the future speed is a linear combination of the current and several past speeds in several areas: directly in the area under consideration as well as in neighboring ones.

2.4 Model Calculation

Once a day the model is trained. Based on the development of events for a certain period of time in the past model builds a prediction. This prediction is compared with subsequent developments. The model has selected coefficients that minimize the discrepancy between the prediction and the actual result. For each time of day, the system calculates its own coefficients, since the nature of traffic changes during the day. Every few minutes a prediction for the next hour is calculated based on the finished model and fresh tracks. When training a model, a huge amount of data is used. The road graph occupies more than 10 gigabytes – information about the topology which edge is connected to which as well as geometry – GPS coordinates of the beginning end and turning points of each road segment.

A new prediction is calculated every 10 min. First, fresh data is processed: they are linked to the schedule the speeds of cars passing one segment are averaged and a prediction is calculated based on the model built for the current day. The data is then prepared for mapping traffic. Sharp changes in speed are smoothed out; for each speed a color is selected taking into account the category of roads and the scale of the map view.

2.5 Quality Control

The quality metric should reflect how much the prediction helps people. In the absence of a prediction, when trying to predict the speed of traffic on a certain road in an hour necessary uses current traffic information. It is clear that traffic jams will change in an hour and the estimate will be incorrect. The estimate obtained using the prediction is also not ideal, but should be closer to reality.

The real situation can be assessed in different ways. It's easiest to rely on custom tracks. As a result, the assessment was 15%. However, there may be systematic deviations in the data. Then the quality metric will give preference to a prediction that diligently reproduces these deviations. To avoid this, measurements of assessors – special people who help measure the quality of service. They drove cars with high-precision GPS navigators to the most important highways and built the tracks. When comparing the prediction with the assessors' tracks a quality indicator of 12% was obtained.

3 Conclusion

In this document an overview of modern intelligent methods for predictive analysis of traffic flows was carried out the Viterbi algorithm used in the construction and display of road traffic on graphs was described in detail. The most relevant prediction methods were investigated and used to predict future road occupancy. The results obtained were compared with the actual results of the accessors and a quality score of 12% was obtained.

The result of the work satisfies the initially set goals and was tested as online service with an incoming data stream of more than 1000 RPS (request per second). The system showed acceptable results for further algorithm and data processing improvement.

References

1. Zhiqiu, H., Shao, F., Sun, R.: A new perspective on traffic flow prediction: a graph spatial-temporal network with complex network information. Electronics **11**, 2432 (2022)
2. Oumaima, E.J.: Ben Othman jalel, veque veronique: a stochastic mobility model for traffic forecasting in urban environments. J. Parallel Distrib. Comput. **165**, 142–155 (2022)
3. Andriyanov, N.A., Dementiev, V.E., Tashlinskiy, A.G.: Development of a productive transport detection system using convolutional neural networks. Pattern Recognit. Image Anal. **32**, 495–500 (2022)
4. Yuan, Y., Zhang, W., Yang, X., Liu, Y., Liu, Z., Wang, W.: Traffic state classification and prediction based on trajectory data. J. Intell. Transp. Syst. (2022)
5. Shaygan, M., Meese, C., Li, W., Zhao, X.G., Nejad, M.: Traffic prediction using artificial intelligence: review of recent advances and emerging opportunities. Transp. Res. Part C: Emerg. Technol. **145**, 103921 (2022)
6. Shinde, S.V., Lakshmi, S.V., Sabeenian, R.S., Bhavani, K.D., Murthy, K.V.S.R.: Traffic optimization algorithms in optical networks for real time traffic analysis. Optik, 170418 (2022)

Application of Machine Learning Methods for the Analysis of X-ray Images of Luggage and Hand Luggage

Nikita Andriyanov(✉)

Financial University Under the Government of the Russian Federation, Leningradsky av. 49, 125167 Moscow, Russia
nikita-and-nov@mail.ru

Abstract. The paper is a scientific solution to the applied problem of detecting dangerous objects using high-precision computer vision models and classical deep learning methods. An approach based on the use of ensemble models in the classification of luggage items is presented. The effectiveness of this approach in terms of the proportion of correct recognitions (accuracy) has been studied. A comparison with neural networks is made. Based on the developed recognition ensembles, the quality of solving the problem of detecting prohibited objects of various types has been improved using the VGG-16 convolutional neural network. The comparison showed that the algorithm provides an increase in the quality of recognition even in comparison with transformer models for computer vision.

Keywords: Prohibited Luggage · Pattern Recognition · Convolutional Neural Networks · Ensemble Approach · X-ray Images · Aviation Security

1 Introduction

Recently, the tasks of ensuring safety in crowded places have become increasingly important. There is talk about the introduction of football fan identification systems, which, based on facial recognition algorithms [1], will identify offenders in the stands.

Also of particular importance are the tasks of monitoring drivers [2, 3]. The fact is that transport security in general is of increased interest with the advent of modern technologies, providing both traffic control and monitoring of the situation on the roads [4, 5]. However, these tasks are solved for images obtained in the optical range by special registration devices, and the detection, for example, of prohibited objects at airports must be automated in a system capable of working with X-ray images.

It was shown in [6, 7] that the use of artificial neural networks on X-ray images in classification tasks can be very effective. However, there is a problem with the high confidence of network responses even with erroneous responses. In addition, detection problems for X-ray images are presented in the literature much less than other computer vision problems [8].

I. Czarnowski et al. (Eds.): KESIDT 2023, SIST 352, pp. 310–316, 2023.
https://doi.org/10.1007/978-981-99-2969-6_28

Overcoming these problems is possible based on the use of image augmentations [9]. However, this does not guarantee that the overfitting effect will be completely eliminated. This is because the images will still be quite similar to each other.

Another solution is the use of ensemble models of neural networks. This approach, firstly, will make it possible to have alternative predictions for different images, and secondly, if correctly integrated, it will reduce the probability of error, and, consequently, increase the proportion of correct recognitions.

Note that the use of the ensemble approach, when several models are combined, is also widespread in classical machine learning [10].

To solve the problem of detecting objects in X-ray images, by analogy with optical images, specialized neural network detectors can be used [11]. Then it is possible to use pre-trained models, including ensemble ones, to refine the detection results.

2 Ensemble Approach for Luggage Image Recognition

As noted above, providing truthful answers is possible by combining the predictions of several artificial neural networks for recognition tasks. In particular, let there be two models. The first recognizes the image as belonging to class A with 90+% confidence, and the second classifies the image as class B with approximately the same confidence. Thus, averaging these probabilities, we get much less confidence for disputable objects. Such thresholds can subsequently be used to organize additional verification.

An actual task in computer vision for the purposes of aviation security is image recognition into 2 classes. These classes are following: prohibited (dangerous) and allowed (safe). Let such a problem be solved using computer vision methods on dataset of X-ray images of luggage. It should be noted that the division into 2 classes quite strongly generalizes different items, since objects of completely different shapes and sizes can belong to both prohibited and permitted items. Figure 1 shows examples of a preprocessed sample, where the images are pure objects of different classes.

On Fig. 1, the top row shows safe items and the bottom row shows dangerous items. Note that the original data set was provided by the Ulyanovsk Institute of Civil Aviation as a result of the interaction with the Barataevka airport named after N.M. Karamzin (Ulyanovsk, Russia).

The total sample size is 3,700 images. At the same time, almost two thirds of images containing permitted items, namely there are 2,500 images for safe items, and only 1,200 images with prohibited items.

Considering the uneven distribution of images by classes, let's balance the dataset. We left 900 images of each class for training, and 300 images of each class for testing. Thus, the ratio of the training and test samples was 3 to 1. Taking into account the balancing of the dataset, the proportion of correct recognitions, in other words, accuracy, was chosen as the main metric.

Machine learning models require a table of features and objects, so you need to get a one-dimensional vector. Then each feature will be characterized by the position of the pixel in the image and the color channel.

Thus, from three-channel images with a size of 125×125 pixels, we get 46,875 features for training. For training and recognition, we will use random forest and categorical boosting models.

Deep learning was performed using transfer learning. ResNet-50, Inception-3, VGG-16 were studied as basic models. All of them are convolutional neural networks. The output of the models was a vector of probabilities of belonging to each class. The final characteristics were obtained by adding such vectors and dividing by the number of models. Then the class with the highest probability was selected.

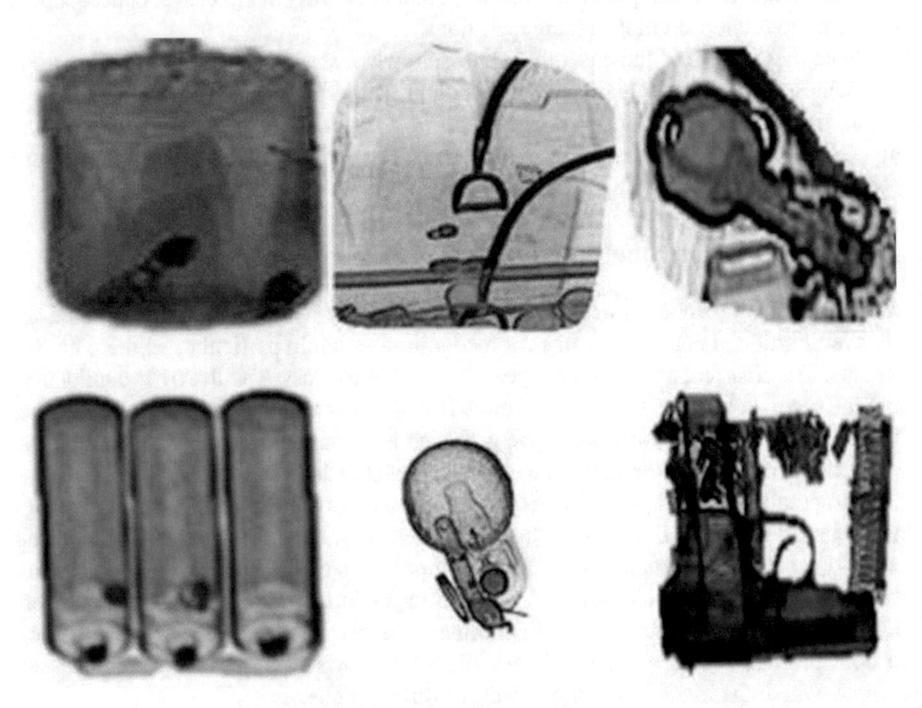

Fig. 1. Examples of luggage images.

Table 1 shows the comparative performance of the models on the test sample.

Table 1. Comparison of recognition models

№	Model	Accuracy, %
1	Random Forest (10 Decision Trees)	67.8
2	CatBoost (categorical boosting)	70.0
3	VGG-16	93.8
4	ResNet-50	97.0
5	Inception-3	94.8
6	VGG + ResNet + Inception	98.2

From the data presented in Table 1, it can be seen that classical machine learning is significantly inferior to methods based on complex architectures of convolutional neural networks. The results are worse by 20–30%. This is because simple models do not use spatial image relationships, unlike convolutional networks. At the same time, the use of the ensemble model provided the best result. It is 98.2%. It should be noted that the recall for prohibited objects in this case was also the highest and amounted to 97.6%.

Table 2 shows precision and recall metrics.

Table 2. Precision and recall of all models

№	Model	Precision, %	Recall, %
1	Random Forest (10 Decision Trees)	69.4	66.3
2	CatBoost (categorical boosting)	68.3	71.2
3	VGG-16	94.2	91.9
4	ResNet-50	96.8	97.0
5	Inception-3	93.5	95.2
6	VGG + ResNet + Inception	98.9	97.6

The study showed that deep learning is able to analyze information from X-ray images. Moreover, the use of ensemble models of neural networks provides extremely high recognition efficiency.

Next, let consider the problem of detecting objects on X-ray images of luggage.

3 Detection of Prohibited Items of Luggage

It is impossible not to note the growing popularity of algorithms for processing multimodal data [12, 13]. The detection of some important features occurs on the basis of different channels of information. In our case, another trend is being investigated, namely, image processing in ranges other than optical. As noted earlier, the formulation of the detection problem in such images is rare. The proposed algorithms are used for either classification or segmentation.

If it is necessary to automatically detect dangerous items of luggage in crowded places, during screening at the airport, it is impossible without a specialized neural network model.

One of the options for a quick solution to this problem is the use of the YOLO family model [14]. Transformer architectures [15] can also be used, but they are too slow.

Data for the detection task was also prepared jointly with the Ulyanovsk Institute of Civil Aviation (Ulyanovsk, Russia) and the local airport. Figure 2 shows an example of an image from the training sample without markup.

The sample size in the discovery task was smaller than in the classification. The sample size was 100 images containing 58 objects of the prohibited class.

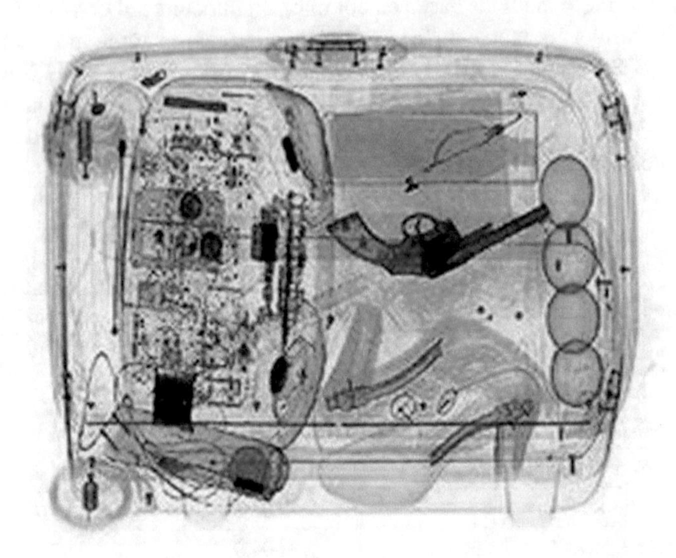

Fig. 2. An example of the image of luggage in the X-ray spectrum.

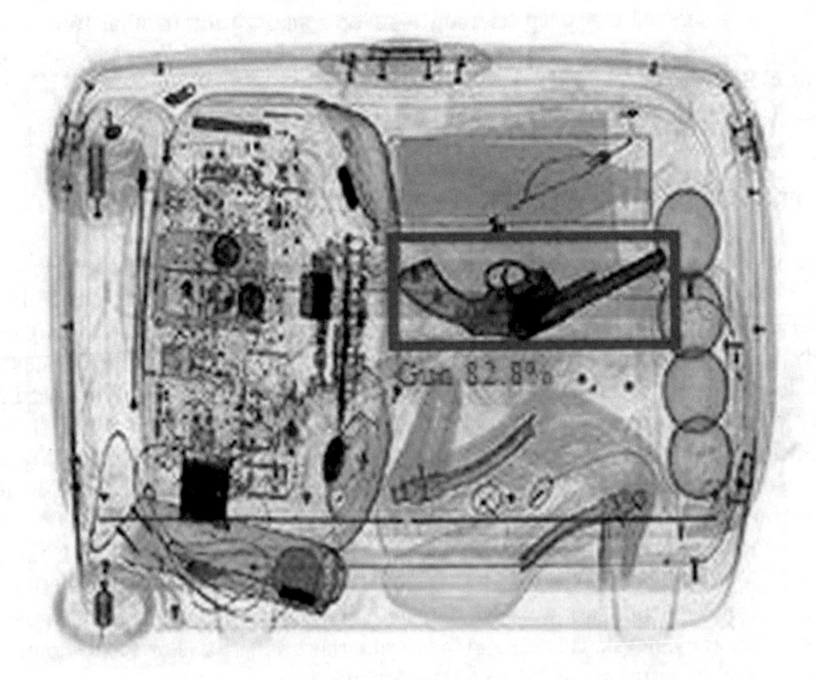

Fig. 3. Detection results.

It also should be noted that 80 images were used for training. Another 20 images, including 12 prohibited items, were used for testing. The cloud service Computer Vision Annotation Tool (CVAT) was used for data markup.

So Fig. 3 shows an example of a labeled image after processing by the YOLO neural network.

The mAP (mean average precision) metric was used to evaluate the effectiveness. However, it should be noted that it was applied to single-class detection.

Table 3 summarizes the results for various models.

Table 3. Comparison of detection models

Model	mAP
YOLO	81.8%
Vision Transformer	76.9%
YOLO + Ensemble	83.3%

From Table 2 we can see that the use of an ensemble for additional recognition leads to an increase in accuracy.

For example, for the YOLO model on test data, mAP = 81.8%. This result is obtained by detecting 9 valid prohibited items and 2 false positives on the test base.

The study showed that models of convolutional and transformer architectures for object detection can be transferred to X-ray images.

4 Conclusions

The article demonstrates the possibility of applying machine and deep learning approaches of computer vision of the optical range to solving problems of detecting and recognizing luggage objects in the X-ray range. It is shown that the ensemble approach, due to the integration of the results of the work of several neural networks, provides the highest recognition accuracy. In the detection problem, the best accuracy was shown by the model based on the YOLO neural network using additional recognition based on the ensemble model.

Acknowledgements. The study was supported by the Russian Science Foundation grant No. 23–21-00249.

References

1. Andriyanov, N., Dementev, V., Tashlinskiy, A., Vasiliev, K.: The study of improving the accuracy of convolutional neural networks in face recognition tasks. In: Pattern Recognition. ICPR International Workshops and Challenges. ICPR 2021. LNCS, vol. 12665, pp. 5–14. Springer, Cham (2021). https://doi.org/10.1007/978-3-030-68821-9_1

2. Andriyanov, N.A., Lutfullina, A.A.: Eye recognition system to prevent accidents on the road. Int. Arch. Photogramm. Remote Sens. Spatial Inf. Sci. **XLIV-2/W1-2021**, 1–5 (2021). https://doi.org/10.5194/isprs-archives-XLIV-2-W1-2021-1-2021
3. Campos-Ferreira, A.E., et al.: Vehicle and driver monitoring system using on-board and remote sensors. Sensors **23**, 814 (2023). https://doi.org/10.3390/s23020814
4. Ahmad, A.B., Tsuji, T.: Traffic monitoring system based on deep learning and seismometer data. Appl. Sci. **11**, 4590 (2021). https://doi.org/10.3390/app11104590
5. Andriyanov, N.A., Dementiev, V.E., Tashlinskiy, A.G.: Development of a productive transport detection system using convolutional neural networks. Pattern Recognit. Image Anal. **32**, 495–500 (2022). https://doi.org/10.1134/S1054661822030038
6. Sorić, M., Pongrac, D., Inza, I.: Using convolutional neural network for chest X-ray image classification. In: Proceedings of 43rd International Convention on Information, Communication and Electronic Technology (MIPRO), Opatija, Croatia, 2020, pp. 1771–1776 (2020). https://doi.org/10.23919/MIPRO48935.2020.9245376
7. Andriyanov, N.A., Gladkikh, A.A., Volkov, A.K.: Research of recognition accuracy of dangerous and safe X-ray baggage images using neural network transfer learning. In: IOP Conference Series: Materials Science and Engineering, vol. 012002, pp. 1–6 (2021)
8. Mustafa, W.A., Salleh, N.M., Idrus, S.Z., Jamlos, M.A., Rohani, M.N.: Overview of segmentation X-Ray medical images using image processing technique. J. Phys. Conf. Ser. **1529**(042017), 1–10 (2020). https://doi.org/10.1088/1742-6596/1529/4/042017
9. Buslaev, A., Iglovikov, V.I., Khvedchenya, E., Parinov, A., Druzhinin, M., Kalinin, A.A.: Albumentations: fast and flexible image augmentations. Information **11**, 125 (2020). https://doi.org/10.3390/info11020125
10. Magidi, J., Nhamo, L., Mpandeli, S., Mabhaudhi, T.: Application of the random forest classifier to map irrigated areas using google earth engine. Remote Sens. **13**, 876 (2021). https://doi.org/10.3390/rs13050876
11. Andriyanov, N.A., Dementiev, V.E., Tashlinskii, A.G.: Detection of objects in the images: from likelihood relationships towards scalable and efficient neural networks. Comput. Opt. **46**(1), 139–159 (2022). https://doi.org/10.18287/2412-6179-CO-922
12. Andriyanov N.: Estimating object coordinates using convolutional neural networks and intel real sense D415/D455 depth maps. In: Proceedings of 2022 VIII International Conference on Information Technology and Nanotechnology (ITNT), IEEE Xplore, pp. 1–4 (2022). https://doi.org/10.1109/ITNT55410.2022.9848700
13. Kim, J.-C., Chung, K.: Recurrent neural network-based multimodal deep learning for estimating missing values in healthcare. Appl. Sci. **12**, 7477 (2022). https://doi.org/10.3390/app12157477
14. Andriyanov, N., et al.: Intelligent system for estimation of the spatial position of apples based on YOLOv3 and real sense depth camera D415. Symmetry **14**, 148 (2022). https://doi.org/10.3390/sym14010148
15. Lee, J., Lee, S., Cho, W., Siddiqui, Z.A., Park, U.: Vision transformer-based tailing detection in videos. Appl. Sci. **11**, 11591 (2021). https://doi.org/10.3390/app112411591

Author Index

A
Abe, Jair Minoro 137, 157
Ahmad, Bashar 109
Aldbs, Heba 237
Andriyanov, Nikita 293, 310
Arima, Sumika 35, 48

B
Babkin, Eduard 250

C
Cabral, José Rodrigo 137
Cheniti-Belcadhi, Lilia 74
Czarnowski, Ireneusz 62
Czyżewski, Andrzej 13

D
de Oliveira, Cristina Corrêa 157
Dementiev, Vitaly 293
Demidovskiy, Alexander 250
do Nascimento, Samira Sestari 157
Duarte, Aparecido Carlos 137

E
Ekhlakov, Roman 302

F
Favorskaya, Margarita N. 147
Fukui, Keisuke 173

G
Gayed, Sana 128
Große, Christine 25

H
Hadyaoui, Asma 74
Hardt, Wolfram 97
Hasan, Samer 109
Hijikawa, Yuta 173
Horikawa, Keito 203

I
Itoh, Yoshimichi 183
Itou, Hiroki 35, 48

J
Jafar, Kamel 109
Jrad, Fayez 237

K
Kanyagha, Hellen Elias 87
Kazieva, Victoria 25
Kirishima, Koki 183
Kostek, Bożena 3

L
Larsson, Aron 25

M
Maeda, Takumi 48
Maginga, Theofrida Julius 87
Mallat, Souheyl 128
Massawe, Deogracious Protas 87
Mizuno, Takafumi 282
Mohammad, Ali 109
Monden, Rei 203, 214

N
Nagai, Isamu 173, 203, 214
Nahson, Jackson 87
Naserifar, Pooya 97
Norikumo, Shunei 273
Nsenga, Jimmy 87

O
Oda, Ryoya 193
Ohishi, Mineaki 183, 225
Ohta, Ryo 48
Ohya, Takao 263
Okamura, Kensuke 183

I. Czarnowski et al. (Eds.): KESIDT 2023, SIST 352, pp. 317–318, 2023.
https://doi.org/10.1007/978-981-99-2969-6

P
Pakhirka, Andrey I. 147

S
Sakamoto, Liliam Sayuri 137
Saleh, Hadi 109, 237
Saleh, Shadi 97
Saoud, Lama 237
Sęk, Oskar 62
Solieman, Monaf 109
Suetin, Marat 293

T
Takada, Daisuke 48
Teles, Nilton Cesar França 157

U
Ulitin, Boris 250

W
Watanabe, Kyo 35, 48

Y
Yamamura, Mariko 225
Yanagihara, Hirokazu 173, 183, 203, 214, 225

Z
Zrigui, Mounir 128
Zykov, Sergey V. 250

Printed in the United States
by Baker & Taylor Publisher Services